Secrets of
ANTIGRAVITY
PROPULSION

"Paul LaViolette is one of the most interesting and innovative thinkers probing the limits and horizons of contemporary physics. In this book he takes up a challenge that many of us have thought about but could not document: the possibility of propulsion systems that practically defy gravity. His findings merit earnest consideration, debate, and discussion."

ERVIN LASZLO, AUTHOR OF
SCIENCE AND THE AKASHIC FIELD

"Paul LaViolette's investigations into this most mysterious of subjects are at once fascinating and prescient."

NICK COOK, AUTHOR OF *THE HUNT FOR ZERO POINT: INSIDE THE CLASSIFIED WORLD OF ANTIGRAVITY TECHNOLOGY*

"One of the boldest and most exciting books on gravity control to be put forward in our times. Paul LaViolette is an outstanding scientist and the first to reverse engineer the B-2's highly classified propulsion system."

EUGENE PODKLETNOV, PH.D., PROFESSOR OF
CHEMISTRY, TAMPERE, FINLAND

"Paul LaViolette has once again unearthed advanced knowledge that can change our lives. This is a landmark book to be read and discussed by anyone concerned about humanity's options for the near future."

JEANE MANNING, AUTHOR OF *THE COMING ENERGY
REVOLUTION: THE SEARCH FOR FREE ENERGY*

Secrets of
ANTIGRAVITY
PROPULSION

Tesla, UFOs, and Classified
Aerospace Technology

PAUL A. LaVIOLETTE, PH.D.

Bear & Company
Rochester, Vermont

Bear & Company
One Park Street
Rochester, Vermont 05767
www.BearandCompanyBooks.com

Bear & Company is a division of Inner Traditions International

Library of Congress Cataloging-in-Publication Data
LaViolette, Paul A.
  Secrets of antigravity propulsion : Tesla, UFOs, and classified aerospace technol-
ogy / Paul A. LaViolette.
    p. cm.
  Includes bibliographical references and index.
  Summary: "A complete investigation of the development and suppression of anti-
gravity and field propulsion technologies"—Provided by publisher.
  ISBN: 978-1-59143-078-0
  1. Unidentified flying objects. 2. Space vehicles—Propulsion systems. 3. Anti-
gravity. I. Title.

  TL789.3.L38 2008
  629.47'5—dc22

                                                                        2008004466

Printed and bound in the United States by Lake Book Manufacturing, Inc.

10 9 8 7

Text design by Jon Desautels and layout by Priscilla Baker
This book was typeset in Sabon, with Bank Gothic and Gill Sans used as display
typefaces

To send correspondence to the author of this book, mail a first-class letter to the
author c/o Inner Traditions • Bear & Company, One Park Street, Rochester, VT
05767, and we will forward the communication.

CONTENTS

Appendices

ACKNOWLEDGMENTS

I dedicate this book to my father, Fred LaViolette (1916–2008), who through our years had been a guiding light for me. In particular I am greatly indebted to him for the long hours he spent helping me edit this manuscript. I would also like to thank my sister, Mary, for her editorial assistance as well. Finally I would like to thank Tom Turman, Thomas Chavez, Guy Obolensky, Larry Deavenport, Jean-Louis Naudin, and others for information about their work that they graciously shared.

1

ANTIGRAVITY: FROM
DREAM TO REALITY

1.1 ▪ TRAVELING TO THE STARS

Interstellar space travel has long captivated the imagination and longing of humankind. Indeed, we have penetrated the cosmos and walked on the moon, while breakthroughs in long-range exploration, such as the Hubble Space Telescope, bring the farthest reaches of space tantalizingly close, rekindling our desire to travel beyond our galaxy. As of yet, we are bound by the frustrating limits of conventional propulsion technology. Skeptics remind us that a spacecraft powered by even the most advanced chemical rockets would need to carry so much fuel that travel over interstellar distances would be out of the question. Alternatively, vehicles equipped with nuclear-powered ion thrusters would have a much greater range. However, the fuel requirements would be such as to make a journey of even a few light-years quite impractical—basic physics tells us that a rocket-powered spacecraft would need a fuel mass that would far exceed the mass of the vehicle itself.

Is there a way to free ourselves of this fuel problem, using a totally different means of propulsion, one that does not require large quantities of mass to be jettisoned rearward for the craft to move forward? Imagine a spaceship that could alter the ambient gravitational field,

1

artificially producing a matter-attracting, gravity-potential well that was just beyond the ship's bow. The gravity well's attractive force would tug the ship forward just as if a very massive, planet-sized body had been placed ahead of it. The ship would begin to "fall" forward and, in doing so, would carry its self-generated gravity well along with it. The gravity well would continually draw the ship forward, while always staying ahead. Through such a carrot-and-stick effect, the ship could accelerate to nearly the speed of light, or maybe even beyond, with essentially no expenditure of energy other than that needed to generate the gravity well.

Is such gravity control possible? Would it be possible to construct a spaceship with small enough propulsion power requirements that interstellar travel could be achieved? The answer is yes. For the past several decades, highly classified aerospace programs in the United States and in several other countries have been developing aircraft capable of defying gravity. One form of this technology can loft a craft on matter-repelling energy beams. This exotic technology falls under the relatively obscure field of research known as electrogravitics.

The origins of electrogravitics can be traced back to the turn of the twentieth century, to Nikola Tesla's work with high-voltage shock discharges, and somewhat later to T. Townsend Brown's relatively unpublicized discovery that electrostatic and gravitational fields are closely intertwined. Unfortunately, the electrogravitic effect has for the most part been ignored by mainstream academics, because the phenomenon isn't anticipated by either classical electrostatics or general relativity, effectively preventing it from being taught in university courses such as physics and electrical engineering. Rather, to unlock the secrets of electrogravitics, one must delve into popular science articles, patents, and relatively obscure technical reports that once held a classified status. Perhaps the best place to begin is to review some of Brown's seminal work.

1.2 ▪ THE BIRTH OF ELECTROGRAVITICS

The American physicist and inventor Thomas Townsend Brown was born in 1905 to a well-to-do Zanesville, Ohio, family. At an early age, he

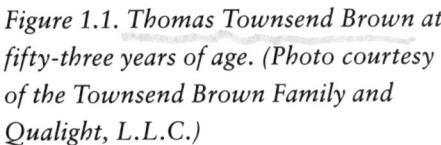

Figure 1.1. Thomas Townsend Brown at fifty-three years of age. (Photo courtesy of the Townsend Brown Family and Qualight, L.L.C.)

displayed a keen interest in space travel and dreamed of one day journeying into space himself. His discovery of the electrogravitic phenomenon occurred during his high school years, when his interest in space travel led him to toying with a Coolidge tube—a high-voltage X-ray-emitting vacuum tube similar to that found in modern dental X-ray machines. Brown had the insight to mount the tube on a delicate balance to investigate whether it might produce any thrust. To his surprise, the tube moved every time it was turned on. Ruling out X-rays as the cause of this mysterious force, he traced the effect to the high voltage he was applying to the tube's plates. He concluded that the tube had moved because its gravity field had somehow become affected by the plate's high-voltage charge.[1-4]

After additional experiments, Brown eventually developed an electric capacitor device that he termed a *gravitator* (or, alternately, *gravitor*). These units were very heavy. One version consisted of a wooden box, 2 feet long and 4 inches square, that contained a series of massive, electrically conductive plates made of lead and separated from one another by electrically insulating sheets of glass, which served as the capacitor's dielectric medium (a *dielectric* is a substance that does not conduct electric current). Another version used a dielectric molded from a mixture of lead monoxide and beeswax encased in Bakelite. The diagram in figure 1.2, which is reproduced from Brown's 1928 patent, shows yet another version made with aluminum plates and paraffin.

When energized with up to 150,000 volts of direct current (DC), Brown's gravitator developed a thrust in the direction of its positively

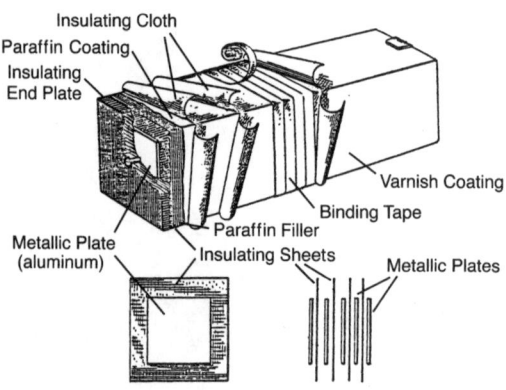

Figure 1.2. A cellular gravitator shown in perspective together with end- and side-view details of its plates. (Brown, 1928)

charged end. One such gravitator, which weighed 10 kilograms, was observed to generate a maximum thrust of 0.1 kilogram (1 newton), a force equal to about 1 percent of its weight.[5,6] When oriented upright on a scale and energized, it proceeded to gain or lose that amount of weight depending upon how the charge polarity was applied. It became lighter when its positive end faced up and heavier when its negative end faced up.

Brown entered the California Institute of Technology in 1922. He spent a good part of his freshman year attempting to win the friendship of his professors and to convince them of his abilities as a first-class "lab man." However, when he began mentioning his ideas about electrogravity, no one would listen. At the end of the year, he had his laboratory equipment shipped from Ohio, set it up in his quarters, and sent invitations to several of his professors, including the renowned Dr. Robert Millikan, to witness a demonstration of the new force he had discovered. No one came. Some time later, one of Brown's friends tested Millikan by asking him whether he knew of anyone who had ever found a way of modifying or influencing the force of gravity. Millikan is said to have replied brusquely, "Of course not; such a thing is impossible and out of the question."

His feelings deeply hurt by the incident, Brown transferred to Kenyon College, in Gambier, Ohio, and the following year he transferred to Dennison University, in Granville, Ohio. One of his physics professors at Dennison, Dr. Paul A. Biefeld, had also been interested in the movement of electric capacitors. Brown had frequent conversations with Biefeld and came to refer to the electrogravitic phenomenon as the

Biefeld-Brown effect, perhaps in respect to Biefeld's own interest in the subject. Still, it is not clear that Biefeld actively collaborated with Brown on his research.

For one of his experiments, Brown arranged a pair of gravitators, one at each end of an arm that was suspended from the laboratory ceiling by a long cord attached to the arm's central fulcrum (figure 1.3). When energized with between 75,000 and 300,000 volts DC, the connecting arm rotated as each gravitator moved in the direction of its positive pole. This force occurred in the same fashion even when the capacitor was immersed in a tank of oil, thereby ruling out the possibility that the effect was produced by a wind of electric ions. Brown's gravitators could produce this motion with a power input of just 1 watt. With each gravitator generating 100 grams of thrust, for a total thrust of 2 newtons, the thrust-to-power ratio of Brown's electrogravitic thrusters calculates to 2,000 newtons per kilowatt. This is 130 times the thrust-to-power ratio of a jet engine, or 10,000 times the thrust-to-power ratio of the space shuttle main engine.

Brown determined that the electrogravitic effect he observed depended on the amount of charge stored in his capacitors. As the applied voltage was increased and a greater amount of charge was stored, the capacitors would respond with a greater amount of electrogravitic force. Moreover, because the intensity of the effect depended upon the

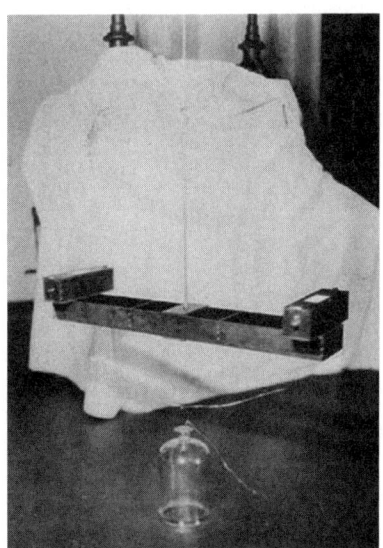

Figure 1.3. An experimental setup designed to measure the thrust produced by Thomas Townsend Brown's gravitators. (Photo courtesy of the Townsend Brown Family and Qualight, L.L.C.)

capacitor's mass, he concluded that the induced motion must be due to the capacitor's ability to generate a localized gravitational field.

After he left Dennison, Brown conducted astrophysics research for four years, from 1926 to 1930, working at a private laboratory in his hometown of Zanesville and also at Ohio's Swazey Observatory, where he was in contact with Dr. Biefeld. In a variation of his rotating gravitator experiment, Brown suspended a single gravitator from his laboratory ceiling by two wires (figure 1.4). The gravitator was hung so that it would stay immersed in a tank of oil, so as to reduce the production of ions. When energized, the pendulum would swing toward the gravitator's positive pole. Brown characterized this electrogravitic phenomenon as an impulse.[7] He noted that less than five seconds was required for the pendulum to reach the maximum amplitude of its swing, but then, even while he maintained the high-voltage potential, his pendulum would gradually return to its plumb position, taking from 30 to 80 seconds to return. He noted further that on its return from maximum deflection, his pendulum would hesitate at definite levels or steps, but repeated trials showed that there were no consistent positions to these steps.

Brown also noted that he would have to give his gravitator a rest after each test to see the effect repeat once again. He had to remove his charging potential for at least five minutes to allow his gravitator sufficient time to "recharge" itself so that it might regain its "former gravitic condition." He did not mention what might have been happening during this recharging process, probably because at that time he had no clear idea himself. He saw that when the duration of the gravitic impulse had been greater, more time was needed off-line to allow the gravitator to refresh itself.

We may gain an understanding of why his gravitator would not

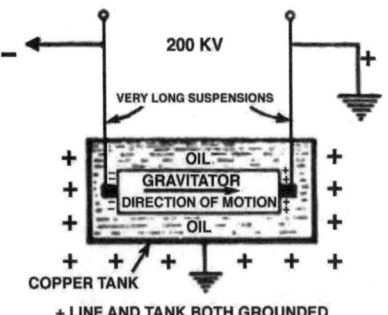

Figure 1.4. Thomas Townsend Brown's gravitator hung in pendulum fashion and was submersed in a tank of oil. (Brown, 1929)

hold its initial gravitic force by analyzing what was happening inside its dielectric. Initially, before high voltage was applied, the dielectric would reside in an unpolarized state. With the application of voltage, current would begin to flow and the gravitator's plates would progressively charge up. The electric field between the plates would exert an electrostatic force on the dielectric's molecules, causing them to displace slightly—the positive molecular charges being tugged in the direction of the gravitator's negative pole and the negative molecular charges being tugged toward its positive pole. As a result, the dielectric would become polarized (see figure 1.5), its electric dipole moment pointing in a direction opposite to the direction of the applied electric field.

The dielectric does not polarize instantaneously in response to the applied voltage; it takes some time to reach full polarization. This time lag is a common property of dielectrics known as *dielectric relaxation.* It is analogous to the property of hysteresis observed when a transformer core is magnetically energized. Most capacitor dielectrics used today have very short dielectric relaxation times—less than microseconds. However, Brown's capacitor must have had a very slow relaxation time, probably because it was rather long from end to end and because of the nature of the wax-litharge mixture of which it was composed. The 30 to 80 seconds or so that the gravitator took to gradually return to its plumb position from its maximum deflection was likely the duration of its dielectric relaxation, the time required for its dielectric to become fully polarized.

During the first few seconds that the voltage was applied, the slowly responding dielectric, for the most part, would have remained unpolarized. Hence the applied electric field, along with its associated gravitic field effect, would have extended with full intensity throughout the gravitator, exerting a maximal gravitic thrust on the dielectric in the

Figure 1.5. The polarized charge arrangement in the gravitator's dielectric when voltage is applied to the gravitator plates. Arrows indicate the direction of the electrogravitic force.

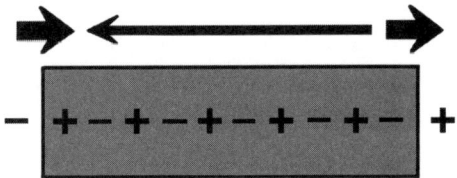

direction of the gravitator's positive pole. However, as the dielectric became increasingly polarized, its oppositely directed electric dipole moment field arising within the dielectric would have progressively increased in strength, progressively canceling out the gravitic effects induced by the externally applied electric field. Thus the thrust pushing the gravitator in the direction of its positive pole would have progressively subsided. Moreover, when the dielectric reached its fully polarized state with its opposed dipole moment field at its maximum, this thrust would have become almost entirely canceled out, leaving the gravitator to return to its plumb position.

As the dielectric became progressively polarized, the gravitator capacitor plates would have been able to hold an increasing amount of electric charge as an increasing number of polarized molecular charges moved adjacent to the plates to attract additional charges. As a result, throughout this polarization interval the gravitator would have been charging up and a current would have been flowing to its plates. Charge would have been accumulating most rapidly in the beginning and the charging rate would have progressively dropped off as the full charged state was approached. Similarly, the reverse gravitic thrust generated by the polarizing dielectric would have caused the overall gravitic thrust to decline most rapidly at the beginning of the pendulum's swing and to subside more slowly as the fully charged state was approached. The observation that the gravitic force subsided in steps may be an indication that the dielectric experienced a succession of abrupt mechanical shifts in its approach to the fully polarized state.

The need to recycle the gravitator between test runs, to discharge it and let it rest so as to "regain its former gravitic condition," is understandable if we realize that it was necessary to allow a sufficiently long rest period for the dielectric to completely depolarize. After the DC voltage supply is shut off, a residual charge will initially remain on the capacitor plates, kept there by the dielectric's residual polarization. Engineers refer to this remnant charge as *dielectric absorption*. It is particularly important in capacitors that are capable of storing a lot of charge. As the dielectric progressively relaxes, this charge is gradually released. Once the gravitator dielectric had relaxed to an unpolarized state, new charges would be able to rapidly accumulate on its electrodes during the next charging cycle. Once again, a steep gravity potential gradient would have been able to form across the gravitator and temporarily exert a net thrust on its massive dielectric until it was again opposed by the dielectric's progressively increasing dipole moment field.

1.3 ▪ A THEORY OF ELECTROGRAVITICS

In August 1927, Brown filed for a British patent on his gravitator idea, which was issued to him in November 1928 (British patent 300,311). In the patent's text, Brown clearly proclaims that the propelling force he has discovered is of an unconventional nature:

> The invention also relates to machines or apparatus requiring electrical energy that control or influence the gravitational field or the energy of gravitation; also to machines or apparatus requiring electrical energy that exhibit a linear force or motion which is believed to be independent of all frames of reference save that which is at rest relative to the universe taken as a whole, and said linear force or motion is furthermore believed to have no equal and opposite reaction that can be observed by any method commonly known and accepted by the physical science to date.[8]

Here he describes his belief that electrogravitic force operates relative to a unique reference frame that is at rest in relation to the universe, an idea that challenges special relativity's notion that a force should operate in the same manner relative to any frame of reference. Moreover, he suggests that this force is reactionless when producing its forward thrust—that is, it produces its forward thrust without any back-directed recoil. He is in effect suggesting that it violates Newton's third law of motion—that every action should produce an equal and opposite reaction. Dr. Patrick Cornille, who repeated Brown's high-voltage pendulum experiment, came to the similar conclusion that Newton's third law of motion was indeed violated (see chapter 12).

On October 28, 1928, just prior to receiving his patent, Townsend submitted to the physics journal *Physical Review* a paper titled "Tapping Cosmic Energy," which described his gravitator experiments.

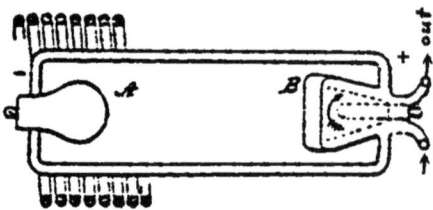

Figure 1.6. A gravitator configured within an evacuated envelope reproduced from Brown's patent. In this version, the negative electrode or cathode (left) is heated to incandescence, thereby encouraging the thermionic emission of electrons, whereas the positive electrode or anode (right) is cooled by circulating air or water. This configuration mimics many of the design features of an X-ray tube (or Coolidge tube), like the ones that Brown used when he first observed the electrogravitic phenomenon. (Brown, 1928)

Unfortunately, the journal rejected his paper, apparently because of its unconventional nature. For one thing, his ideas challenged Einstein's theory of gravitation, which had by then become staunchly accepted by the physics community. One year later, Brown published a less technical version of his findings in *Science and Invention Magazine*[9] and succeeded in impressing a large number of people with his work.

In 1930 one of Brown's colleagues wrote about the gravitator to Colonel Edward Deeds, who was one of Brown's longtime acquaintances. In his letter he wrote, "I have had a number of scientists view the gravitator and they have all been absolutely amazed at its action, frankly stating that whereas they see the results and the movements of the gravitator, it is absolutely unexplainable by any laws of physics that they know."[10]

At that time, Brown had no theory to explain electrogravity. It would not be until twenty years later that he sketched out a theory of sorts, which he made notes about in one of his lab notebooks. But a theoretical methodology that actually predicted charge-mass coupling and that could begin to make some sense out of electrogravitics in a unified-field-theory context did not begin to emerge until the late 1970s with the development of subquantum kinetics.[11-13] It is useful to review a bit about this theory here, as it will help us interpret the novel results that Brown was getting.

Subquantum kinetics offers an explanation for gravity that is substantially different from Einstein's relativity theory. Whereas general relativity postulates that masses exert an attractive gravitational force on other bodies by warping the space-time dimensional fabric around themselves, subquantum kinetics proposes that masses have no such effect on the geometry of space or time. Subquantum kinetics assumes that space is geometrically flat, or Euclidean; hence, it conforms to the geometrical rules most everyone learns in high school math class. It predicts that a mass creates a classical gravity potential field and that a gradient in such a field exerts a force on a remote body by affecting how that body's constituent subatomic particles regenerate their physical form. (Details of how potential fields are generated and how they accelerate material particles through form regeneration are further discussed in chapter 4.)

Subquantum kinetics also differs from general relativity in its prediction of gravitational field polarity. According to general relativity, masses only attract other masses, never repel them. Although Einstein did introduce the notion of a matter-repelling effect whose magnitude he symbolically represented by a quantity called the cosmological constant, this was not part of his general relativity theory, but was an ad hoc correction factor added to his field equations so that they would not predict a universe that was spontaneously contracting due to self-gravitation.

Einstein had attempted to expand his relativity theory to encompass both electromagnetism and gravitation, but he was unsuccessful. Relativity was unable to predict any connection between charge polarity and gravitational field polarity.

Subquantum kinetics, on the other hand, predicts that gravity should have two polarities. It permits the creation of either a matter-attracting gravity potential well or a matter-repelling gravity potential hill and predicts that these two gravity polarities should be directly correlated with electric charge polarity. That is, positively charged particles such as protons would generate gravity wells, whereas negatively charged particles such as electrons would generate gravity hills. When protons and electrons combine to compose electrically neutral atoms, the gravitational polarities of the protons and electrons for the most part would neutralize one another. However, because a proton's gravity well is theorized to marginally exceed an electron's gravity hill, electrically neutral matter would produce a small, residual matter-attracting gravity potential well, thereby generating the gravity we commonly experience pulling us to Earth.

Subquantum kinetics predicts that a matter-repelling gravity potential hill should form on the negatively charged side of a capacitor and that a matter-attracting gravity potential well forms on the positively charged side. The intervening gravity potential gradient would produce a gravitational force on the capacitor's massive dielectric that would act to pull it in the direction of the positively charged plate (figure 1.7). The more prominent the gravity hill and well, the steeper the gravity potential gradient and the stronger the produced gravitational thrust. While this force was present, the capacitor would behave as if it was being tugged forward by a very strong gravitational field emanating from an invisible planetary mass situated ahead of its positive pole and as if it was being pushed forward by an equally strong repulsive gravitational force emanating from behind its negative pole. If the capacitor was placed with its positive pole facing up and was energized such that it generated a sufficiently steep vertical gravity gradient, theoretically the downward pull of gravity could be entirely overcome. (For a more detailed mathematical analysis of how this electrogravitic force might be quantified, see the text box on pages 13–16.)

At present there is no easy way to check the prediction that an individual electron might have negative gravitational mass because any matter-repelling gravitational force it might produce would be greatly

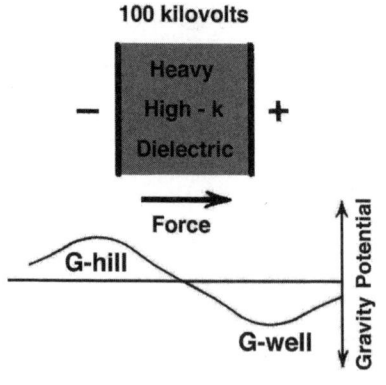

Figure 1.7. The electro-gravitational force effect produced by charging a capacitor to a high voltage. (P. LaViolette, © 1994)

overpowered by its electrostatic force interactions with surrounding matter. That is, no one has found a way to screen out these electrostatic forces sufficiently to allow an accurate measurement of a single particle's gravitational mass. However, when large numbers of electrons and protons are differentially accumulated, as at the opposite poles of a charged capacitor, the cumulative effect of the negative gravitational potentials of the electrons appears to be great enough to produce an observable macroscopic force. That force is the electrogravitic effect that Brown observed.

Quantifying the Electrogravitic Effect

Subquantum kinetics, then, predicts that a charged body should generate a gravitational mass, m_g, that scales directly with the magnitude of its electrical charge. Their proportional equivalence is expressed by the following electrogravitic coupling relation:

[gravitational mass] IS PROPORTIONAL TO [electric charge]

or with symbols:

$$m_g \propto q, \tag{1}$$

Thus, a body that has a fourfold increase in positive electric charge should produce a fourfold-greater positive gravitational mass. Also, a fourfold increase in negative electric charge should produce a fourfold-greater negative (mass-repelling) gravitational mass. Moreover, because electric charge comes in either a positive or negative polarity, $\pm q$, gravitational mass would similarly be induced in either of two polarities correlated with the charge polarity.

The same electrogravitic rule holds when expressed in terms of electric charge *density*, ρ_e, and gravitational mass *density*, ρ_m, quantities that refer to the amount of charge or gravitational mass per unit volume. Their proportional equivalence is expressed as:

[gravitational mass density] IS PROPORTIONAL TO [electric charge density]

or with symbols:

$$\rho_m \propto \rho_e \tag{2}$$

We may also express this charge–mass correspondence in terms of energy potentials or, to use another phrase, in terms of field potentials. For example, a positively charged body that is characterized by a positive charge density, ρ_e, would create a positive electric potential within itself. This elevated potential would create an electric potential field, $\varphi_e(r)$, that would appear as an electric potential hill having its maximum centered on the charged body and a magnitude that progressively declined with increasing radial distance r from that body. The parenthetical expression, (r), indicates that the field magnitude varies with distance r.

As noted in relation 2, a body having a positive electric charge density would produce a proportionate positive gravitational mass density, ρ_m, that would supplement its inherent natural mass density. This in turn would create a proportional negative gravity potential within the body supplementing its naturally produced negative gravity potential, which in turn would generate an extended gravity potential field $-\varphi_g(r)$. This gravity field would be configured as a gravity potential well centered on the charged body, its gravity potential progressively rising to more positive values with increasing radial distance r from that body.

In the case of a negative charge density, these field polarities would be reversed, resulting in an electric potential well centered on the body that in turn would produce a gravity potential hill. Note that when speaking of gravity fields, what we term a "positive mass" by convention is one that produces a matter-attracting gravity potential well. In the case of electric charge, on the other hand, by convention a positive charge would produce a positive electric potential hill.

The electrogravitic relations presented in (1) and (2) may be expressed in terms of field potentials as:

[gravity potential] IS PROPORTIONAL TO [negative electric potential]

or with symbols:

$$\varphi_g(r) \propto -\varphi_e(r). \tag{3}$$

Hence, an electric potential field gradient extending between the positive and negative plates of a capacitor would produce a proportional gravity potential field gradient of opposite sign across the capacitor's intervening dielectric; recall figure 1.5.

Also, Newton's second law tells us that a gravity potential field will generate a force on a body that is proportional to the magnitude of the field gradient multiplied by the body's inertial mass. This may be expressed mathematically by the equation:

$$\mathbf{F}_g(r) = -Gm_o\, \nabla\varphi_g(r), \qquad (4)$$

where $\mathbf{F}_g(r)$ is the gravitational force acting on a body, G is the gravitational constant, m_o is the inertial mass of the affected body, and $\nabla\varphi_g(r)$ is the local gravity potential gradient that is sometimes alternatively symbolized as **grad** $\varphi_g(r)$. The bold type on the force and gradient symbols indicates that they are vector quantities having direction as well as magnitude. Basically this equation states that the steeper the gravity field gradient, the greater the produced force, as was mentioned earlier in connection with figure 1.7. Or, alternatively, the greater the magnitude of $\nabla\varphi_g(r)$, the greater the produced force.

The quantity $-G\nabla\varphi_g(r)$ in equation 4 is termed the *gravitational acceleration* and is sometimes symbolized as g(r). Thus equation 4 may be rewritten to yield the more condensed expression for gravitational force: $\mathbf{F}_g(r) = m_o\, g(r)$. Often the magnitude of a gravitational accelerating force is measured in terms of "g's," or multiples of Earth's gravitational acceleration pulling us toward Earth, which at Earth's surface has a value of about 980 cm/s^2. This should not be confused with the inertial "g" symbol, which quantifies the magnitude of a mechanical accelerating force experienced by a jet pilot or rocket astronaut as inertial force resisting acceleration. Thus, an electrogravitic acceleration of 10 g's would signify a gravitational acceleration ten times that produced naturally at Earth's surface. Depending on the polarity and orientation of the applied electric field, this artificially induced gravitational acceleration may be engineered either to supplement or to counter that produced by Earth's field.

Equation 4 may be combined with proportionality relation 3 to express the gravitational force $\mathbf{F}_g$ acting on a body (or dielectric) in terms of the product of the inertial mass m_o of that body (or dielectric) and the voltage gradient, $\nabla\varphi_e(r)$, that spans it:

$$\mathbf{F}_g(r) = k\, m_o\, \nabla\varphi_e(r). \qquad (5)$$

> The constant k added in here is an experimentally determined electrogravitic proportionality constant that quantifies the charge-to-mass coupling relationship. Hopefully, future experimentation will provide a value for this constant. Equation 5, then, mathematically expresses the electrical induction of a gravitational force.

1.4 ■ ELECTROGRAVITIC MOTORS

In his 1928 British patent, Brown also introduced his invention of a gravitator motor. This involved a series of gravitator cells arranged in a circle (figure 1.8). By ensuring that the cells were spaced sufficiently far apart from one another and that the spacing medium was less dense than the dielectric medium within each cell, the cells would collectively generate unbalanced forces and hence produce rotation. He noted that this motor may either be "independently excited," that is, run by an external source of electric power, or be "self-excited," that is, energized from electric power that it generates itself.

A later version of his gravitator motor was described in U.S. patent 1,974,483, filed in February 1930 and which was issued to Brown in September 1934. This used a rotor made from alternating sectors of marble and varnished wood, separated by copper-plate electrodes across which a high-voltage charge was applied (see figure 1.9). In another variation, he used alternating sectors of lead oxide and paraffin wax; essentially he alternated a high-density dielectric with a low-density dielectric.

In his 1928 patent, where he discussed the possibility of powering his motor from electric power that the motor itself would produce, he

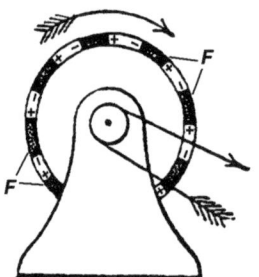

Figure 1.8. A gravitator motor composed of gravitator cells (F) positioned around the circumference of a wheel. (Brown, 1928)

pointed out that the electric power output generated by the motor could far exceed the electric input needed to run it. He stated:

> Here it will be understood that the energy created by the operation of the motor may at times be vastly in excess of the energy required to operate the motor. In some instances the ratio may be even as high as a million to one. . . . In said self-excited motors the energy necessary to overcome the friction or other resistance in the physical structure of the apparatus, and even to accelerate the motors against such resistance is believed to be derived solely from the gravitational field or the energy of gravitation.[14]

In effect, Brown boldly states that his motor is a perpetuum mobile. There is a question as to whether he was overstating this motor's over-unity capability, for he makes no reference to experimental data. Also, there is no evidence of anyone having reproduced this design and having obtained such high electrical or mechanical outputs. Nevertheless, such a blatant violation of the first law of thermodynamics in principle is possible in cases in which a gravitational field is made to follow a circular path, as in Brown's gravitator motor. That is, because the gravitators mounted on the wheel's periphery would generate a circumferentially oriented gravity field and carry this field along as the wheel turns, regardless of the wheel's position, the induced gravity field would always cause further rotation. In effect, the wheel would rotate in a state of circular free fall. Just as a mass is able to fall forever in an infinitely deep pit, so too would this rotor be able to turn indefinitely without reaching the end of its potential energy supply. All the while, power could be extracted from the wheel's shaft at no cost, save that needed to power the gravitators.

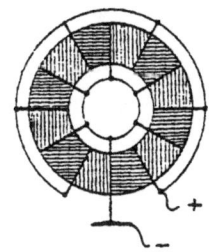

Figure 1.9. A rotor component for an electrostatic motor built and patented by Thomas Townsend Brown that used dielectric sectors of alternating high and low density. (Brown, 1930)

Such vortical gravity fields are rarely observed in nature, because Earth's field is for the most part directed radially with respect to Earth's center. However, there may be marginal exceptions to this rule, as is the case in the vicinity of Argostoli Bay, on the island of Cephalonia, located off the northwest coast of Greece. Several kilometers northwest of the coastal town of Argostoli, there is a place where water from the bay flows inland, runs downhill from sea level, and, after a few hundred meters, disappears into a fissure in the rock. To find where this water goes, Austrian geologists added 350 pounds of a tracer dye to this inflow and, using sensitive equipment, two weeks later detected this same dye on the other side of the island fourteen kilometers to the east in a spring issuing from a subterranean cavern. Curiously, the water in this cavern is situated several meters above sea level and eventually flows downhill, emptying back into the bay. Thus, the water makes a complete circle! One hundred years ago, local residents fashioned a channel for this inflowing water and built a waterwheel to harness its energy to produce electric power (see figure 1.10).

What causes water on the western side of this bay to flow downhill, below sea level, and then flow uphill toward the eastern side, returning once again to the bay? Some have suggested that geothermal, subterranean hydrostatic pressures may be responsible for siphoning the water upward. Because of the existence of several other unusual phenomena in the region, the Greek physicist Panagiotis Pappas believes that a gravitational field anomaly may instead be responsible. For one thing, the water flow in Argostoli Bay changes its direction about every quarter of an hour. This is most easily seen from the vantage point of the one-kilometer-long bridge that spans the shallow southern end of the bay. There, one can see water flowing briskly under the bridge and passing through its arches at speeds of up to one meter per second, but after some minutes it slows to a stop, reverses, and begins to pick up speed in the opposite direction. This effect is not at all related to lunar tides, which occur on a much longer, twelve-hour cycle.

Across the bay from Argostoli, near the village of Loukouri, lies a huge boulder that for many years was observed to very slowly sway back and forth. Because of its motion it came to be called Kounopetra, meaning "rocking rock." If a sheet of paper was placed under one end of

Figure 1.10. Waterwheel on Cephalonia Island built over a sluiceway to generate electricity from inflowing water. The water level drops about 2 meters below sea level by the time it reaches the waterwheel and thereafter drops several more meters before entering a fissure. (Photo by the author)

this rock, some time later we would find that the sheet was caught under the rock and could not be removed. Later still, however, the rock's center of gravity would shift and once again the paper could be removed. Perhaps the boulder's slow rocking, the gradual change in water-flow direction in the adjoining bay, and the gravitational anomaly responsible for propelling the subterranean flow of seawater uphill to its spring outlet all arise from the same cause—a vortical instability in the local gravitational field that causes motion tangential to Earth's surface. If so, the waterwheel at Argostoli may have been the first gravitational perpetual-motion machine built in modern times.

1.5 ▪ BROWN'S GRAVITO-ELECTRIC DISCOVERIES

Brown kept a sharp eye on the daily operation of his electrogravitic motor. In the course of his studies, he found that the rate of rotation of his motor was not constant; it varied depending on the time of day. Further observation revealed that its torque rose and fell according to the lunar and solar cycles. A diurnal sidereal cycle was also present in which the gravitic torque changed as a result of the Earth's rotation relative to a fixed point in space lying in the general direction of the galactic center. He observed similar cyclic influences in his gravitator pendulum experiments in which the total duration of the pendulum's developed impulse was seen to vary with cosmic conditions, such as the pendulum's alignment with the sun and moon at times of conjunction or opposition. Ruling out factors such as changes in temperature and supplied voltage, he concluded that the impulse was governed solely by the condition of the ambient gravity field potential. He found that any number of different kinds of gravitators, operating simultaneously at very different voltages, revealed the same impulse duration at any given instant and underwent equal variations over extended periods of time. The cause of these variations greatly intrigued him and became a focus of his gravity research throughout his life.

In 1930, Brown left Swazey Observatory and began working at the Naval Research Laboratory in Washington, D.C., as a specialist in radiation, field physics, and spectroscopy. From 1931 to 1933, the Naval Research Laboratory placed him in charge of a project whose stated purpose was to investigate certain unusual "electric" effects found in fluids and in massive high-K dielectrics. Brown found that such massive high-K dielectrics exhibited the strongest electrogravitic coupling. Again, he found that the magnitude of the electrogravitic thrust varied with the time of day.

Explaining the Dielectric Constant, K

Often the permittivity of a dielectric is expressed in terms of the dielectric constant K of the material, which is the ratio of its permittivity to the permittivity of empty space, $\varepsilon_o = 8.85 \times 10^{-12}$ farads per meter: that is, $K = \varepsilon/\varepsilon_o$. So if two capacitors are compared, one having a dielectric between its electrodes with a tenfold-higher K value, and if both capacitors are charged to the same voltage, the capacitor with the higher K dielectric will be able to store ten times as much electric charge. K values can range from near unity, such as the value for air, to more than 20,000 for certain ceramic compounds. When Brown was conducting his first tests, he used lead monoxide as a dielectric for one of his gravitators, which has a K of 26. Some ceramic compounds, such as barium titanate, not only can have a very high dielectric constant, ranging from 2,000 to 10,000 K, but they also happen to be quite heavy. More recently, a ceramic compound called barium zirconium titanate (also known as BZT), which also is quite massive, has been found to have a dielectric constant of 23,000.

Brown constructed expensive recording instruments, some of which resembled the electrostatically energized multisegmented rotor he had developed in the 1920s but which used massive dielectrics with much higher K values. He called these *sidereal electrometers*. For several years, he took continuous readings with them under carefully controlled conditions, keeping voltage and temperature constant and shielding his units from magnetic and electrostatic fields in the environment. His sidereal electrometer rotor was typically 12 inches in diameter and was suspended from its center by a thin wire that allowed it to rotate under torque in a horizontal orientation. A sequencer applied 11,000 volts for thirty seconds across the rotor segments, causing the rotor to turn by several degrees. The power was then shut off for three minutes to allow the rotor to return to its relaxed, untorqued position. The cycle would then repeat. The rotor's energized and relaxed angular positions were automatically recorded on a slowly advancing paper strip, and later the trends were statistically processed to check for possible cyclic correlations. In 1973, Brown wrote the following about his findings:

There were pronounced correlations with mean solar time, sidereal time and lunar hour angle. This seemed to prove beyond a doubt that the thrust of the "gravitators" varied with time in a way that related to solar and lunar tides and a sidereal correlation of unknown origin. These automatic records, acquired in so many different locations over such a long period of time, appear to indicate that the electrogravitic coupling is subject to an extraterrestrial factor, possibly related to the universal gravitational potention or some other (as yet) unidentified cosmic variable.[15]

In addition, Brown's Naval Research Laboratory investigations unexpectedly revealed that the electric resistivities of certain high-density dielectrics also undergo cyclic changes correlated with solar and sidereal time. He devised a resistance-sensing device that was able to measure these changes. Unlike his sidereal electrometer, it had no moving parts. He made observations with these two types of detectors, both in Washington and at sea on the Navy-Princeton International Gravity Expedition to the West Indies conducted on board the U.S. submarine S-48. Interestingly, Admiral Hyman Rickover, who was then a lieutenant, served as the executive officer (second in command) for this expedition. Brown's laboratory findings were summarized in a study titled "Anomalous Behavior of Massive High-K Dielectrics," which, it seems, has yet to be declassified. A Freedom of Information Act request was made to the Naval Research Laboratory in May 1995 to retrieve a copy of this document. However, the response came back that the library had no record of it. Either they did not do a thorough search or it was relocated and its existence and whereabouts are presently classified.

The results of these gravito-electric measurements were so encouraging that in 1937 a decision was made to extend the investigation and to establish another naval field station some distance west of Washington. Measuring equipment was set up in a constant-temperature vault in the basement of Brown's home in Zanesville, with provisions made for automation of the data-recording process. These new measurements confirmed the Naval Research Laboratory findings. The field station was moved the next year to the University of Pennsylvania, in Philadelphia. The investigation was interrupted during World War II but was resumed again from 1944 to 1949 in California, at Laguna Beach and Los Angeles. The project was sponsored by the Townsend Brown Foundation, a scientific research organization established by Brown's parents in the mid-1920s.

In a letter he wrote in 1968 to the researcher Thomas Turman, Brown commented about the observed variations in electrogravitic force:

> There are a number of mysteries concerning the nature of the [electrogravitic] force, largely the variations which it undergoes. There appear to be at least three semi-diurnal cycles:

1) relating to mean solar time (with maxima at 4 AM and 4 PM)

2) relating to lunar hour angle with maxima approximately 2 hours after the upper and lower meridian transit of the moon, and

3) relating to sidereal time with a sharp peak at 16h S.T. [Greenwich sidereal time] and a minor maximum at 4h S.T. The reasons for these variations as well as the reasons for the almost continuous secular variations [are] completely unknown.[16]

At sixteen hours Greenwich sidereal time, the western end of the Scorpius constellation was reaching its zenith, a sky position lying within 25 degrees of the galactic center. Consequently, Brown theorized that the sidereal effect he was observing was due to some kind of radiation emanating from the center of our galaxy. He concluded that these "sidereal rays" were not electromagnetic in nature and did not resemble cosmic rays. They had no known ionizing power, were not disturbed by

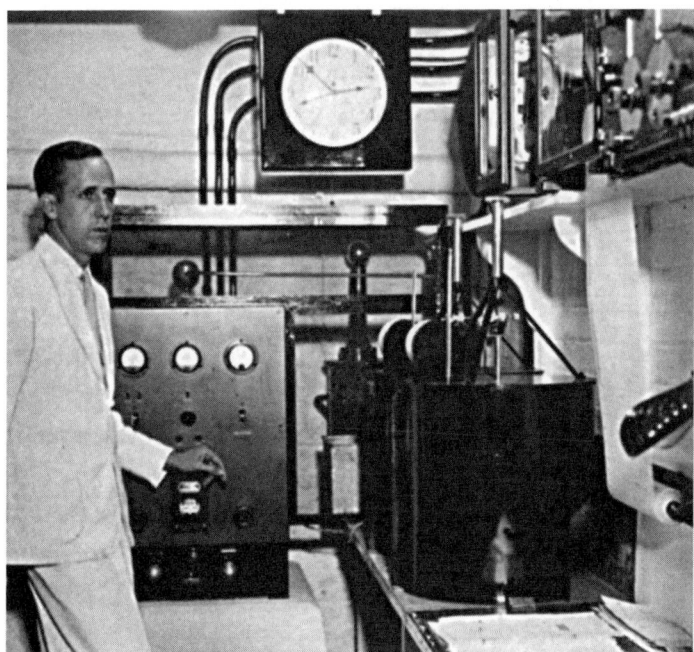

Figure 1.11. Thomas Townsend Brown in his underground gravito-electric monitoring station at his home in Zanesville, Ohio. (Photo courtesy of the Townsend Brown Family and Qualight, L.L.C., 1937)

Earth's magnetic field, and were highly penetrating. He eventually came to feel that they must be high-frequency gravitational waves.

Brown resumed his sidereal measurements in 1970 from an isolated site on Catalina, an island situated off the coast of Southern California. Around this time, he discovered a new correlated effect. He found that certain materials, including massive high-K ceramic dielectrics, certain kinds of resistors, complex silicates, and natural igneous rocks and clays, spontaneously generate DC electric potentials, with some materials producing as much as 0.7 volt. Moreover, he found that this generated DC potential varied from hour to hour and from day to day in much the same way as the resistance variations he had observed in the Naval Research Laboratory experiments.[17] In a paper about his findings, he commented:

> It has been found that certain basaltic and granitic rocks exhibit a self-potential which undergoes large cyclic variation not related to temperature, pressure, humidity or other local variables. Long-time monitoring has revealed periods of the year when the self-potential correlates consistently with sidereal time, reaching maximum and minimum values vectoring on the Galactic center (17h 43m RA). At other times, solar cycles predominate and [the] sidereal component disappears. Even so, a circadian pattern nearly always exists which cannot be correlated with ambient laboratory conditions. Hence, it is of interest not so much that a self-potential exists, but that it varies with a cosmic pattern.[18]

Brown's discovery that these variations were registered on two different kinds of detectors helped to support his hypothesis that the sidereal effect was due to an energy flux, as opposed to simply a potential gradient. Whatever it was, this phenomenon apparently had the ability to input electric energy into certain dielectric materials, substances that he named *petrovoltaics*. Because his measurements indicated that this flux could penetrate even to subterranean vaults, he concluded that it might be reasonably identified with high-frequency gravity wave radiation. He found that, in addition to their DC voltage, petrovoltaics also generate alternating current (AC) electric noise, spanning a broad radio frequency band. He theorized that this AC component may arise from cosmic gravity waves that constantly pass through the substance and impart some of their energy to it. He speculated that the rock might act as a rectifier, converting a portion of these energy fluctuations into DC potential.

If electric energy is spontaneously generated in petrovoltaics, it is reasonable to expect that they would also be evolving heat. In fact, in the 1920s, the American inventor and industrialist Charles Brush took measurements on petrovoltaics and demonstrated that they spontaneously gave off heat even though they were not radioactive.[19] He reported his findings in a *Physical Review* paper titled "Retardation of Gravitational Acceleration and the Spontaneous Evolution of Heat in Complex Silicates, Lavas, and Clays." His calorimetric results were subsequently confirmed by Dr. Elmer Harrington, of the National Bureau of Standards.[20] Probably because it was not well understood, the phenomenon received little attention from the scientific community. If such heat evolution indeed exists, it is reasonable to speculate that a substantial portion of the geothermal flux originating from the Earth's crust arises in this fashion.

In 1974, Brown set up his automated recording equipment at the Haleakala Observatory on Maui for high-altitude observations (10,000 feet), and in 1975 he moved his laboratory to an underground vault at the University of Hawaii in Honolulu. Later, he also took measurements at the bottom of a 300-foot mine shaft in Berkeley, California. His collection of measuring instruments now included a sidereal electrometer, a dielectric resistance sensor, a petrovoltaic self-potential detector, and a "K-wave" detector. All the instruments registered variations that showed sidereal correlations. In this way, he established that this sidereal phenomenon influenced electrogravitic coupling in a bidirectional fashion. It affected both the *electrogravitic* conversion of electrostatic potential into gravitational force and the *gravito-electric* conversion of gravitational wave energy into electric power.

Brown's K-wave detector could measure very small changes in a capacitor's dielectric constant, thereby monitoring small changes in the local electric permittivity of space—the ability of space to store electric charge. A capacitor's electric permittivity—ε—is equal to its dielectric constant K times ε_0, the electric permittivity of matter-free space, that is, $\varepsilon = K\varepsilon_0$. The K-wave detector registered changes believed to be caused by slight variations in ε_0. Brown felt that long-term changes in ε_0 could account for historical variations in the measured value of the speed of light.

The circuits Brown used for the K-wave detector and the dielectric resistance detector are shown in figure 1.12. Another version of his K-wave detector used a spent nickel-cadmium battery cell in place of a high-K capacitor. Figure 1.13 presents portions of a nine-day strip chart recording the voltage (in millivolts) spontaneously generated by a piece of Koolau basalt in August 1978.[21] The voltage varied cyclically with time of day and reached a maximum at times when the galactic center reached the zenith. He also found that detectors separated by distances of up to eighty kilometers occasionally registered concurrent events, or "bursts," indicating that they had been triggered by a common external source.[22]

1.7 ■ THE PHILADELPHIA EXPERIMENT

Another interesting episode in Brown's career that should be mentioned, but for which documentation is very sparse and contradictory, concerns his work with the Navy on the Philadelphia Experiment. This was a highly classified research project reportedly conducted in the Philadelphia Navy Yard in October 1943 whose alleged objective was to render a naval vessel invisible both to radar and to the naked eye.

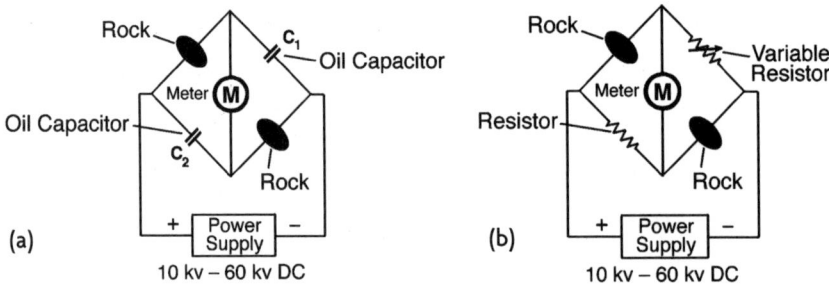

Figure 1.12. Bridge circuits that Brown used for his K-wave detector (a) and for his dielectric resistance detector (b). (Taken from entries in Thomas Townsend Brown's 1974 laboratory notebook)

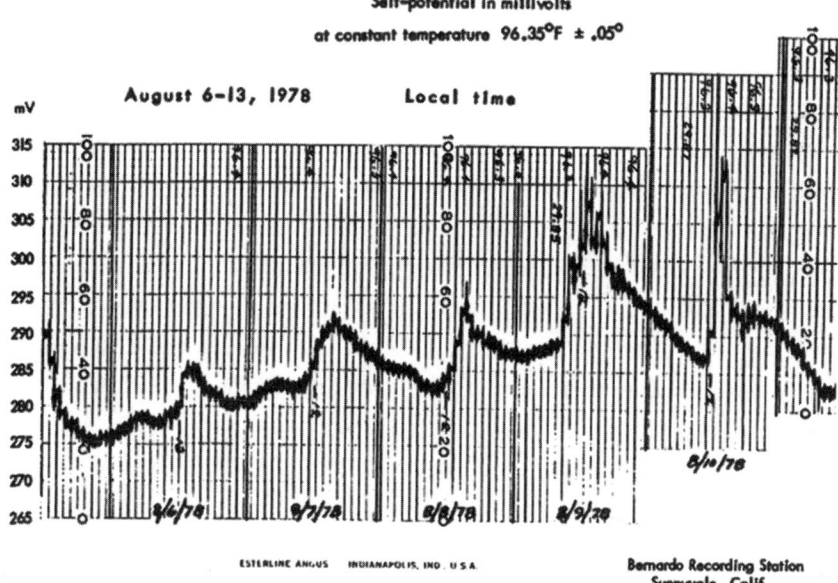

KOOLAU BASALT

Self-potential in millivolts
at constant temperature 96.35°F ± .05°

Figure 1.13. A portion of a nine-day strip chart recording of the voltage generated by a piece of Koolau basalt. Voltage maxima occur at times when the Galactic center reached its zenith. (Diagram courtesy of the Townsend Brown Family and Qualight, L.L.C.)

The list of scientists said to have worked on the project includes Albert Einstein, Vannevar Bush, John von Neumann, and Nikola Tesla. Before describing this further, it is worth reviewing what Brown was doing in the years leading up to the project.

Early in 1933, while working at the Naval Research Laboratory in Washington, D.C., Brown was given temporary leave to serve as a physicist on a geophysical expedition to the Caribbean sponsored by the Smithsonian Institute and financed by businessman Eldridge Johnson, cofounder of the Victor Talking Machine Company, which was the forerunner of RCA.[23] The Johnson-Smithsonian Expedition, which was conducted on board Johnson's immense yacht the *Caroline,* involved mapping the locations of underwater rifts.

However, there was much more to this expedition than just science. While on this cruise, Brown had the opportunity to meet Johnson and several of his associates, who included his wealthy business partner Leon Douglass and the British master spy William S. Stephenson, who, years later during World War II, earned the title "the man called Intrepid." In his Internet-published biography about Brown, Paul Schatzkin states that he learned from one of Brown's former close acquaintances that Johnson and Douglass were members of Stephenson's international intelligence network and that while on board the *Caroline,* Brown himself became recruited into its ranks.[24] Schatzkin dubbed this network the Caroline Group and said that it was to play a significant role in the course of Brown's life. Much of Schatzkin's inside information came from an individual he code-named Morgan, who at that time held a high-ranking position in one of the U.S. intelligence agencies and in earlier years had worked closely with Brown.

In the years that followed, Brown held a number of jobs. One particularly worth mentioning is his assignment in 1938 to serve as an assistant engineering officer on the maiden voyage of the USS *Nashville.* On its return trip from Europe, this ship ferried across the Atlantic $50 million in gold bullion that was being transferred from the Bank of England to the Chase Manhattan Bank in New York. While Brown was on that voyage, an electrogravitic research laboratory was established for him at the University of Pennsylvania. Schatzkin wrote that Johnson was involved in the construction of this laboratory, whose operation was funded from part of the money that the *Nashville* was transferring.[25]

In 1939, Brown left the University of Pennsylvania to work as a material and process engineer with the Glenn Martin Company in Baltimore, an early forerunner of the Lockheed Martin aerospace corporation. Shortly afterward, in 1940, the Navy called on him to head up a "mine sweeping research and development project" under the Bureau of Ships in Washington, D.C. William Moore wrote that Brown directed a staff of fifteen Ph.D.'s and was allotted a research budget of nearly $50 million for the project.[26] One might suspect that the funding money came from the very same stash that had been transferred to the Chase Manhattan two years earlier. This was a significant sum of money, about 5 percent of the U.S. Navy's 1940 budget! We are left to speculate whether the Caroline Group was somehow involved.

Whatever the case, this project must have been very important, and one wonders whether it was dealing with just "mine sweeping."

Following the attack on Pearl Harbor and the beginning of America's direct involvement in World War II, Brown was assigned to the Naval Operating Base in Norfolk, Virginia, as officer in charge of the Atlantic Fleet Radar Materiel School and Gyro-Compass School. In the summer of 1942, he was assigned to return to Philadelphia to disassemble the scientific equipment kept at his laboratory at the University of Pennsylvania and ship it to Norfolk. He continued his work there, at the Atlantic Fleet Radar School, until retiring from the Navy near the end of 1943.

From a brief entry Brown made in one of his autobiographies, we find that after his assignment to the Bureau of Ships in Washington, D.C., and before his assignment to the Atlantic Fleet Radar School, he was assigned to the Philadelphia Navy Yard as an assistant machinery superintendent for "outfitting new ships." In the autobiography, Brown wrote:

> My activities during the war were largely as follows:
>   1. Acoustic and Magnetic Mine Sweeping (Officer-in-Charge) Bureau of Ships, Washington, D.C.
>   2. Assistant Machinery Superintendent (outfitting new ships) Philadelphia Navy Yard
>   3. Naval Research Laboratory–Radar Refresher
>   4. Atlantic Fleet Radar School (Commanding Officer) Naval Operating Base, Norfolk, VA. Advanced teaching and writing of textbooks, Officer and Librarian.[27]

Curiously, the navy yard assignment as well as his "radar refresher" assignment at the Naval Research Laboratory in Washington are omitted in other biographies of Brown. His autobiography does not give dates for these assignments. However, his biography in *Who's Who in American Science* lists him as finishing his work at the Bureau of Ships in 1941.[28] Also, Moore's article lists Brown as beginning his work at the radar school shortly after the December 1941 bombing of Pearl Harbor, hence in 1942.[29] His assignment to the Philadelphia Navy Yard, then, would likely have been sometime during 1941. This would have placed him in the very location where the USS *Eldridge* DE 173 destroyer escort is said to have been outfitted in preparation for the Philadelphia Experiment and where the invisibility experiment was alleged to have been conducted in October 1943. Considering Brown's technical caliber as a research scientist, without further information one is left to wonder whether the nautical machinery that he was in charge of outfitting at the Philadelphia Navy Yard might have been equipment for a research experiment to be conducted aboard a ship, lending credence to claims that he had worked on the Philadelphia Experiment.

Later in his life, Brown was privately asked by family friend and business associate Josh Reynolds about his involvement in the Philadelphia Experiment. Brown answered that he "was not permitted to talk about that part of his work"; however, he did comment that "much of what has been written about the project is grossly exaggerated."[30] Here, he was probably referring to claims some have made that the ship had been made to travel through time or that it had teleported itself to Norfolk Harbor, where it was alleged to have reappeared for a few minutes before disappearing and reappearing once again in the Philadelphia Navy Yard. Yet the fact that he did not flatly deny his involvement in the project leads one to suspect that the rumors of his involvement are true.

Moore, coauthor of the book *The Philadelphia Experiment,* once asked Brown to edit a rough draft of an article he was writing on Brown's life. Moore had planted a paragraph describing a series of experiments, sponsored by the Navy, that were based on the effects and equipment later associated with the Philadelphia Experiment. He had done this intentionally to see Brown's reaction. Although Brown made other corrections and notes for changes to the manuscript, he allowed the entire test paragraph on the Philadelphia Experiment to remain intact. Thus, we are left to conjecture that tales of the existence of this project may be true and that Brown had somehow been involved in this project, although what his involvement was is open to speculation.

In their book *The Philadelphia Experiment,* Moore and Charles Berlitz cite letters attributed to a former sailor, Carlos Allende (a.k.a. Carl Allen), that suggest that the USS *Eldridge* was made invisible on

October 28, 1943, when it enveloped itself in a very strong magnetic field.[31] They said a large amount of electric power from onboard generators was used to resonantly excite large degaussing coils that were wrapped around the inside of the ship's steel hull. The resonant excitation would have set up a pulsating magnetic field, turning the ship into a giant electromagnet. This intense field was said to make the ship invisible both to radar and by sight!

According to Allende, the crew of the ship experienced physical and mental side effects so horrendous that the project was immediately terminated. He alleged that most of the crew were found to be violently sick after the field had been shut off, some were missing, and some had gone crazy. Most unusual, five men were found fused to the metal of the ship's structure, some crew members being stuck in steel bulkheads, others within the ship's deck, and another with the ship's railing stuck through his body. Allende also claimed that for a period of time, ranging from minutes to, in some cases, months, men would spontaneously become invisible and unable to move, speak, or interact with other people. Such people were said to have become "caught in the Flow" or "stuck in a freeze." Depending on the duration of the mishap, recovered victims were said to be left with symptoms ranging from psychological trauma to insanity. Allende maintained that those who lived were discharged from the Navy as "mentally unfit" for duty regardless of their condition.

Although it is difficult to sort out fact from fiction when trying to understand what had been done in the Philadelphia Experiment, laboratory research has shown that a metal object can be made radar invisible by high-intensity magnetic fields. At the 1994 Tesla Symposium in Colorado Springs, K. Corum, J. Corum, and J. Daum described an experiment in which they wound a high-amperage coil around a 2-inch-thick, 14-inch-diameter steel torus.[32] They found that when the coil was electrified with a sudden surge of current of several thousand amps or more discharged from a large high-voltage capacitor, the high-gauss magnetic field produced around the torus caused a fivefold reduction in radar reflection from the steel core. Some term this the Corum-Daum effect. The production of optical invisibility, however, has yet to be reported by scientists working outside of the classified world.

Electromagnetic wave experiments conducted by the independent researcher John Hutchison lend some credibility to the report that sailors had been found fused with the vessel's metal structure. Beginning in 1979, Hutchison experimented with high-voltage, high-frequency longitudinal wave emissions similar to those Tesla was producing. Employing a Van de Graaff generator and two or more Tesla coils, he was able to create wave interference zones in which a number of strange phenomena were observed. These included the fusion of dissimilar materials such as wood and metal, cold liquefaction or fragmentation of metal, invisibility, and levitation. Examples of metal splitting and fusion of dissimilar materials are shown in figures 1.14 and 1.15. In the fusion phenomenon, the substances do not dissociate; they retain their individual compositions. A piece of wood, for example, could sink into a metal bar with neither the wood nor the bar coming apart.

Interestingly, Brown's work on magnetic minesweeping would have made him a prime candidate for work on this version of the Philadelphia Experiment. In his autobiography, he describes how he had developed a new technique for blowing up magnetic mines—submerged explosive devices that are triggered when a steel-hulled vessel passes over them. A detector in a mine senses the temporary alteration in the Earth's mag-

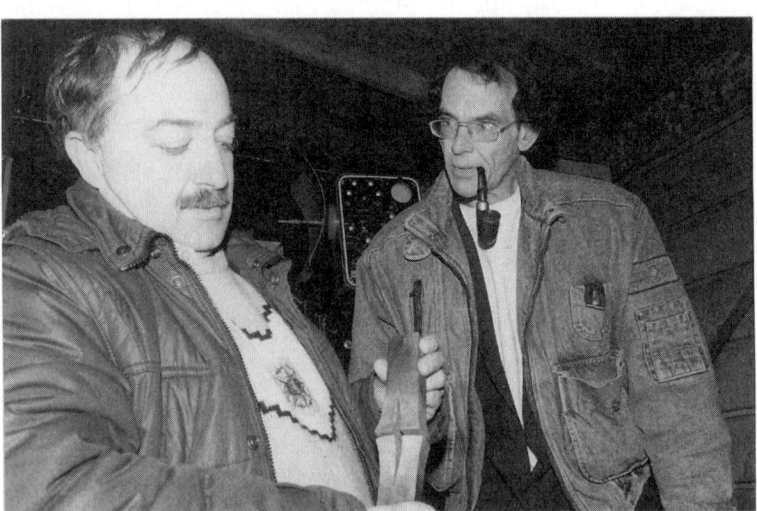

Figure 1.14. Professor Panos Pappas (left) holding a 2-inch-wide brass bar that was split by the Hutchison Effect. John Hutchison is shown standing to the right. (Photo courtesy of P. Pappas)

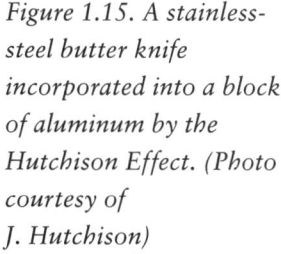

Figure 1.15. A stainless-steel butter knife incorporated into a block of aluminum by the Hutchison Effect. (Photo courtesy of J. Hutchison)

netic field intensity produced by the steel hull and detonates the mine's explosive. Brown had devised a method of exploding these mines by floating a loop of degaussing cable on the water's surface and passing 300 amperes of current through it, producing a magnetic field that triggered the mines to explode. The cable, which typically measured 3.5 inches in diameter, could easily carry a current of several hundred thousand amperes or more. Such a cable would have been ideal to generate an extremely high-intensity magnetic field around a ship. If so, Brown's work in Washington at the Bureau of Ships and later at the Philadelphia Navy Yard may have involved more than just research on magnetic minesweeping. The $50 million research project he was heading, which involved a team of fifteen Ph.D.'s and reportedly had occasional input from Einstein himself, was most likely directly connected with the fabled Philadelphia Experiment.

Another, very different, account of the story, presented by Gerry Vassilatos in his book *Lost Science,* claims that the cloaking effect was instead brought about by enveloping the ship in a very intense pulsing electrostatic field and makes no mention of magnetic fields.[33] Vassilatos's account is not as documented as Moore's, as he gives no indication of what sources he used for the rather detailed information he gives. One is left with the impression that portions of his story have been improvised. Vassilatos writes that the invisibility effect was first serendipitously noticed at a classified military arc-welding facility that had developed a new spot-welding technique for fabricating very durable armor-plated vessel hulls. The process employed a very intense, high-amperage discharge supplied from an immense bank of high-voltage capacitors. When the titanic, lightning-like discharge was applied to the hull, the resulting shock wave reportedly rocked the entire welding facility. Vassilatos writes that during the discharge, an optical blackout region was seen around the arc, and tools left in the vicinity of the discharge were displaced or found to have vanished. Scientists from the Naval Research Laboratory who were called in to investigate determined that the blackout was not a neural retinal bleaching phenomenon and that the tool disappearance was not due to thermal vaporization. Something far more exotic was going on. They eventually concluded that the momentary buildup of high electric field potentials in the vicinity of the arc in some unexplained manner induced a state of invisibility and even caused local dematerialization of objects. The project was code-named Project Rainbow, and Vassilatos says that at one point Brown was brought in to consult on the project.

Vassilatos writes that after conducting a series of further experiments, researchers devised an experiment that attempted to render an armored tank invisible. Capacitors of very large capacitance were arranged in a ring, and the tank was placed at the ring's center. The capacitors were oriented so that their plates were parallel to the circle's circumference, that is, with their polarization axes directed toward the circle's center. They were synchronously energized with high-voltage, high-amperage pulses conducted in phase along a spokelike array of cables extending from the center of the ring out to each capacitor. In this way, the capacitors acting together were able to build up a very high electric field potential, presumably with a negative potential in the ring's interior. Tesla had done years of research with high-voltage shock discharges, which could explain why he was allegedly called in to consult on the project.

According to Vassilatos, as a next step they scaled up the cloaking experiment to attempt to make an entire ship invisible. He says that they sought to control the effect by adjusting the electric field's intensity to a moderate level so that a state of invisibility might be produced without inducing complete dematerialization. He claims that Brown bowed out of the project prior to the test on the *Eldridge,* which reportedly ended in tragedy.

While the Navy claims that the story of the Philadelphia Experiment is entirely myth, Brown's hesitation to speak about the subject suggests that something very important and highly classified was going on in Philadelphia during his wartime service. One's suspicions are piqued about the significance of the whole affair because of the tremendous amount of disinformation that has apparently been circulated to purposefully cause confusion. Conflicts emerge even in Brown's own biographical records spanning this period. It is as if these years of Brown's life are shroudeded in a blurry haze. Conflicting accounts give the impression of there being two Townsend Browns, one account placing him at the naval base in Norfolk, Virginia, during 1942 and 1943, the other account having him working at Lockheed Vega Aircraft in Burbank, California, during this same period. This duplicity leaves us asking whether it had been Brown and not the *Eldridge* that had been teleported in space and time during that mysterious 1943 experiment.

According to the version that Moore published in 1978, Brown retired from the Navy in December 1943 after having suffered a nervous collapse.[34] He says that Brown subsequently took six months off to recover at his home following the recommendation of a team of naval physicians.

He began employment in June 1944 at the Advanced Projects Unit of Lockheed Vega Aircraft in Burbank. This was the forerunner of Lockheed's modern Skunk Works. We are led to believe that Moore's account should be accurate, because prior to its publication he gave Brown the opportunity to check over the draft of his article to make any necessary corrections. The Lockheed Vega employment date that Moore gives is consistent with that listed in the *Who's Who* biography published after Brown's death, which states that Brown was employed at Lockheed Vega as a radar consultant from 1944 to 1945.[35]

A. L. Kitselman, a mathematician who worked at this Lockheed facility, met Brown there and became his longtime friend. In an essay he wrote in 1962, Kitselman describes Brown as "a quiet, modest, retiring man—exactly the sort one expects to find in important research installations. He was a brilliant solver of engineering problems, and I soon found that he was more familiar with fundamental physical laws than anyone I had met. So many of us are strictly textbook scientists that it is stimulating to find someone who has first-hand knowledge."[36]

In this essay, Kitselman comments that Brown had previously suffered a collapse after working too long and too hard at the Norfolk radar school, was subsequently retired from the Navy, and then, after a six-month rest at home, came to work at Lockheed Vega. Hence Kitselman's account corroborates portions of Moore's story.

According to this timeline, Brown would have been working for the Navy during the critical period when the Philadelphia Experiment was conducted and would have had his nervous collapse around the time of the disastrous failure of this invisibility experiment. In fact, in their book *The Philadelphia Experiment,* Moore and Berlitz quote Riley Crabb, founder of Borderland Sciences Research Foundation, as saying that the cause of Brown's breakdown was directly related to the Philadelphia Experiment. Crabb noted that if such a disaster had happened to the crew of the ship, it is not too difficult to imagine the mental pressures that those in charge would have experienced.

Schatzkin has come to entirely different conclusions about Brown's whereabouts during this key period. At Morgan's suggestion, he obtained from the Navy a copy of Brown's resignation letter, which is dated September 30, 1942, and which states, "I herewith submit my resignation from the Navy for the good of the naval service in order to escape trial by General Court Martial."[37] If we are to believe this date, this was to have occurred just two months after Brown had shipped his equipment from the University of Pennsylvania to Norfolk.

Schatzkin also obtained an official copy of Brown's Navy fitness report dated October 5, 1942. Describing this report, he wrote:

> The final fitness report is almost completely blank. Instead of the usual details, the page is struck through with a single pen-stroke, above which is hand-written "See remarks." And on the second page, in the "remarks" section that in previous reports had displayed so many glowing assessments of Lieutenant Brown's character and service, Captain Hinkamp writes, "In view of the circumstances under which this officer was detached, I desire to make no comment."[38]

We know something is amiss in the Navy's records because they contain no reference to Brown's assignment in 1942 to the Atlantic Fleet radar schools in Norfolk. However, trusting that the naval records or Brown's discharge papers had not been altered by covert operatives in the interest of protecting any top-secret naval research projects from exposure, Schatzkin accepted October 5, 1942, as the date of Brown's detachment. He then suggested that within two weeks of the date Brown left the Navy, he began working at Lockheed Vega. Schatzkin proposed that Brown had neither a nervous collapse following his discharge nor a subsequent six-month recuperation period. Schatzkin's version of Brown's history then conflicts with both that given by Moore and that given by Kitselman, both of whose accounts he maintains are seriously flawed. The suggestion that Kitselman's account might be flawed, however, comes as somewhat of a surprise, seeing as he was one of Brown's close friends. In writing his essay, he should have had easy access to input from Brown as well as an interest to ensure that he got his facts straight about Brown's departure from the Navy. Also, Brown himself had checked over Moore's story prior to its publication, so if there was such a major error as the date and circumstances of his departure from the Navy, why did Brown not catch it? Considering that there is no record of Brown having expressed any doubts about the accuracy of Moore's or Kitselman's account, one is surprised by the allegation that they were in error.

Furthermore, there is the inconsistency of the date when Brown began working at Lockheed Vega. Schatzkin places his arrival at the end of October 1942, while Moore states the arrival date was more than one and a half years later, in June 1944—a start date that is also corroborated by the account given in the *Who's Who* biography. So which version is correct, the revised timeline based on Navy records or the preexisting biographic timeline that was developed with Brown's full knowledge? Unfortunately, Brown is no longer around to comment, having passed away in 1985.

To support his argument for Brown's early departure, Schatzkin cites a Federal Bureau of Investigation (FBI) report that claims to have been filed in March 1943. This report states that by that date, Brown had resigned from the Navy and returned to his home in Los Angeles, as told by an anonymous informant (name blacked out). But should this anonymous informant be relied upon? Schatzkin himself admits that much of the information the report provides about Brown is inaccurate and contradictory. The filing date given on the report appears to be among the fabrications. The report states, "He [Brown] had his own laboratory and had purchased equipment from his own funds for use in his experimental work, and this equipment was taken by Subject when he was detached from the Fleet Service School." This equipment included gravito-electric sensor equipment, which was among the apparatus that had earlier been transported from Brown's University of Pennsylvania laboratory to Norfolk. According to Schatzkin's revised timeline, this equipment would have then been transported from Norfolk to Los Angeles around October 1942, when he claims that Brown was discharged.

However, the revised timeline does not jibe well with Brown's account of the dates and locations at which he was conducting gravito-electric measurements. In his March 1975 paper titled "Anomalous Diurnal and Secular Variations in the Self-Potential of Certain Rocks," Brown discusses dates and locations at which he conducted gravito-electric measurements, mentioning his work at the Naval Research Laboratory (1931–1933) and his research at the University of Pennsylvania (1939). Then he writes, "The investigation was interrupted by World War II but was resumed in 1944 in California by the Townsend Brown Foundation (an Ohio non-profit corporation) and was carried forward in two locations in especially constructed shielded rooms at constant temperature."[39]

If we accept the traditional timeline in which Brown is discharged from the Navy in December 1943 and transports his equipment to California around that same time, then his stated 1944 date for resuming his gravito-electric measurements in California makes sense. This implies that he wasted no time in setting up his equipment to start collecting data once again. On the other hand, if we accept the Navy-FBI timeline that has Brown being discharged in October 1942, we would have to conclude that he shipped his equipment to California at the end of 1942 and left it sitting boxed up for more than a year before setting it up. However, it seems unlikely that Brown would have tolerated having his detectors "off the air" for such a long time period. Could it be that the FBI report was actually filed in 1944 and its date was at some later point changed to 1943 in an effort to rewrite Brown's official history?

To support his 1942 date for Brown's Navy discharge, Schatzkin refers to a bound laboratory notebook that he believes Brown had used while at Lockheed Vega.[40] The ledger's notes are written in Brown's handwriting and contain occasional dates that also appear in Brown's handwriting, the oldest date near the beginning of the book being December 1, 1942, and the most recent date near the end of the book being May 2, 1944. The notebook's cover page is neatly hand printed and reads:*

T. T. BROWN

VEGA AIRCRAFT CORP.

BURBANK, CALIF.

NOTES

We are left to consider the possibility that the notebook contains lecture notes that Brown began writing while teaching at the Atlantic Fleet schools in Norfolk. The last dated entry in the notebook would have been made after Brown had left the Navy and had moved to California, prior to going to work at Lockheed. He may have labeled the notebook as "Vega Aircraft Corp." because he wanted his notes with him at his new job, or he may have purposely mislabeled the notebook in this way so that naval intelligence would not squirrel it away in some classified storage room.

If we instead accept that Brown actually wrote these notes while he was at Lockheed Vega and that he began working there as early as October 1942, then we are confronted with the inconsistency of this date with those given in Brown's autobiographies and with the question of why his gravito-electric sensor equipment would supposedly have been stored unused for more than a year. Also, with this early-departure scenario, it is difficult to understand why Brown wished to resign from the Navy at the height of World War II, just nine months after Japan

---

*Information from photocopied sample pages sent to me courtesy of Paul Schatzkin. Being hand printed, it is more difficult to tell whether the cover page is also in Brown's handwriting; however, members of his family affirm that it is.

had bombed Pearl Harbor and at a time when his Navy career looked so promising. According to the FBI report, Brown was "reported to know more about Radar detection than any individual in the U.S. Navy." So why would the Navy let him go at such a crucial time of need? If, on the other hand, Brown's decision to leave the Navy arose as a result of a nervous breakdown brought about by the great weight of guilt he felt from being associated with a project that had suffered an immensely tragic outcome, as Moore and Vassilatos suggest, then his departure at the later date of December 1943 becomes more understandable. The Navy administrators who had knowledge of this classified project and who themselves shared the guilt of its outcome would have sympathized with Brown's wish for departure and released him from service, even knowing how indispensable he was.

According to Schatzkin, "Starting in the fall of 1942 there is virtually no documentation available that might shed some light on just what Brown was doing during those crucial years."[41] He notes that the Brown family files are devoid of any correspondence or documentation from roughly that time until the end of World War II and that they have very little information about his activities at Lockheed Vega.

So considering the absence of information from both the Navy records and the Brown family files, we are left only to speculate. Had some military intelligence organization gone out of its way to ensure that any record of Brown's activities during this period was either erased or classified to keep a tight cover on Brown's wartime research activities? Despite its official denial, did the Navy conduct a highly secret project on ship invisibility and was Brown involved in it? Perhaps the adage "Where there's smoke, there's fire" applies here. One suspects that something very strange and clandestine was under way in the Philadelphia–D.C.–Norfolk area during the 1942 to 1943 time period.

In July 1946, the *Eldridge* was decommissioned and placed in the Reserve Fleet. In 1951, the United States transferred her to the Greek navy, in which she served as the HS *Leon* until the 1990s. One Greek engineering professor related that he formerly served on the *Leon* as a naval officer specializing in electrical engineering.[42] While on board, he noted several odd things about the ship. One was that he saw numerous remnants on the inside of its hull of heavy-duty cables that once ran along the length of the ship. These were in the form of insulated metal bars measuring 10 to 15 centimeters in width that had been cut in between their points of attachment to the hull. Other large-diameter cables were also present fully intact that were presumably part of the electric wiring for the ship's propulsion system. The *Eldridge* was a Cannon class electric drive ship, meaning that instead of having a shaft running from its engine directly to its propeller, as most ships do, it had a diesel-powered electrical generator whose power was conveyed through heavy-duty cables to a huge electric motor at the ship's stern that drove the propeller. The *Eldridge*'s ability to produce large amounts of electric power with an onboard generator would have made it ideal to use in conducting the Philadelphia Experiment.

The other unusual thing that the professor noted was that one room adjacent to the ship's hull was barred from access, its hatch having been welded shut. The commanding officer had instructed the ship's crew that it was forbidden for anyone to try to enter the sealed room. What this forbidden zone hid will perhaps never be known, for the ship was decommissioned and sold as scrap sometime after 1992.

2

BEYOND ROCKET PROPULSION

2.1 ▪ BROWN'S ELECTRIFIED FLYING DISCS

During the years following World War II, Brown continued to improve his gravitator device in his spare time, financing his efforts through the Townsend Brown Foundation. By 1950 he had built a test apparatus to demonstrate the electrogravitic propulsion concept in a pair of disc airfoils. He set a 6-foot-long horizontal beam on a pivot so that it could rotate about its midpoint, and from each end of the beam he suspended two lightweight saucer discs by means of 7-foot-long tethers (figure

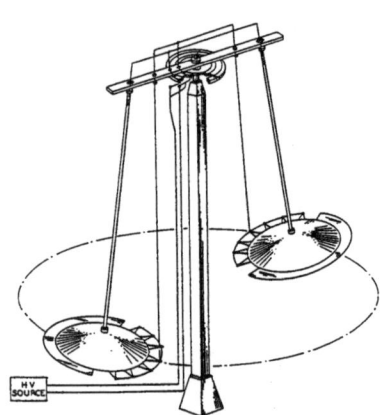

Figure 2.1. Thomas Townsend Brown's flying disc setup.

Figure 2.2. Thomas Townsend Brown's 2-foot-diameter experimental disc airfoil. (From Project Winterhaven, plate 1; photo courtesy of the Townsend Brown Family and Qualight, L.L.C.)

2.1). When the saucers were in flight, rotating tethers extended sideways and expanded the diameter of the flight course to as much as 20 feet. In one version, each disc was made of two curved aluminum shells, measuring 1.5 feet in diameter, fixed on either side of a 2-foot-diameter Plexiglas sheet (figure 2.2).[1] High-voltage power of up to 50,000 volts was supplied through feed wires to positively charge a fine outboard wire running along each disc's leading edge and to negatively charge the aluminum disc body. When electrified with approximately 50 watts of this high-tension power, the discs traveled around their 20-foot-diameter course at speeds of up to twelve miles per hour.[2,3]

The wire electrodes ionized the surrounding air, forming a cloud of positive ions around the leading wire and a cloud of negative ions around the disc body. Although ions would continuously leave these clouds as a result of being attracted to the oppositely charged electrodes, the electrodes would resupply ions at a sufficiently fast rate so as to maintain a positive-ion space charge at the front of the disc and a negative-ion space charge on the disc body (see figure 2.3).

As to how the disc generates its propulsive force, two possibilities present themselves. One is that the ion clouds it emits produce electrostatic fields that act on charges attached to the disc's leading-edge wire and to its main body, producing a net forward thrust. The other possibility is that an electrogravitic thrust may be present whereby the

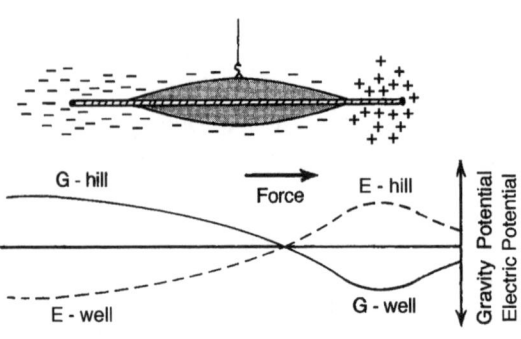

G - hill

Force

E - hill

E - well

G - well

Gravity Potential
Electric Potential

Figure 2.3. A side view of
one of Thomas Townsend
Brown's flying discs,
as normally energized,
showing the location
of its ion-space charges
and induced gravity field
gradient. (P. LaViolette,
© 1994)

positive- and negative-ion clouds would create, respectively, a gravity potential well and a gravity potential hill in their vicinity. As new positive charges are continuously added to the cloud, they replace charges that leave the cloud through attraction to the disc's negative pole. As a result, the cloud will maintain a moderately deep gravity well at its bow through a kind of dynamic equilibrium. The same will hold for the disc's rearward negative charges. Despite the mobility of the individual negative ions, the negative-ion cloud as a whole will persist and create a net gravity hill. Consequently, the gravity potential gradient established across the disc's body between this hill and the well propels the disc forward in the direction of its positive-ion cloud.

By accumulating charges in the air in the form of fore and aft ion clouds, large quantities of charge may build up, comparable to the quantity of charge on the plates of a high-K dielectric capacitor. But because these charges are freshly created, there is little time for them to polarize the ambient air. Furthermore, due to the disc's forward motion, the air dielectric around the disc is continuously replaced by new, unpolarized air, and this also contributes to maintaining the air dielectric in a relatively unpolarized state. Consequently, the electric and gravity potential fields are able to extend between the oppositely charged fore and aft clouds unopposed by any electric dipole moment in the intervening air. Hence a substantial gravity field gradient could span the disc and exert a maximal forward thrust.

As the disc moves forward, its associated positive- and negative-ion clouds also move forward, transporting their generated electrostatic and gravity field gradients along with them. Consequently, each disc rides its advancing wave much like a surfer riding an ocean wave. Dr. Mason Rose, one of Townsend's colleagues, describes the disc's gravitic principle of operation:

> The saucers made by Brown have no propellers, no jets, no moving parts at all. They create a modification of the gravitational field around themselves, which is analogous to putting them on the incline of a hill. They act like a surfboard on a wave. . . . The electro-gravitational saucer creates its own "hill," which is a local distortion of the gravitational field, then it takes this "hill" with it in any chosen direction and at any rate.[4]

A full-scale version of Brown's vehicle was thought to be able to accelerate to thousands of miles per hour, change direction, or stop merely by altering the intensity, polarity, and direction of its electric charge. Because the wavelike distortion of the local gravitational field would pull with an equal force on all particles of matter, the ship, its occupants, and its load would all respond equally to these maneuvers. The occupants would feel no stress at all, no matter how sharp the turn or how great the acceleration. A turbo-jet airplane, by comparison, must produce a twentyfold increase in thrust just to attain a twofold gain in speed. Whereas jets and rockets attempt to combat the force of gravity through the application of opposed brute force, electrogravitics instead attempts to directly control gravity so that this longtime adversary is made to work for the craft rather than against it.

Partly with the help of his friend Kitselman, who was then teaching calculus in Pearl Harbor, Brown's discs came to the attention of Admiral Arthur Radford, commander in chief of the U.S. Pacific Fleet at the Pearl Harbor Navy Yard. In 1950, Brown was hired as a consulting physicist to stage a demonstration. Nothing immediately came of this. However, two years later, on March 21, 1952, Brown was visited at his Los Angeles laboratory by Vic Bertrandias, a well-connected Air Force major general. He dropped in unexpectedly, just when Brown was about to demonstrate his flying discs to a group of colleagues. Once there, Bertrandias demanded that he be included in the demonstration. Having formerly served as vice president of Douglas Aircraft, he was well informed on the state of the art in aviation technology and knew that Brown's discs could have important military applications. Shaken by what he saw, Bertrandias urgently telephoned Lieutenant General H. A. Craig the following morning to voice his concerns. An excerpt from a declassified transcript of their conversation reads as follows:

> Bertrandias: the thing frightened me—for the fact that it is being held or conducted by a private group. I was in there from about 1:30 until 5:00 in the afternoon and I saw these two models that fly and the thing has such a terrific impact that I thought we ought to find out something about it—who the people are and whether the thing is legitimate . . . if it ever gets away, I say it is in the stage in which the atomic development was in the early days.
>
> Craig: I see.
>
> Bertrandias: It was quite frightening. I made the inquiry whether the Air Force or the Navy knew anything about it and I was told—no. But I tell you, after hearing it and all the other things that I had heard, I was quite concerned about it. . . . I am of the opinion that if all I heard the other day—if it ever comes true, and somebody occupies space with that instrument, it is a bad deal for somebody.
>
> Craig: Well, we will look into it, Vic.

Craig subsequently initiated a background check on the Townsend Brown Foundation.

Bertrandias was also a close friend of General Albert Boyd, director of Air Force Systems Command at Wright Air Development Center. It was under Boyd that Air Force Systems Command carried out most of its early, super-secret research projects on antigravity propulsion.[5] Brown's work may have been encroaching into an area in which the Air Force had established a substantial lead.

Perhaps Brown sensed Bertrandias's fearful reaction and was concerned that he might initiate formal military classification of Brown's electrogravitic work, for just two weeks after Bertrandias's visit, Brown and his two associates, Mason Rose and Bradford Shank, held a press conference to publicize the fantastic possibilities of this electrogravitic propulsion technology. In this way, they got the word out before things got too hushed up. Reporters from the *Los Angeles Times* were invited to view Brown's flying discs in operation and had a chance to read a paper prepared by Rose that explained the Biefeld-Brown antigravity effect and how it could be used to propel a full-scale antigravity spacecraft. The next day the *Times* carried a story about Brown's discs and how flying saucers (also popularly known as UFOs, short for "unidentified or unconventional flying objects") might function on a similar principle.[6] It quoted Rose as saying that details about Brown's work had been given to some Navy admirals and that there was military interest, although no censorship had yet been imposed. Like the Air Force, the Navy had an active interest in advanced aviation technology.

About two months after the L.A. press conference, in June 1952, the Office of Naval Research (ONR) sent Will Cady to investigate a number of Brown's inventions, including his flying discs. The ONR data indicate that Cady witnessed a pair of 1.5-foot-diameter discs achieve a top speed of three miles per hour with a propulsion efficiency of 1.5 percent while drawing 15 watts of power at 47 kilovolts.[7] This was about one-fourth the speed and efficiency obtained with the 2-foot-diameter model. Did Brown stage this more modest performance with the intention to reveal just enough to get the military interested but not enough to make the demonstration so astounding that they might demand classification of his work? One alternative suggested by Paul Schatzkin is that there had been a security breach during Brown's Pearl Harbor demonstration and that Brown had been asked to purposely downplay the performance of his device in order to mislead foreign intelligence agents into thinking that his invention was not worth pursuing.[8]

Presumably, Cady was unable to see the discs perform at higher voltages because of the limitations of the power supply that Brown chose for the occasion. That is, the ONR data indicate that the output voltage progressively leveled off to 47 kilovolts as the control panel voltage dial was turned up to increasingly higher settings. This indicates that the 0.7 milliamp that the test rig was drawing was more current than the high-voltage power supply was designed to provide.* For his own research,

---

*The power supply's upper-limit current would be determined by the impedance of the secondary winding of its high-voltage transformer.

Brown probably used a transformer that had a slightly higher current rating, perhaps 2 milliamps. Cady conducted a test in which he removed one of the saucers from its carousel and suspended it from the ceiling to measure its static propelling thrust at various applied voltages. At 47 kilovolts, he observed that the discs delivered only 8 grams of thrust.

Cady concluded that the technology was impractical for aviation because the discs were propelled with an efficiency over an order of magnitude less than the efficiency of a jet engine. He had failed to realize that the trends in his own data showed that the speed and propulsion efficiency of the discs increased exponentially with increasing voltage and that he had been observing their behavior in a very unfavorable voltage range. A logarithmic plot of the ONR data (see figure 2.4) reveals that above 38 kilovolts, the velocity of the discs increased according to the 5.5 power of voltage and that their propulsion efficiency increased according to the 4.5 power of voltage.[9] These projections may be somewhat optimistic, as most of Brown's writings state that thrust increases according to the second or third power of voltage. Nevertheless, enormous speeds and efficiencies would undoubtedly have been attained at higher voltages.

Cady maintained that it was unnecessary to introduce exotic ideas such as electrically induced gravity fields because the behavior of the discs could be explained entirely in terms of the conventional ion-wind effect. That is, he believed that the discs obtained their thrust because ionized particles impacting the disc electrodes imparted more of their momentum in the forward direction than the reverse direction. On the contrary, although ion-wind forces would have been present, such forces would have been too small to account for the thrust. Furthermore, vacuum chamber tests that Brown later carried out on electrostatically charged rotors and saucers showed that thrust persisted even in the absence of ion discharge.

Cady also suggested that the discs may have been propelled forward by unbalanced electrostatic forces operating between the discharged ions and the disc that was discharging them. This is a more likely possibility than ion wind. For example, positive ions emitted by the leading wire electrode would move toward the negatively charged disc body, setting up a positive-ion space charge behind the wire and ahead of the

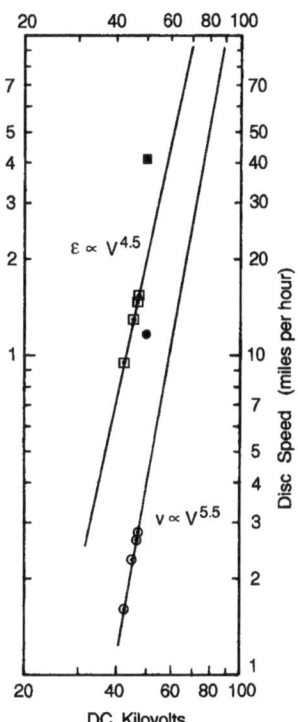

Figure 2.4. A logarithmic plot showing how the velocity (right line) and efficiency (left line) of Thomas Townsend Brown's electrogravitic discs increased with voltage. Empty squares and circles indicate the ONR measurements of the 1.5-foot-diameter discs, while solid squares and circles indicate the performance of an improved 2-foot-diameter model. (P. LaViolette, © 1997)

disc (see figure 2.3). These charges would repel forward the positively charged wire and attract forward the negatively charged disc body. As the saucer's speed increases, the airflow would assist in displacing the positive ions behind the wire, thereby improving the forward propulsive force. Also, the positive-ion wind and the airflow passing the disc would blow the negative ions toward the rear of the disc, and as a result, their space charge would electrostatically repel forward the negatively charged rear wire and disc body. As a result, both the negative and positive ions would work together to create a forward thrust on the saucer.

Brown referred to a mass effect (gravitational force effect) operating in the electrokinetic movement of massive high-K dielectrics but did not similarly report a mass effect operating in the case of his electrokinetic disc experiments. Thus it is not clear how much of the thrust he was attributing to gravitic forces and how much to electrostatic forces. Nevertheless, his research colleagues did seem to think that a new electrogravitic principle was needed to account for the propulsion. In his 1952 write-up, Rose stated that "anyone wanting to understand electrogravitation and its application to astronautics must dismiss the principles of electromagnetics in order to grasp the essentially different principles of electrogravitation. . . . Electrogravitation must be understood as an entirely new field of scientific investigation and technical development."[10]

As we can see from entries he made in 1943 in one of his laboratory notebooks, Brown was exploring an "ether" theory interpretation of the electrogravitic phenomenon, one that has many similarities to subquantum kinetics. More will be said about this in chapter 4.

The ONR researcher's skeptical reaction was typical of individuals who were used to thinking in conventional scientific terms. In one of his articles on Brown, the journalist Gaston Burridge wrote that many scientists and engineers had watched the discs fly and most of them concluded that the discs were propelled by the well-known "electric wind" phenomenon and not by some new principle of physics. One engineer blurted out to him, "The whole thing is so screwball I don't want to even talk about it!" Other engineers reportedly objected to the lack of mathematical substantiation. Burridge explained, "To engineers and scientists one equation is worth a thousand words!" But even an equation is of little use unless it has values assigned to at least some of its main parts. When these were not forthcoming, from a technical point of view, it appeared Brown was walking on straw legs.[11]

2.2 ■ THE SECOND PEARL HARBOR DISC DEMONSTRATION

Some years later, around 1953 or 1954, in the hope of renewing the Navy's interest, Brown again staged a demonstration at Pearl Harbor for a number of admirals. This time his demonstration was on a much larger scale. From a gymnasium ceiling, at a height of 50 feet, he suspended a revolving horizontal beam that tethered a pair of 3-foot-diameter discs (see figure 2.5). Powered by a potential of 150 kilovolts, the discs flew around a 50-foot-diameter course at such an impressive speed that the subject became highly classified. The speed may have been in excess of 100 miles per hour, because the May 1956 issue of the Swiss aeronau-

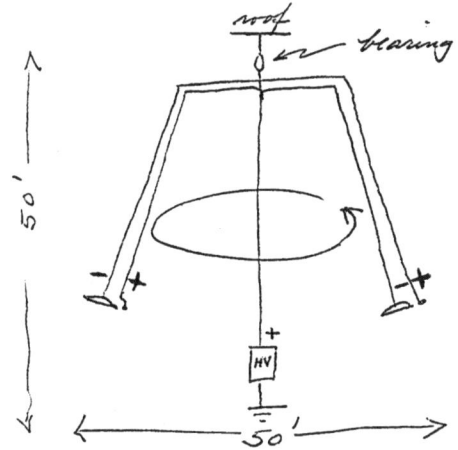

Figure 2.5. Sketch made by Thomas Townsend Brown showing the test setup used for demonstrating his 3-foot-diameter flying discs. (From Brown's November 1, 1971, letter to T. Turman; see appendix A)

tics magazine *Interavia* stated that the discs were capable of attaining speeds of several hundred miles per hour when charged with several hundred kilovolts![12] Such high velocities are not surprising considering that the ONR test data show that the speed of the discs increased exponentially with voltage.

Brown used a different disc design for this later demonstration. During a telephone conversation he had in the early 1970s with electrical engineer Tom Turman, Brown disclosed that the airfoil disc design depicted in his 1960 patent (no. 2,949,550) was an inferior one. The cross-sectional view presented in that patent shows the spun aluminum discs having a knife-thin edge at their periphery (as shown in figure 2.3). The discs used in the 1952 ONR test were of a similar design. On the other hand, the discs that Brown flew in his gymnasium demonstration had a blunt profile, as shown in figures 2.6 and 2.7. This design consisted of two spun aluminum discs cupped on either side of a Plexiglas sheet, but the upper disc had a "triarcuate" cross-sectional profile—a convex central bulge that turned concave farther out and that finally terminated in a convexly curved rim with a radius of curvature of ½ inch or more. The outer rim of the lower half of the disc had a flat profile, but its outer edge was curved to make a smooth transition to the edge of the upper disc.

Also, the leading-edge electrode used in the disc flown in Brown's gymnasium demonstration was of much smaller diameter. In a letter he wrote to Turman in 1971, Brown noted in a sketch that he used an

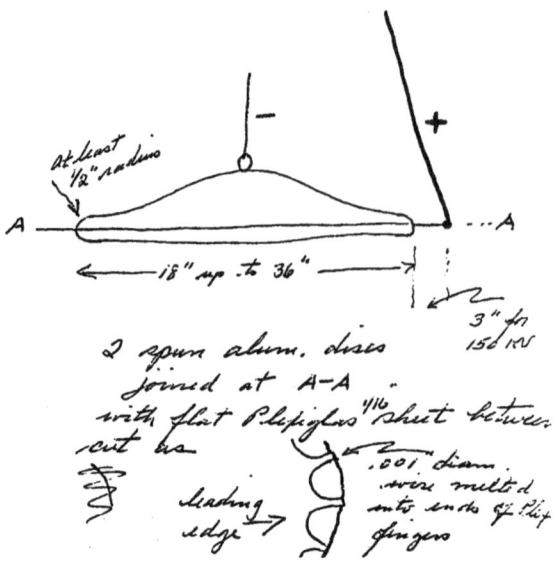

Figure 2.6. Sketch made by Thomas Townsend Brown showing the design of the 3-foot-diameter disc airfoils he demonstrated to the military. (From Brown's November 1, 1971, letter to T. Turman)

electrode that had a diameter of only 1 mil (0.001 inch). This is five times smaller in diameter than the wire he used on the discs flown in his ONR test in Los Angeles. Moreover, it is far smaller than the diameter he had specified in his 1960 patent. His patent states that saucers designed to be energized at voltages greater than 125 kilovolts would preferably have leading-edge electrodes of large cross-section made of rods or hollow pipes having diameters measuring from ¼ to ½ inch (e.g., 250 to 500 mils) to ensure that their surface potential gradient was below the threshold required to produce a visible corona. He maintained that energy losses associated with coronal ionization reduced the achievable thrust, but, as he acknowledged in his 1971 letter, the design specified in the patent was inferior to what he used in his Pearl Harbor gymnasium demonstration. The leading-edge electrodes of the discs he flew in that demonstration would have had a much steeper field gradient at their surface, which would have allowed them to emit ions more effectively.

Burridge has commented that the discs emanated a slight humming sound as they flew.[13] This implies that Brown may have been applying a nonreversing high-voltage AC potential across his disc that, on average, established a DC potential across its electrodes. He may have used a rectifier bridge circuit to convert the 60-cycle AC output from

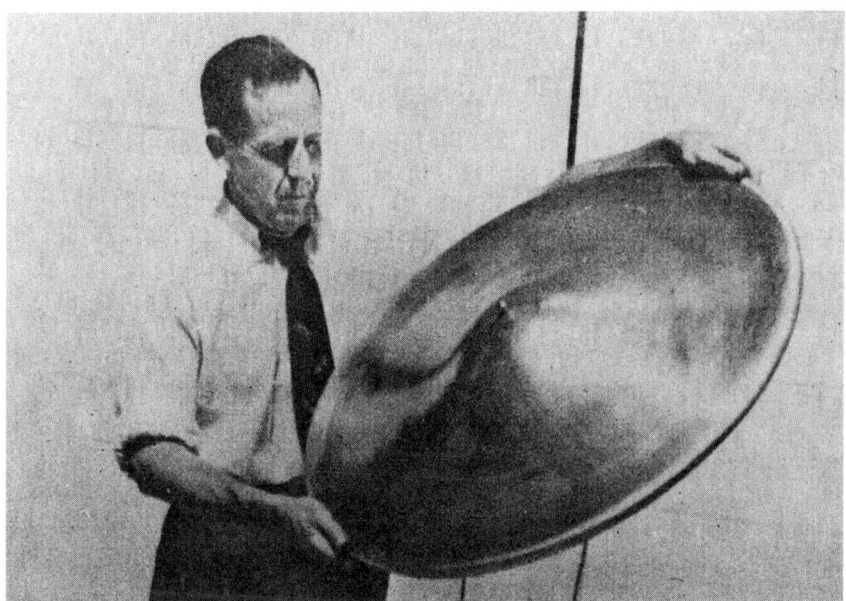

Figure 2.7. Thomas Townsend Brown holding one of his 3-foot-diameter discs, which he referred to as an experimental triarcuate ballistic electrode. (Courtesy of the Townsend Brown Family and Qualight, L.L.C.)

his high-voltage transformer into a series of unfiltered half-wave voltage cycles oscillating at 120 hertz. His use of a blunt profile for the edge of the negatively charged disc body would have helped to suppress negative-ion coronal discharge toward the front of the craft. This would especially have become an important issue in this gymnasium demonstration because he flew his discs at much higher voltages than used in flying his earlier model. The asymmetrical profile of his saucer, with its curved upper surface and flat lower surface, would also have been beneficial since airflow over the surfaces would have given the saucer aerodynamic lift during flight.

After his Pearl Harbor demonstration, Brown traveled to the mainland. Upon returning to Hawaii, he found that his room had been broken into and that some government agency had confiscated his models and notebooks and sealed his laboratory.[14] A day later, the Navy informed him that they had his notebooks and that he could have them back. A few days after that, they said that they were not interested in his work. They claimed that the discs must be powered by ion wind and, hence, that they would not work in outer space.[15] So here we find that the Navy had the opportunity to make changes, in the interest of national security, to any of Brown's laboratory notebooks, including his Vega notebook.

2.3 ▪ PROJECT WINTERHAVEN

The negative evaluation that came out of the 1952 ONR investigation of Brown's electrokinetic discs temporarily slowed down the Pentagon's endorsement of his work, but it did not halt the eventual implementation of his electrogravitic technology. In an effort to secure government funding, Brown wrote up a proposal in 1952 urging the Navy to initiate a highly secret project to develop a manned flying saucer as the basis of an interceptor aircraft with Mach 3 capability and proposed that this might follow along the same lines as the Manhattan District Project, which developed the atomic bomb at the end of World War II. This confidential January 1953 submittal was code-named Project Winterhaven.[16] Extrapolating the numbers from the performance charts for Brown's laboratory-model flying discs, Project Winterhaven estimated that larger discs operating at 5 million volts, rather than 50,000, should be able to develop speeds of 1,150 miles per hour (Mach 1.5) in the presence of atmospheric resistance and in excess of 1,800 miles per hour (Mach 2.5) in the upper atmosphere.

This rather conservative speed estimate was based on the assumption that disc speed extrapolated linearly with voltage, when in fact the evidence suggested a nonlinear relationship in which disc speed would rise exponentially with applied voltage. Thus, considering that Brown's 1½-foot-diameter discs had achieved speeds of twelve miles per hour when energized at 50 kilovolts, the report conservatively estimated that larger versions should be able to achieve speeds at a hundred times greater when energized at 5,000 kilovolts. The Pearl Harbor Navy Yard demonstration, which Brown performed shortly after this proposal was submitted and was intended as a demonstration of the aviation technology he was proposing in Project Winterhaven, indicated that speeds far higher than this would be possible. Because his 3-foot-diameter airfoils were shown to achieve speeds of several hundred miles per hour when energized at just 150 kilovolts, even a simple linear voltage-speed extrapolation would indicate that speeds of Mach 13 (10,000 miles per hour) would have been more reasonable for a 5-megavolt disc. However, since the jet speed record at the beginning of 1953 was only Mach 1.88, Brown could guarantee Navy interest even with his more conservative speed estimates.

Project Winterhaven proposed a five- to ten-year-long research and development (R&D, or RAND) program that was to be carried out in stages. Beginning with 2-foot-diameter discs powered at 50 kilovolts, it would proceed to 4-foot-diameter discs powered at 150 kilovolts, and finally to a 10-foot-diameter disc powered at 500 kilovolts. The proposal also suggested making a 10-foot demonstration model capable of vertical levitation as well as horizontal thrust.

The aircraft that Brown had in mind for military development probably looked similar to the version described in his U.S. patent 3,022,430, which was filed in July 1957 (figure 2.8).[17] Like his small-scale flying disc models, this craft produced a cloud of positive ions at its bow and a cloud of negative ions at its stern. Brown did not discuss gravity field effects in his patents, probably because he felt that such unconventional concepts might jeopardize a patent's ultimate acceptance. So although his patent alludes to ion thrust as being the craft's means of propulsion,

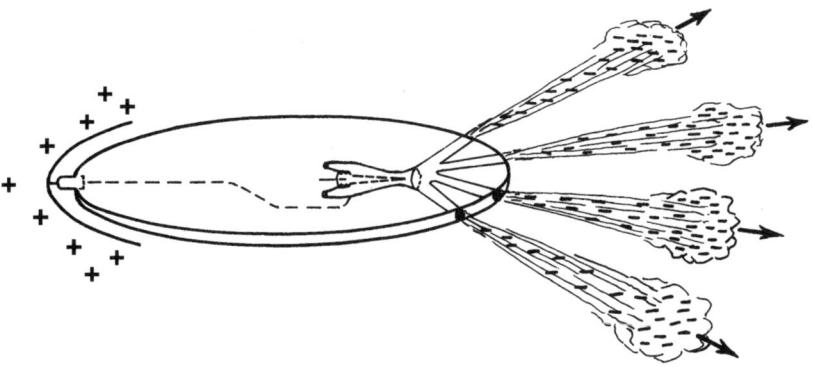

Figure 2.8. The flying disc Thomas Townsend Brown proposed for development under Project Winterhaven may have looked like this. (After Brown, U.S. patent 3,022,430, figure 1)

on the basis of his own research Brown was convinced that the propulsion also involved electrogravitic effects.

In parallel with this effort to develop an electrogravitic craft, Project Winterhaven also planned to investigate various methods to generate the high voltages required on board the aerospace vehicle. The discs that Brown had flown in his carousel demonstration were energized from a heavy, high-voltage laboratory transformer powered by wall current. A full-size airborne craft, however, would need to carry its own energy source, one capable of delivering far more power at far higher voltages than were used for this demo model. However, a conventional turbine generator and transformer unit capable of delivering the required amount of power would have been prohibitively heavy. Consequently, Brown took a very different approach. He recommended investigating a device that he called a "flame-jet" electrostatic generator. This was essentially a jet engine modified to electrify its exhaust stream, turning it into a powerful electro-hydrodynamic generator (figure 2.9, taken from Brown's 1965 patent).

The jet engine's exhaust nozzle was to be fitted with a negatively charged needle electrode and a positively charged plate electrode. A 50,000-volt starter transformer located on the craft would cause the needle to emit a stream of negative ions into the jet exhaust. The ions, however, would never succeed in reaching the nozzle's positive electrode because they would be whisked out of the nozzle throat and away from the craft by the high-velocity exhaust gases. The departing negative ions would acquire a very large negative voltage potential relative to the electrodes in the jet's nozzle. The farther they would be forced from the

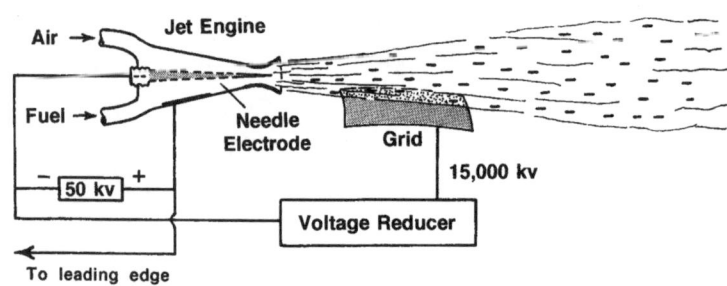

Figure 2.9. A high-voltage flame-jet generator design patented by Brown. (After Brown, U.S. patent 3,022,430, figure 3)

craft, the greater would be their potential difference. A Van de Graaff generator, a common fixture in most high school physics laboratories, operates on a similar principle. In that case, a rapidly running conveyor belt takes the place of the engine exhaust for transporting and separating the negative ions from the positive. The Winterhaven proposal indicated that such an engine would be capable of generating up to 15 million volts, three hundred times higher than the 50-kilovolt input potential initially used to electrify its needle ionizer.

By positioning a metallic grid in the exhaust stream just beyond the generator's nozzle, some of the exiting exhaust ions could be collected at an intermediate voltage of around 1 million and, after being stepped down in voltage, could be recycled to drive the generator's needle ionizer. Thus, once the generator got going, the 50-kilovolt starter transformer that supplied the initial power could be turned off, leaving the generator to run itself from its own electrical output. In the alternate design, shown in figure 2.10, the downstream exhaust grid is replaced by a series of conical baffles that become charged to successively negative potentials, with the outermost baffle being the most negative and attaining a charge of many millions of volts.

Furthermore, Brown proposed that the flame-jet generator's positive electrode be connected to an ionizer wire running along the craft's leading edge (see figure 2.8). As a result, when the generator was in operation, a positive-ion space charge would build up in front of the craft, counterbalancing the negative-ion space charge built up in the exhaust plume trailing the craft. The gravity gradient generated between these oppositely polarized ion clouds would induce a forward-pulling gravitational force. The craft could be steered to one side or the other by diverting its exhaust through one or the other of its side nozzles, thereby producing a corresponding shift in its gravitational propulsion field.

Brown found that his flame-jet generator conveyed charges most effectively when its flame was adjusted to an orange-red color, indicating incomplete combustion of its fuel.[18] Incomplete combustion would produce large numbers of charged submicron-size particles (0.003 to 0.03 micron in diameter) that, upon being ionized, would grow in size to form Langevin ion smoke particles (> 0.03 micron). Being much more massive than air ions, Langevin ions would move considerably slower

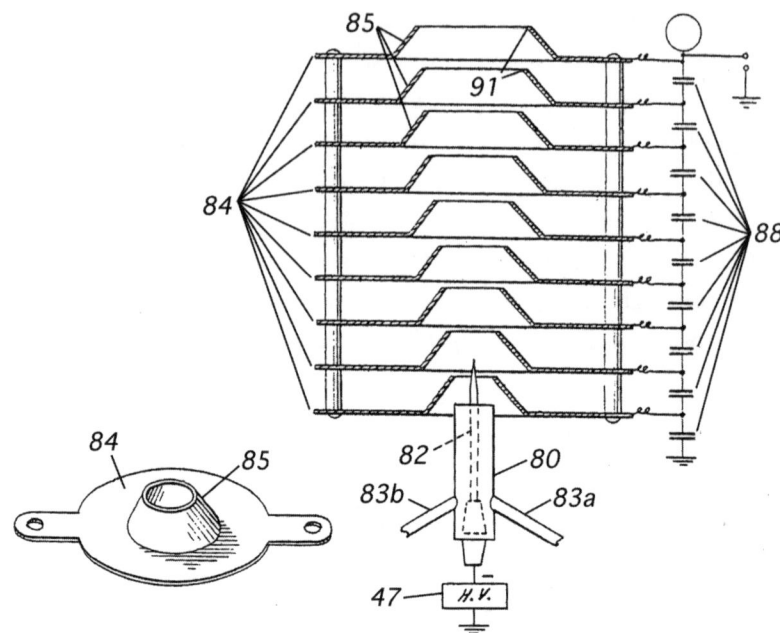

Figure 2.10. A high-voltage flame-jet generator, a second design depicted in Brown's patent. Numbers indicate the following: (47), a 50-kilovolt power supply; (80), exhaust nozzle; (82), needle electrode; (83a and 83b), air and fuel inlets; (84 and 85), baffles; (88), capacitors; (91), jagged electron emitter edges. (After Brown, U.S. patent 3,022,430, figure 4)

under the influence of an electric field. For example, in a field of 10,000 volts per centimeter, the kind Brown often worked with, Langevin ions would travel about 1 centimeter per second, as compared with 450 miles per hour for air ions. Hence, once they are ejected from the craft, the negative exhaust ions would not readily return to the front of the aircraft to neutralize the positive ions. Consequently, a much larger negative-ion space charge would build up behind the craft that, in turn, would substantially increase the forward-directed gravitic propulsion force. Although the positively charged air ions produced at the front of the craft would have a much higher ionic mobility, the bow shock front that would form under high-speed flight would tend to deflect these ions away from the body of the disc, thereby retarding their rearward flight toward the negative-ion space charge cloud.

The Winterhaven proposal stated that the electrogravitic motor would be essentially soundless, vibrationless, and heatless and that its internal resistance losses would be almost negligible and its speeds enormous. The motor's thrust could be controlled by regulating the applied voltage, and its flight velocity could be braked or even reversed simply by reversing its electric polarity. The proposal commented that past laboratory research had indicated that an electrogravitic motor would set up a gravitational field independent of that of Earth. Hence, an electrogravitic spacecraft could maintain sustained acceleration even after leaving the gravitational influence of Earth. It predicted top speeds far beyond those of jet propulsion or rocket drive, with the possibility of approaching the speed of light in free space.

A 1960 report titled "Electrohydrodynamics," issued by the Electrokinetics Corporation of Bala Cynwyd, Pennsylvania, presented an idea for a vertical takeoff aero-marine vehicle that was a variation of the one proposed in Project Winterhaven (figure 2.11).[19] The proposed vehicle was to be 24 feet high and 70 feet in diameter. Through experimentation, Brown found that a disc whose upper surface had a triarcuate-shaped profile, a helmet-shape similar to that shown in figure 2.11, produced the best vertical thrust.

As described in the report, the vehicle would use spherical pontoons to rest on the water surface (see figure 2.12, Landing Position). To

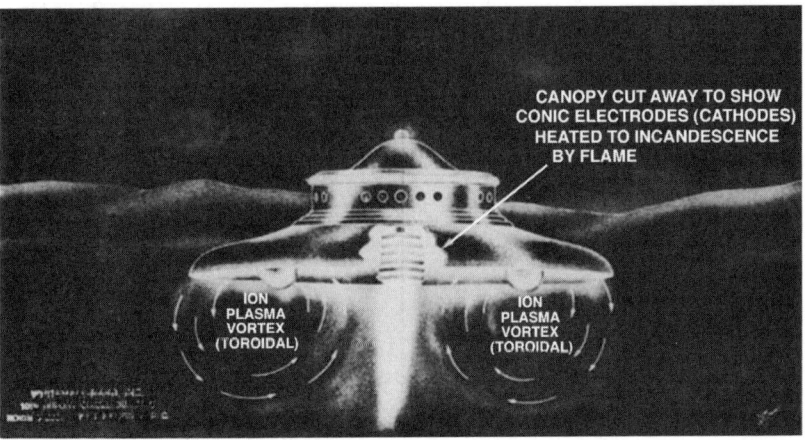

Figure 2.11. Prototype aero-marine vehicle powered by a high-voltage flame-jet generator. (Courtesy of the Townsend Brown Family and Qualight, L.L.C.)

take off, the craft would ignite its flame jet without electrical excitation (figure 2.12, Stage 1). The jet would be powered by either kerosene or solid rocket propellants. The exhaust would set up a positive pressure beneath the canopy that would loft the craft about ten feet above the water surface, allowing the pontoons to be partially retracted. A gyro wheel in the cabin dome would provide horizontal stability and orientation control.

Once the vehicle was airborne, the flame would be electrically energized by applying high voltage to the incandescent cathode (figure 2.12, Stage 2). An electric gradient would establish itself along the length of the exhaust plume, with the voltage progressively increasing downstream to reach a potential difference of several million volts. A potential gradient would similarly build up on the nozzle's conical collector electrodes, which collect power for the craft. The exhaust would set

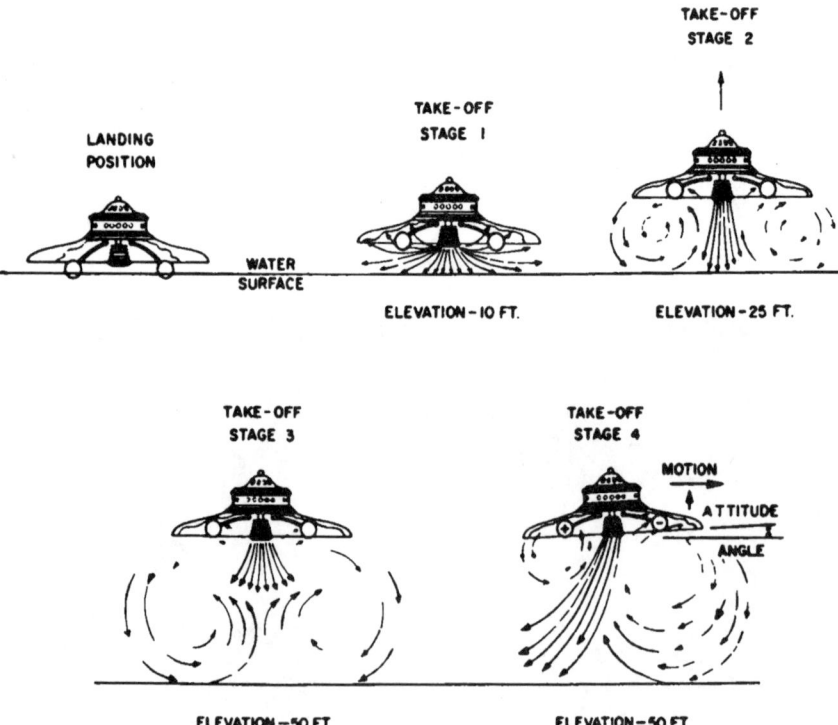

Figure 2.12. Stages in the takeoff of Brown's electrohydrodynamic propulsion vehicle. (Courtesy of the Townsend Brown Family and Qualight, L.L.C.)

up a negative-ion space charge below the craft, and ionizer electrodes on the canopy would create positive ions above the craft. The positive charges emitted from the canopy would create an ion wind that would move downward and radially inward toward the central cathodic axis, setting up a toroidal ion vortex. The flow would receive negative charges from the exhaust, which would cause it to accelerate upward toward the canopy. The large volume of upward-moving plasma would impart upward momentum to the undersurface of the positively charged canopy, thereby helping to loft the vehicle. Also, the inflowing ion plasma would buck the flame jet's outgoing gas flow, confining its flow and increasing the lift pressure beneath the canopy. At this stage, the vortex would be slightly larger than the diameter of the craft. The craft would have risen to about a 25-foot elevation and would continue to rise as subcanopy pressures induce further lift. At this point, the pontoons would have completely retracted. Smoke tests, which Brown conducted on an 18-inch-diameter model, showed that this electrode geometry indeed generates a toroidal vortex (see chapter 3, figure 3.3).

The report states that as the craft rises, the toroidal vortex would expand to about three (craft) diameters. The axial upwelling would continue to buck the downward jetstream to add to the jet's upward thrust (figure 2.12, Stage 3). At about a 50-foot elevation, the subcanopy pressure would diminish to equal the weight of the craft and equilibrium would be established. In this fashion, the craft would be able to hover at about this height, its canopy riding upon the vortex, whose aerodynamic pressure would act at all points against the craft's underside, providing lift. Electrogravitic forces would also contribute to this effect, as would electrostatic forces between the craft and charges in the surrounding air and charged plasma.

Horizontal thrust in any direction would be provided by altering the symmetry of the electric field and of the resulting vortex pattern (figure 2.12, Stage 4). This would allow flame gas to escape to the rear of the craft and the forward side of the vortex to exert traction upon the water surface. The unbalanced canopy pressures would provide forward thrust, and the additional lift of the leading edge would cause a change in the altitude of the craft.

Project Winterhaven also requested funds to develop solid-state electrogravitic motors similar to Brown's early gravitators. It stressed the importance of improvements that had been achieved in developing new dielectric materials, noting that the available dielectric constant K-values had progressively increased from 6 to 100 to 6,000 to 30,000 and beyond. It proposed engineering a 500-pound high-K dielectric motor for propelling a model ship and envisioned that this would presage the development of much larger motors for ships weighing thousands of tons.

Yet another Winterhaven project concerned the investigation of electrogravitic communication equipment that would transmit and receive electrogravitic waves. Early in 1952, Brown conducted a demonstration in Los Angeles of one such transmitting and receiving system, through which he successfully transmitted a signal over a distance of 35 feet to a receiver located within an electrically grounded metallic enclosure.[20] He used a relaxation oscillator as his transmitter. This consisted of a high-voltage power supply that continuously charged a capacitor, which, in turn, periodically discharged itself through a small spark gap when its voltage had reached a certain value (figure 2.13). Just as in his early capacitor experiments, Brown reasoned that electrogravitic coupling

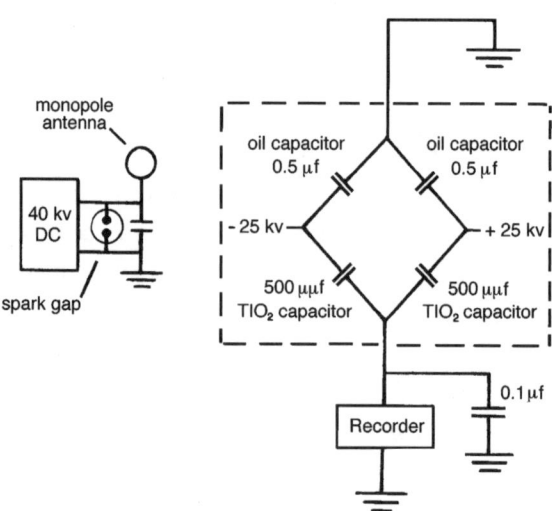

Figure 2.13. Schematic circuit diagrams of Brown's electrogravitic wave transmitter (left) and receiver (right).

would cause the capacitor to radiate gravitational waves. His receiver antenna consisted of a charged high-voltage capacitor bridge circuit similar to the one he used for his gravito-electric measurements. Because the grounded Faraday shield that surrounded his receiver antenna was able to prevent the entry of ordinary electromagnetic waves, he concluded that the signal being conveyed was gravitic rather than electromagnetic and that his capacitor bridge was able to detect gravitational disturbances.

In a patent disclosure Brown wrote in September 1953, he described another version of this communication device, which was designed to send a signal from an audio oscillator to an electrically shielded radio receiver. However, instead of capacitors, this device used heavy spherical masses for the transmitter and receiver antennae.[21] His Project Winterhaven proposal envisioned that electrogravitic wave transmission and reception could be developed into a fundamentally new communication technology. It noted that because of the extreme penetrating ability of these waves, messages could be transmitted to submarines and to underground shelters and military installations, locations inaccessible to normal radio-wave communication. Brown's spherical antenna gravitational wave generator bore a close resemblance to devices developed by scientist and inventor Nikola Tesla in the early twentieth century. How such devices generate gravity waves may be understood in terms of the subquantum kinetics ether methodology (see box).[22,23]

A Gravity Wave Model

As mentioned earlier, subquantum kinetics predicts that a positive charge should induce the formation of a gravity potential well and a negative charge should induce the formation of a gravity potential hill. Hence, a spherical conductor that is alternately charged and discharged should radiate both an electric potential wave and a gravity potential wave. These waves would be scalar waves rather than force field vector waves because they would consist of changes in energy potential (i.e., ether concentration), which is a scalar quantity. That is, a potential wave would have a measurable magnitude at a given point in space but no associated direction. Electric charges on a spherical monopole antenna would not be laterally displaced to any appreciable extent, as they would be in a conventional dipole antenna; hence, they would induce a minimal magnetic field. So unlike Hertzian electromagnetic waves, which oscillate transverse to their direction of travel, these scalar waves would induce no forces transverse to a wave's direction of travel. They would instead produce longitudinal electric and gravity potential field gradients that would induce longitudinal forces, that is, forces aligned with the wave's direction of propagation.

Tesla's ether sound-wave model for describing radiant energy waves fits this energy potential description quite well because sinusoidal variations in ether concentration may be visualized as alternate compressions and rarefactions of the ether medium, analogous to the compressions and rarefactions of air molecules in a sound wave. A high ether concentration would correspond to a high energy potential, and a low ether concentration would correspond to a low energy potential. So according to this model, the antenna would be radiating spherical waves of alternating ether concentration. In the event that the waves were made to repeatedly reflect back and forth under resonance, they could reinforce one another to produce stationary wave patterns characterized by very strong longitudinal forces.

The Project Winterhaven proposal requested that its projects be carried out under a Department of Defense R&D contract administered by a prime contractor. It advocated a cooperative participation of four commercial corporations engaged in applied research and four academic institutions engaged in pure research. The four corporations were to comprise Lear Inc. (for gravimetric field measurements), Jansky & Bailey (for electrogravitic wave communication research), Brush Development Company (for development of high-K dielectric thrust motors), and Hancock Manufacturing Company (for development of the disc airfoils). The academic institutions included Stanford Research Institute, the University of Chicago, and the Franklin Institute.

The proposal acknowledged that the Pentagon had classified some of Brown's past electrogravitic research. It noted that additional data confirming the existence of the electrogravitic coupling effect have been "associated with government research projects of a highly classified status, and publication has been precluded." Nevertheless, Brown's desire in proposing Project Winterhaven was not to keep this technology secret but to accelerate its development so that it could benefit humanity.

Unfortunately, his dream of unrestricted access to electrogravitic technology was not to be realized. Perhaps unknown to him at that time, highly classified work on electrogravitics then in progress had proceeded to a relatively advanced stage of development. In particular, out of all the ideas proposed in Project Winterhaven, the electrogravitic wave communicator device came closest of all to this ongoing sensitive propulsion research.

2.4 ◾ ANTIGRAVITY RESEARCH: TOP SECRET

Brown's effort to promote his electrogravitic propulsion concept received its greatest boost from Aviation Studies (International) Ltd., a privately owned, London-based aviation intelligence consulting firm.* Since its formation in 1950, Aviation Studies has marketed reports to aerospace companies and government defense departments on a wide variety of subjects, giving information about various kinds of aircraft, rockets, and missiles (e.g., their design features, prices, production run sizes, foreign arsenal sizes); data on nuclear, thermal, and directed-energy weapons; assessments of foreign government military intelligence capabilities (e.g., organization missions, manpower, intelligence advisories); and much more. Their price lists for publications available between 1957 and 1960 and for publications issued in 1993 are reproduced in appendix B.

Richard Worcester, the director of Aviation Studies, had become convinced that Brown had discovered something that could radically revolutionize aviation technology. So beginning in August 1954, his think tank began an effort to promote Brown's ideas to the aerospace industry, indicating that the rewards of success in developing electrogravitics technology were too far-reaching to be overlooked. They

---

*The firm's address was given as Aviation Studies (International) Ltd., Sussex House, 37A Parkside, Wimbledon, London SW19 5NB, UK. However, currently no address for this organization is available.

began including news items about electrogravitics technology in their weekly newsletter, *Aviation Report,** and by 1956 also began sponsoring research into high-K dielectric materials for use in electrogravitic aircraft. Their catalytic effort proved successful because industry involvement in electrogravitics expanded exponentially from 1954 onward. Around the late 1950s, antigravity propulsion research went underground and little was heard about it, although today the effort continues secretly on a scale rivaling the Manhattan Project's effort to develop the atomic bomb.

In February and December 1956, Aviation Studies published two summary reports on electrogravitics. The February report, titled "Electrogravitics Systems: An Examination of Electrostatic Motion, Dynamic Counterbary and Barycentric Control," presents an illuminating survey of government and industry's early involvement in the field of antigravitics R&D.[24] Its cover page lists its origin as the Gravity Research Group, a subdivision of the Aviation Studies Special Weapons Study Unit, but it is now known to have been written by Worcester. The December report, titled "The Gravitics Situation," also written by Worcester, was issued as being produced by Gravity Rand Limited, an affiliate of Aviation Studies.[25] It provides additional information about aviation industry progress in developing electrogravitic antigravity technology. Gravity Rand had no affiliation with the Rand Corporation. Rand, an acronym for Research and Development, is often included in the names of companies that are involved in R&D work.

It is relatively difficult to obtain original copies of these documents. Although the card catalog at the U.S. Library of Congress in Washington, D.C., has a card on file for the February 1956 "Electrogravitics Systems" report (LOC no. 3,1401,00034,5879; call no. TL565.A9), when I tried to check out the report in 1985, the librarian found that it was missing from its shelves. A subsequent check of the Library of Congress computer database showed that one other library in the United States kept a copy of the report. That was the Wright-Patterson Air Force Base Technical Library in Dayton, Ohio. Originally marked "confidential,"

---

*A 1993 circular states that this newsletter covers aerospace policy and contains equipment and intelligence data, economic/technical acquisition needs, and government news, analysis, and trends. I do not know if it continues to be published.

the report had been declassified sometime prior to when I obtained it in 1985 and is currently available for public scrutiny. It may be borrowed from Wright-Patterson through the interlibrary loan system, but doing so may require some persistence on the part of the requester since the document is not cataloged on all Wright-Patterson computer databases, so Air Force librarians might overlook its presence. The February 1956 report is reproduced in appendix C.

An original copy of the December 1956 report was more difficult to locate. Librarians at Wright-Patterson could not find a copy in their stacks and an attempt to obtain the document from Aviation Studies was also unsuccessful. The director responded that copies of both 1956 electrogravitics studies could not be found in its files. Currently, however, it is possible to download these documents from the Internet. Aircraft and missile companies that have been purchasing the Aviation Studies reports may still keep copies of these older issues in their technical libraries, but company officials may be hesitant to share them with outsiders. There is an apparent effort to keep this subject very quiet. For example, Loral Vought Systems Corporation of Grand Prairie, Texas, a company that is heavily involved in missile R&D and that served as a subcontractor to Northrop Grumman Corporation on development of the B-2 stealth bomber, had listed the December 1956 report on its library database, but in a 1993 telephone conversation, one of their librarians told me that the document was marked as "destroyed." Although she mentioned that three or four other Aviation Studies reports were listed, she was unwilling to divulge their titles and cited company policy that prevented the documents from being loaned out.

The subtitle of the February 1956 report—An Examination of Electrostatic Motion, Dynamic Counterbary, and Barycentric Control—blatantly indicates that it deals with gravity control. The words *dynamic counterbary* and *barycentric control* translate to mean antigravity propulsion and the control of gravity, the root word *bary* coming from the Greek βαρη, meaning weight. More specifically, page 19 of this report defines *counterbary* as "the manipulation of gravitational force lines" and *barycentric control* as "the adjustment to such manipulative capability to produce a stable type of motion suitable for transportation."

The glossary of the December 1956 Aviation Studies report defines *counterbary* as "another name for lofting . . . the action of levitation where gravity's force is more than overcome by electrostatic or other propulsion." It defines *barycentric control* as "the environment for regulation of lofting processes in a vehicle."

The term "dynamic counterbary," or contrabary, was coined by the renowned German scientist Burkhard Heim, who used it publicly for the first time in 1952 in a Stuttgart lecture titled "Dynamic Contrabary and the Solution of Astronautical Problems." Heim had been heavily engaged in gravity force field research in the early 1950s and claimed to have discovered what he called the "contrabaric effect," a way of inducing a gravitational force field by electromagnetic means.

The February 1956 report briefly reviews Brown's seminal work and mentions his 1952 Project Winterhaven proposal to develop an electrogravitic interceptor disc of approximately 35 feet in diameter that would be capable of attaining Mach 3 (2,250 miles per hour) and executing sharp-edged changes of direction. The authors of the report were quite convinced that electrogravitics involves a nonconventional method for artificially altering a vehicle's gravity field, because the report begins by stating:

Electrogravitics might be described as a synthesis of electrostatic energy used for propulsion . . . and gravitics (or dynamic counterbary) in which energy is also used to set up a local gravitational force independent of the earth's . . .

The essence of electrogravitics thrust is the use of a very strong positive charge on one side of the vehicle and a negative on the other. The core of the motor is a condenser and the ability of the condenser to hold its charge (the K number) is the yardstick of performance . . .

. . . The electrogravitics saucer can perform the function of a classic lifting surface—it produces a pushing effect on the under surface and a suction effect on the upper, but, unlike the airfoil, it does not require a flow of air to produce the effect.[26]

The report summarizes electrogravitics work that was being done in the United States and in Great Britain and even indicates that several antigravity test rigs were in operation at that time. It also includes extracts from various issues of *Aviation Report* dated between August 1954 and December 1955. These give an illuminating view of how interest in electrogravitics progressively expanded over this eighteen-month period. Some interesting excerpts are quoted here in chronological order:

ANTI-GRAVITATION RESEARCH

The basic research and technology behind electro-antigravitation is so much in its infancy that this is perhaps one field of development where not only the methods but the ideas are secret. Nothing therefore can be discussed freely at the moment. Very few papers on the subject have been prepared so far, and the only schemes that have seen the light of day are for pure research into rigs designed to make objects float around freely in a box . . . long term aims . . . envisage equipment that can defeat gravity.

Aviation Report, August 20, 1954

MANAGERIAL POLICY FOR ANTI-GRAVITICS

Anti-gravitics work is . . . likely to go to companies with the biggest electrical laboratories and facilities. It is also apparent that anti-gravitics, like other advanced sciences, will be initially sponsored for its weapon capabilities. There are perhaps two broad ways of using the science—one is to postulate the design of advanced type projectiles . . . The other, which is a longer term plan, is to create an entirely new environment with devices operating entirely under an anti-gravitic envelope.

Aviation Report, August 24, 1954

GRAVITICS FORMULATIONS

Some extremely ambitious theoretical programs have been submitted and work towards realization of a manned [anti-gravitic] vehicle has begun. On the evidence, there are far more definite indications that the incredible claims are realizable than there was, for instance, in supposing that uranium fission would result in a bomb.

Aviation Report, September 7, 1954[27]

An October 1954 report refers to Brown's 1952 Winterhaven Project proposal and indicates that the Pentagon was about to begin funding the development of electrogravitic aircraft:

ELECTRO-GRAVITIC PROPULSION SITUATION

Under the terms of Project Winterhaven [1952] the proposals to develop electro-gravitics to the point of realizing a Mach 3 combat type disc were not far short of the extensive effort that was planned for the Manhattan District. Indeed the drive to develop the new prime mover is in some respects rather similar to the experiments that led to the release of nuclear energy in the sense that both involve fantastic mathematical capacity and both are sciences so new that other allied sciences cannot be of very much guide. In the past two years since the principle of motion by means of massive-K was first demonstrated on a test rig, progress has been slow. But the indications are now that the Pentagon is ready to sponsor a range of devices to help further knowledge. . . . Tentative targets now being set anticipate that the first disc should be complete before 1960 and it would take the whole of the "sixties" to develop it properly, even though some combat things might be available ten years from now. . . . The frame incidentally is indivisible from the "engine." If there is to be any division of responsibility it would be that the engine industry might become responsible for providing the electrostatic energy (by, it is thought, a kind of flame) and the frame maker for the condenser assembly which is the core of the main structure.

Aviation Report, October 12, 1954[28]

Note that the October report mentions Brown's flame-jet high-voltage generator concept as a means for generating electrostatic energy and suggests that such a device would be developed by the jet engine industry. A November 1954 report describes the Air Force's first attempts to draw up specifications for the development of an electrogravitic vehicle and notes that the goal of a Mach 3 combat saucer would be possible through an extrapolation of technology available at that time:

GRAVITICS STEPS

Specification writers seem to be still rather stumped to know what to ask for in the very hazy science of electrogravitic propelled vehicles. They are at present faced with having to plan the first family of things—first of these is the most realistic type of operational test rig, and second, the first type of test vehicle. In turn this would lead to sponsoring of a combat disc. The preliminary test rigs which gave only feeble propulsion have been somewhat improved, but of course the speeds reached so far are only those more associated with what is attained on the roads rather than in the air. But propulsion is now known to be possible, so it is a matter of feeding enough KVA [kilovolt-amperes] into condensers with better K figures. 50,000 is a magic figure for the combat saucer—it is this amount of KVA and this amount of K that can be translated into Mach 3 speeds.

Aviation Report, November 19, 1954[29]

The term "KVA," which stands for kilovolt-amperes, is used exclusively in referring to power consumption in which the power source is AC. Its use in the above passage suggests that the disc being discussed was to use high-K capacitors powered with AC, rather than DC.[30] If the disc's capacitors were instead powered exclusively with DC, then it would have been more proper to refer to kilowatts of power, or KW.* Also, this same report later states, "Perhaps the main thing for management to bear in mind in recruiting men is that essentially electrogravitics is a branch of wave technology and much of it starts with Planck's dimensions of action, energy, and time, and some of this is among the most firm and least controversial sections of modern atomic physics." So here is further acknowledgment that researchers were actively investigating the use of *time-varying* electric fields for electrogravitic propulsion.

Although Brown's early demonstration discs were powered with high-voltage DC power, later demonstrations, such as the one given in a

---

*When a capacitor is energized with AC, there is a phase shift between the current and voltage sine waves such that the current change through the capacitor tends to lead the voltage change across it. Consequently, the product of average current times average voltage (or kilowatts) is not a good measure for the average power of a half cycle. In such cases, electrical engineers instead refer to kilovolt-amperes.

gymnasium at Pearl Harbor, appear to have instead used rectified AC. Also, in one of his patents, Brown briefly alludes to energizing high-K dielectrics with high-frequency AC, but he kept fairly quiet about this part of his work. How an AC-energized capacitor might be used to produce an amplified electrogravitic thrust is described in the next chapter.

During a January 25, 1955, meeting of aviation leaders held in New York, George S. Trimble, vice president of advanced design for Glenn Martin Aircraft in Baltimore, was quoted by the Associated Press as saying, "Unlimited power, freedom from gravitational attraction, and infinitely short travel time are now becoming feasible."[31] He then added that eventually all commercial air transportation would be in vehicles operating on these fantastic principles. Recall that Brown had briefly worked at the Baltimore Glenn Martin plant sixteen years earlier, before the beginning of World War II. Undoubtedly, he had planted the seeds about electrogravitics at that early date.

At the same meeting, Dr. Walter R. Dornberger, a guided missile consultant for the Bell Aircraft Corporation, predicted that airliners would eventually travel at 10,000 miles per hour (Mach 13). This would make possible a trip from New York to Sidney, Australia, in approximately one hour. Two weeks later, Aviation Studies issued a report disclosing that many aircraft companies were aware of the existence of this antigravity technology:

MANAGEMENT NOTE FOR ELECTRO-GRAVITICS

New companies . . . who would like to see themselves as major defence prime contractors in ten or fifteen years time are the ones most likely to stimulate development. Several typical companies in Britain and the U.S. come to mind—outfits like AiResearch, Raytheon, Plessey in England, Rotax and others. But the companies have to face a decade of costly research into theoretical physics and it means a great deal of trust. Companies are mostly overloaded already and they cannot afford it, but when they sit down and think about the matter they can scarcely avoid the conclusion that they cannot afford not to be in at the beginning.

Aviation Report, February 8, 1955[32]

In July 1955, *Aviation Report* quoted Lawrence D. Bell, founder of Bell Aircraft, as saying that the tempo of development leading to the use of antigravitational vehicles would accelerate and that breakthroughs that had become feasible at that time would advance the introduction of such vehicles ahead of the time it had taken to develop the turbojet.[33] That same issue predicted that government procurement would open up "because the capabilities of such aircraft are immeasurably greater than those envisaged with any known form of engine."

On October 15, 1955, the Department of Defense issued a news release informing the public that some government aircraft under development could resemble flying saucers. Secretary of the Air Force Donald A. Quarles stated:

> . . . we are now entering a period of aviation technology in which aircraft of unusual configuration and flight characteristics will begin to appear . . . The Air Force and other Armed Services have under development several vertical-rising, high performance aircraft . . . Vertical-rising aircraft capable of transition to supersonic horizontal flight will be a new phenomenon in our skies, and under certain conditions could give the illusion of the so-called flying saucer.[34]

Although Quarles did not refer to any unconventional propulsion technology, it may be no coincidence that just one year earlier the Pentagon had begun plans to fund the development of Brown's electrogravitics technology. To camouflage the truly exotic nature of the project, the news release called attention to the disc-shaped AVRO car, developed by AVRO Ltd. of Canada. The AVRO car was an ill-conceived vehicle that used a conventional air turbine that was ducted to provide vertical lift. Unfortunately, its design was inherently unstable; it had the persistent tendency to flip over after rising just a few feet off the ground.

The November 1955 issue of *Aviation Report* acknowledges the key role that the Aviation Studies newsletter played in catalyzing the development of the electrogravitics industry:

Electrogravitics Feasibility

The feasibility of a Mach 3 fighter (the present aim in studies) is dependent on a rather large K extrapolation, considering the pair of saucers that have physically demonstrated the principle only achieved a speed of some 30 fps [feet per second]. But, and this is important, they have attained a working velocity using a very inefficient (even by to-day's knowledge) form of condenser complex . . .

It was, by the way, largely due to the early references in *Aviation Report* that work is gathering momentum in the U.S. Similar studies are beginning in France, and in England some men are on the job full time.

Aviation Report, November 15, 1955[35]

Later that month, Ansel Talbert, military and aviation editor for the *New York Herald Tribune,* published a series of articles on the aviation industry's interest in gravity control. On November 20, he wrote:

A number of major, long-established companies in the United States aircrafts and electronics industries also are involved in gravity research. Scientists in general, bracket gravity with life itself as the greatest unsolved mystery in the Universe. But there are increasing numbers who feel that there must be a physical mechanism for its propagation which can be discovered and controlled. Should this mystery be solved it would bring about a greater revolution in power, transportation, and many other fields than even the discovery of atomic power. The influence of such a discovery would be of tremendous import in the field of aircraft design where the problem of fighting gravity's effects has always been basic.[36]

Talbert's article displayed a photo of two General Dynamics Convair Division scientists conducting a research experiment aimed at controlling gravity. It showed them facing an apparatus supported on pillars that was wired with electrical connections. In an article dated November 21, Talbert named six other firms that were involved in such studies:

Aircraft industry firms now participating or actively interested in gravity include the Glenn L. Martin Co. of Baltimore, builders of the nation's first giant jet-powered flying boat; Convair of San Diego, designers and builders of the giant B-36 intercontinental bomber and the world's first successful vertical take-off fighter; Bell Aircraft of Buffalo, builders of the first piloted airplane to fly faster than sound and a current jet "vertical takeoff and landing" airplane, and Sikorsky division of United Aircraft, pioneer helicopter builders. Lear, Inc., of Santa Monica, one of the world's largest builders of automatic pilots for airplanes; Clarke Electronics of Palm Springs, California, a pioneer in its field, and the Sperry Gyroscope Division of Sperry-Rand Corp., of Great Neck, L.I., which is doing important work on guided missiles and earth satellites, also have scientists investigating the gravity problem.[37]

Talbert also named several physicists who were interested in pursuing gravity control research:

> ... current efforts to understand gravity and universal gravitation both at the sub-atomic level and at the level of the Universe have the positive backing today of many of America's outstanding physicists. These include Dr. Edward Teller of the University of California, who received prime credit for developing the hydrogen bomb; Dr. J. Robert Oppenheimer, director of the Institute for Advanced Study at Princeton; Dr. Freeman J. Dyson, theoretical physicist at the institute, and Dr. John A. Wheeler, professor of physics at Princeton University, who made important contributions to America's first nuclear fission project.[38]

Others mentioned to be working on understanding gravity included Dr. Vaclav Hlavaty of the University of Indiana and Drs. Stanley Deser and Richard Arnowitt of Princeton's Institute for Advanced Study. Unlike his colleague Albert Einstein, Hlavaty believed gravity simply to be one aspect of electromagnetism.

In his November 21 article, Talbert further acknowledged the existence of a widespread industry program geared toward gravity control research:

> Many in America's aircraft and electronics industries are excited over the possibility of using its magnetic and gravitational fields as

a medium of support for amazing "flying vehicles" which will not depend on the air for lift. Space ships capable of accelerating in a few seconds to speeds many thousands of miles an hour and making sudden changes of course at these speeds without subjecting their passengers to the so-called "G-forces" caused by gravity's pull also are envisioned. These concepts are part of a new program to solve the secret of gravity and universal gravitation already in progress in many top scientific laboratories and long-established industrial firms of the nation.

William P. Lear, inventor and chairman of the board of Lear, Inc., one of the nation's largest electronics firms specializing in aviation, for months has been going over new developments and theories relating to gravity with his chief scientists and engineers. Mr. Lear in 1950 received the Collier Trophy from the President of the United States "for the greatest achievement in aviation in America" through developing a light-weight automatic pilot and approach control system for jet fighter planes. He is convinced that it will be possible to create artificial "electro-gravitational fields whose polarity can be controlled to cancel out gravity." He told this correspondent: "All the (mass) materials and human beings within these fields will be part of them. They will be adjustable so as to increase or decrease the weight of any object in its surroundings. They won't be affected by the earth's gravity or that of any celestial body. This means that if any person was in an anti-gravitational airplane or space ship that carried along its own gravitational field . . . — no matter how fast you accelerated or changed course—your body wouldn't any more feel it than it now feels the speed of the earth."[39]

It is unlikely that an industrialist as prominent as Lear would make such a strong statement unless he himself had seen concrete evidence that such an electrogravitic effect was possible. Bell, whose company in Buffalo had built the first piloted aircraft in history to fly faster than sound, also was optimistic about the results of gravity research then in progress. The *New York Herald Tribune* quoted him as saying, "Aviation as we know it is on the threshold of amazing new concepts. The United States aircraft industry already is working with nuclear fuels and equipment to cancel out gravity instead of fighting it."[40]

Grover Loening, the first engineer hired by the Wright brothers and whose forty-year career in aircraft design, construction, and consulting had been decorated by the U.S. Air Force, told Talbert, "I firmly believe that before long man will acquire the ability to build an electromagnetic contra-gravity mechanism that works. Much the same line of reasoning that enabled scientists to split up atomic structures also will enable them to learn the nature of gravitational attraction and ways to counter it."[41]

Trimble's company, Glenn Martin, was the first in the United States to investigate electrogravitational propulsion.[42] This is not surprising given that Brown worked for Martin as early as 1939. The *New York Herald Tribune* said that, under Trimble's initiative, Martin Aircraft was building a laboratory between Baltimore and Washington to house the new Research Institute for Advanced Study, which would be committed to investigating the theoretical basis of electrogravitics and to conducting programs in applied research. Regarding the development of this new technology, the *Herald Tribune* quoted Trimble as saying, "I think we could do the job in about the time that it actually required to build the first atomic bomb if enough trained scientific brain-power simultaneously began thinking about and working towards a solution. Actually the biggest deterrent to scientific progress is a refusal of some people, including scientists, to believe that things which seem amazing can really happen."[43]

Dudley Clarke, president of Clarke Electronics, was also reported to be optimistic about gravity control. In an article dated November 22, Talbert stated, "Mr. Clarke notes that the force of gravity is powerful enough to generate many thousand times more electricity than now is generated at Niagara Falls and every other water-power center in the world—if it can be harnessed. This impending event, he maintains, will make heat and power needed by one family for an indefinite period."[44]

Two weeks after the *Herald Tribune* story came out, *Aviation Report* stated:

ELECTRO-GRAVITICS EFFORT WIDENING

Companies studying the implications of gravitics are said in a new statement, to include Glenn Martin, Convair, Sperry-Rand, Sikorsky,

Bell, Lear Inc. and Clark[e] Electronics. Other companies who have previously evinced interest include Lockheed, Douglas and Hiller. The remainder are not disinterested, but have not given public support to the new science—which is widening all the time. The approach in the U.S. is in a sense more ambitious than might have been expected.

Aviation Report, December 9, 1955[45]

Of these companies, Brown had particularly strong ties with Lockheed, having worked there just ten years earlier. The Aviation Studies' "Electrogravitic Systems" report, issued two months after the December 9 article quoted above, noted the increasing number of U.S. aviation companies that were expressing interest in antigravity propulsion technology:

One of the difficulties in 1954 and 1955 was to get aviation to take electrogravitics seriously. The name alone was enough to put people off. However, in the trade much progress has been made and now most major companies in the United States are interested in counterbary. Groups are being organized to study electrostatic and electromagnetic phenomena. Most of the industry's leaders have made some reference to it. Douglas has now stated that it has counterbary on its work agenda but does not expect results yet awhile. Hiller has referred to new forms of flying platform, Glenn Martin say gravity control could be achieved in six years, but they add that it would entail a Manhattan District type of effort to bring it about. Sikorsky, one of the pioneers, more or less agrees with the Douglas verdict and says that gravity is tangible and formidable, but there must be a physical carrier for this immense trans-spatial force. This implies that where a physical manifestation exists, a physical device can be developed for creating a similar force moving in the opposite direction to cancel it. Clarke Electronics state they have a rig, and add that in their view the source of gravity's force will be understood sooner than some people think. General Electric is working on the use of electronic rigs designed to make adjustments to gravity—this line of attack has the advantage of using rigs already in existence for other defence work. Bell also has an experimental rig intended, as the company puts it, to cancel out gravity, and Lawrence Bell has said he is convinced that practical hardware will emerge from current programs. Grover Leoning is certain that what he referred to as an electro-magnetic contra-gravity mechanism will be developed for practical use. Convair is extensively committed to the work with several rigs. Lear Inc., autopilot and electronic engineers have a division of the company working on gravity research and so also has the Sperry division of Sperry-Rand. This list embraces most of the U.S. aircraft industry. The remainder, Curtis-Wright, Lockheed, Boeing and North American have not yet declared themselves, but all these four are known to be in various stages of study with and without rigs.[46]

The report added that a certain amount of antigravity work was also going on in Europe. It mentioned two French companies getting involved and several private ventures developing rigs in Britain. It also mentioned that one Swedish company, two Canadian companies, and several German companies were also making studies. The Airplane Corporation and Gluhareff Helicopter were among the foreign companies that had recently joined the growing gravity research club.

The report extrapolated that it should be possible to produce a Mach 3 fighter disc by electrifying the craft with million-volt potentials and using surface coatings having K-values of more than 10,000. By that time, K figures of 6,000 had been obtained from some ceramic materials, and researchers had demonstrated 30 percent weight reductions in some energized devices. Moreover, there were prospects of synthesizing ceramics with K figures as high as 30,000. Thus it was felt that an operational manned aircraft could be built simply by scaling up what was then already in existence.

The emphasis on using high-K dielectric materials for the craft's hull indicates that its designers planned to achieve gravity control primarily by electrically charging the craft's surface rather than depending entirely on the gravitic effect of external ion clouds. Nevertheless, since this vehicle was to obtain its high-voltage power from a flame-jet generator, its designers probably planned to make beneficial use of such auxiliary ion cloud effects.

One month after the February 1956 Aviation Studies report was released, *Interavia* magazine echoed similar optimism regarding the practical application of Brown's electrogravitic technology:

> Such a [gravitic] force raised exponentially to levels capable of push-ing man-carrying vehicles through the air—or outer space—at ultra-high speeds is now the object of concerted effort in several countries. Once achieved, it will eliminate most of the structural difficulties now encountered in the construction of high-speed aircraft . . . The force is not a physical one acting initially at a specific point in the vehicle that needs then to be translated to all the other parts. It is an electro-gravitic field acting on all parts simultaneously. Changes in direction and speed of flight would be effected by merely altering the intensity, polarity and direction of the charge.[47]

In December 1956, Aviation Studies issued a second progress report that pointed out the military advantages of the technology and men-tioned that government funding was being continued:

> Electrostatic discs can provide lift without speed . . . This could be an important advance over all forms of airfoil which require induced flow; and (lift without airflow) is a development that deserves to be followed up in its own right, and one that for military purposes is already envisaged by the users as applicable to all three services (Army, Navy, & Air Force). This point has been appreciated in the United States and a program in hand may now ensure that develop-ment of large sized disks will be continued. This is backed by the U.S. Government, but it is something that will be pursued on a small scale. This acceptance follows Brown's original suggestion embodied in Project Winterhaven.[48]

The report also made the following revealing assessment of the elec-trogravitics industry situation:

> Already companies are specializing in evolution of particular com-ponents of an electrogravitics disk. This implies that the science is in the same state as the ICBM, namely that no new breakthroughs

are needed, only intensive development engineering. This may be an optimistic reading of the situation; it is true that materials are now available for the condensers giving higher K figures than were postulated in Winterhaven as necessary, and all the ingredients necessary for the disks appear to be available. But industry is still some way from having an adequate power source, and possessing any practical experience of running such equipment.[49]

The report suggests that other companies were duplicating Brown's flying disc experiment and similarly obtaining speeds in the range of hundreds of miles per hour. It states, "High speeds in electrostatic propulsion of small discs will be worth keeping track of (by high speed one means hundreds of mph) and some of these results are beginning to filter through for general evaluation."[50] Interest in the subject of antigravity continued to accelerate in the following years. In January 1957, the Institute of Field Physics at the University of North Carolina in Chapel Hill held a weeklong scientific conference on the role of gravitation in physics. The conference was attended by forty-five physicists from the United States and seven other countries. Brown was undoubtedly among them. Interestingly, the Wright Air Development Center of Wright-Patterson Air Force Base was one of the sponsors of the meeting and was also in charge of publishing the conference proceedings.[51] That same year, J. E. Surrat Jr., vice president of the Society of Aeronautical Weight Engineers, said that Wright-Patterson was equipped with a multimillion-dollar installation designed for the research and study of antigravity forces.[52]

A. V. Cleaver, who worked as assistant chief engineer at the Aero Engine Division of Rolls-Royce, assessed the status of electrogravitics in a February 1957 article published in the prestigious *Journal of the British Interplanetary Society.*[53] He estimated that government and industry in the United States were spending on the order of $5 million annually on fundamental research on electrogravitics and noted that firms in France, Italy, and Japan may also have been researching the phenomenon.

Nevertheless, unknown to the many newcomers being indoctrinated into the field of electrogravitics, this multicompany R&D effort was merely supplementing a highly classified effort that had already been in progress since the end of World War II. This preexisting project, known as Project Skyvault, was actually ahead in achieving the goal of a manned antigravity craft. Yet before examining this project, let us study subsequent developments made by Brown that greatly improved the propulsion force of his technology.

3

ONWARD AND UPWARD

3.1 ■ THE PARIS EXPERIMENTS

Skeptics had claimed that Brown's flying discs were propelled entirely by ion wind pressure and would lose their propulsive force if tested in a vacuum chamber where few air molecules would be present, but in 1955 and 1956, they were proved wrong. Under the sponsorship of the French government, Brown conducted a series of vacuum chamber experiments at facilities made available by Société Nationale du Constructions Aeronautiques du Sud-Ouest, a Paris-based aeronautical corporation. There, he successfully flew a pair of miniature saucer airfoils in a high vacuum of less than one billionth of an atmosphere. Not only did the discs propel themselves more efficiently, but they also sped faster, since, without ion leakage, they could be energized with greater voltages. The tests used a 200-watt power source to supply DC potentials ranging from 70 to 220 kilovolts.[1] Few details are known about these tests because the results were considered a confidential matter. However, it appears the discs measured about 4 to 5 inches in diameter and had a central body made of solid aluminum.[2,3] By comparison, the 1.5- and 3-foot-diameter discs that Brown had tested in his carousel demonstration were made of Plexiglas and lightweight sheet aluminum.

In addition to these miniature airfoils, Brown conducted vacuum chamber tests of a rotor apparatus. (Appendix D presents one of his reports on this experiment.[4]) The apparatus consisted of an arm that rotated about a central bearing and that was fitted at each end with a pair of electrodes (figure 3.1). When the electrodes were oppositely charged, the rotor spun like a pinwheel, revolving around its axis in the negative-to-positive direction. It was found that the torque increased in an exponential fashion with applied voltage. At times when a stream of electrons would discharge from the negative to the positive plate, the rotor would acquire a momentary burst of forward thrust. At around 150 kilovolts, the rate of rotation became so great after four or five discharges that the voltage had to be reduced for fear that the rotor might fly apart and shatter its glass bell-jar enclosure. Moreover, Brown found that the thrust persisted even when the capacitor elements were each surrounded by Plexiglas enclosures in the manner shown in figure 3.1. Since there was no way that ions could escape from the enclosures, ion thrust could be ruled out as a motive force. He also used asymmetrical capacitors having electrodes of differing sizes in which either the positive plate was larger than the negative or the negative plate was larger than the positive. However, neither of these geometries had any appreciable effect on the amount of thrust generated by the discharge events.

In February 1973, Dr. Rolf Schaffranke, who wrote the book *Ether Technology* under the pseudonym Rho Sigma, received a letter from Brown responding to inquiries he had made about Brown's Paris experiments.[5] In that letter Brown disclosed that the thrust on the rotor was several orders of magnitude larger than ion thrust could account for. He also related to Schaffranke that he had obtained a greater thrust when a massive high-K dielectric such as barium titanate was placed between

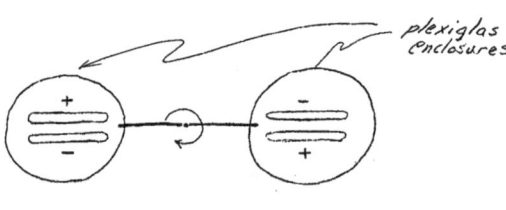

Figure 3.1. Top view of the electrogravitic rotor used in Thomas Townsend Brown's Paris vacuum chamber experiments. (After T. Brown.)

the capacitor plates. He also acknowledged that a residual thrust was obtained even when there were no discharges, clear evidence of the existence of a Biefeld-Brown electrogravitic effect.

Several effects could explain why very large thrusts accompanied each discharge event. As one possibility, the volley of high-energy electrons that formed the spark discharge could have delivered an electrogravitic impulse to the positive electrode. The electron burst would have moved from the negative to the positive electrode of the sparking rotor capacitor at close to the speed of light. That is, in the prevailing hard vacuum, electrons accelerated by a 150-kilovolt potential would have attained a velocity of about 82 percent of the speed of light. Subquantum kinetics predicts that these free electrons would have generated a local gravity potential hill, and as they flew toward the positive electrode, they would have carried this gravity potential hill with them. In the rotor rest frame, this would have appeared as a forward-propagating gravity potential wave. The sharp potential rise at the leading edge of this wave would have had a matter-repelling effect, which would have given a thrust impulse in the negative-to-positive direction as it momentarily passed through the positive electrode mass and any intervening dielectric. This electrogravitic impulse effect is further discussed in chapter 6 in connection with the gravity beam experiments of the Russian materials scientist Eugene Podkletnov.

The nonlinear field gradient associated with each spark could also have contributed to the thrust. Brown observed that the discharges were emitted from a point on his negative electrode and fanned out to produce a broad luminescence on his positive electrode. In other words, regardless of the size of his positive electrode, whether it was a 4-inch disc or a half-inch sphere, the field produced from the discharge would have fanned out from a small high-flux-density region to a larger low-density region. The fanning geometry of the discharge and the sudden onset of the discharge would momentarily have produced a nonlinear electric field between the capacitor plates. This in turn would have generated a large virtual-charge gradient between the capacitor electrodes along with an accompanying gravity potential gradient that would have momentarily induced a large thrust on the capacitor dielectric in the direction of the positive electrode, where the field's flux density would have been lowest. We will defer further discussion of such virtual charge electrogravitic effects until section 4.2 of the next chapter.

Brown's work as a consulting physicist for Société Nationale de Constructions Aeronautiques du Sud-Ouest came to an end in 1956, when the company merged with a larger company, Sud Est, that apparently had no interest in electrogravitics. That summer he returned to the United States with all his papers and took up residence in Washington, D.C.* There, he contacted the Navy, hoping to show them his Paris data, which proved that their earlier ion wind theory was wrong. He was met by Admiral Rickover, but instead of showing interest in Brown's results, the admiral sternly advised him not to take his electrogravitics work any further, that it would be best if he dropped it.[6] Nevertheless, by the end of 1956, the Pentagon had begun sponsoring electrogravitics research that was then in progress at some of the major aerospace firms and had apparently elevated the matter to a top-secret status. Was Brown, the father of this amazing technology, to be excluded from the inner circle of companies chosen to develop his ideas?

Brown was unwilling to give up that easily. He continued his work under the sponsorship of a Delaware company he had formed called Whitehall-Rand Corporation, which had offices in both Washington and London. He probably chose the name Whitehall to allude to the executive branch of the British government, which is based largely on Whitehall Street in London. In July 1957, while serving as its director of research and development, he protected his electric disc and flame-jet-driven aircraft ideas by applying for three U.S. patents (2,949,550; 3,018,394; and 3,022,430). When these were issued, in 1960 and 1962, they were assigned to Whitehall-Rand. Brown was the only contributor of patents to this company.

---

*By the end of August, Brown had founded the National Investigations Committee on Aerial Phenomena (also known as NICAP), a UFO research organization that at the time was the largest and most influential of such organizations. Nevertheless, four months later, as a result of a disagreement over what the organization's objectives should be, he was forced to resign as its director, the position being subsequently filled by Major Donald E. Keyhoe.

3.2 ▪ OVERUNITY LEVITATION

In the fall of 1957, Brown teamed up with Dr. Frank King and Agnew Bahnson Jr., who also had a strong interest in antigravity research. Bahnson, an industrialist from Winston-Salem, North Carolina, had in 1956 played an instrumental role in establishing the Institute of Field Physics at the University of North Carolina, an organization dedicated to the study of gravitation. Having a longtime interest in Brown's electrostatic antigravity propulsion work, Bahnson constructed a well-equipped private laboratory in Winston-Salem and invited Brown down as consultant to work with himself and King. Beginning in November 1957 and continuing for several years, the three carried out electrogravitics research on various kinds of "ballistic electrode" saucer models. Bahnson kept a record of their work in a series of laboratory notebooks, and some of this was reviewed by Charles Yost in the second issue of *Electric Spacecraft Journal.*[7] This work led Brown and Bahnson to file a series of U.S. patent applications in May 1958: an "electrokinetic apparatus" patent awarded to Brown in June 1965 (3,187,206) and two "electrical thrust producing device" patents awarded to Bahnson in 1960 and 1966 (2,958,790 and 3,263,102).

That same year, Brown and his friends organized a company called Rand International Limited, with Brown serving as its president. Together, they carried on electrogravitics experiments and applied for more than seventy-five patents in twelve major countries (the United States, Australia, Canada, France, Belgium, Great Britain, Germany, Holland, Italy, Japan, Sweden, and Switzerland).

It was around this time that Brown succeeded in developing a 15-inch-diameter, dome-shaped saucer that was capable of levitating its own weight! Kitselman, Brown's mathematician friend, related that he had contacted Brown after being out of touch with him for several years and was told, "The lift isn't just 1 percent any longer; the apparatus will now lift 110 percent of its own weight!" Kitselman and his wife immediately flew to Washington and with their own eyes saw a moderately heavy gadget made of metal and Pyrex lift itself right up when 50 kilovolts of electricity were applied and float steadily when a slightly lower voltage was used.[8]

In an April 1973 letter to Schaffranke, Brown confirmed that he had performed this demonstration but indicated that he had conducted experiments throughout the entire voltage range from 50 to 250 kilovolts DC. Illustrating his letter with the sketch depicted in figure 3.2, Brown wrote:

> Mr. Kitselman witnessed an experiment utilizing a 15" circular, dome-shaped aluminum electrode, wired and energized as in the attached sketch. When the high voltage was applied, this device, although tethered by wires from the high voltage equipment, did rise in the air, lifting not only its own weight but also a small balance weight which was attached to it on the underside. It is true that this apparatus would exert a force of upward of 110% of its weight.[9]

In a November 1, 1971, letter written to electrical engineer Tom Turman to respond to some of Turman's questions, Brown described tests on an 18-inch-diameter disc that lifted 125 percent of its weight:

> We used a triarcuate ballistic electrode as the anode and a small electrode underneath as the cathode . . . The large electrode was made of a balsa umbrella-like frame with aluminum foil covering. A thin glass stand-off insulator mounted the cathode as shown in the drawing [drawing shown in appendix A]. The lift of this unit at 170 kv was

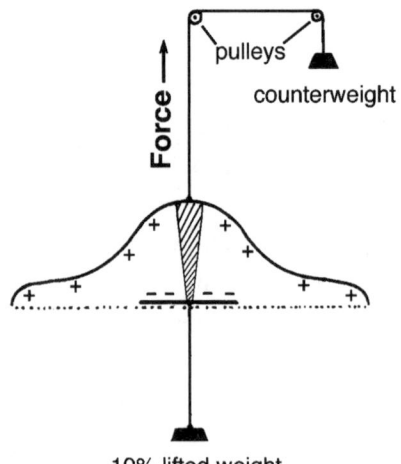

Figure 3.2. A cross-sectional view of a model electrogravitic saucer that was capable of sustained levitation. (Brown, April 5, 1973, letter written to R. Schaffranke)

about 125 grams. The electrode structure itself weighed only about 100 grams, so it was actually self-levitating.[10]

The report titled "Electrohydrodynamics," issued in March 1960 by the Electrokinetics Corporation, presents a diagram of this 18-inch-diameter test model that shows that a toroidal air current vortex was generated beneath the arcuate electrode when the electrode was electrified (figure 3.3).[11] It notes that this vortex was an effective aerodynamic pattern for inducing lift, although vacuum chamber tests that were conducted showed that any momentum that may have been imparted by this ion wind would have been many orders of magnitude too small to account for the observed thrust. The report notes that hydrostatic pressure exerted against the entire inner surface of the large arcuate electrode resulted in a lift force.

In a second letter to Turman, Brown drew cross-sectional views showing how the hydrostatic air pressure is distributed beneath the positively charged electrode for differing inclinations of the negative electrode (see figure 3.4).[12] The pressure was found to be up to 0.25-inch water gauge (~0.64 gram per cm^2) and more positive below the electrode than in the disc's immediate surroundings. Estimating from Brown's sketch, the positive pressure beneath the positive electrode would have averaged about 0.1 inch water gauge, or about 0.25 gram per cm^2. An 18-inch-diameter saucer would have had a cross-sectional area of 1640 cm^2; hence, this pressure would have imparted an upward force of up to 400 grams, more than enough to support the 125-gram weight.

Commenting on various means of steering or stabilizing the flight of such a saucer, Brown wrote:

Figure 3.3. Thomas Townsend Brown's 18-inch-diameter triarcuate ballistic electrode and its generation of a toroidal airflow vortex. (Courtesy of the Townsend Brown Family and Qualight, L.L.C.)

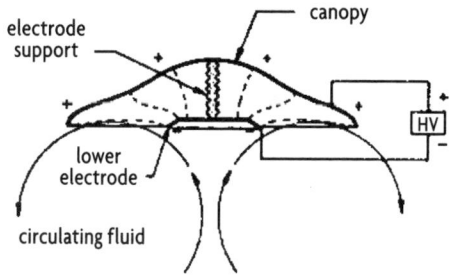

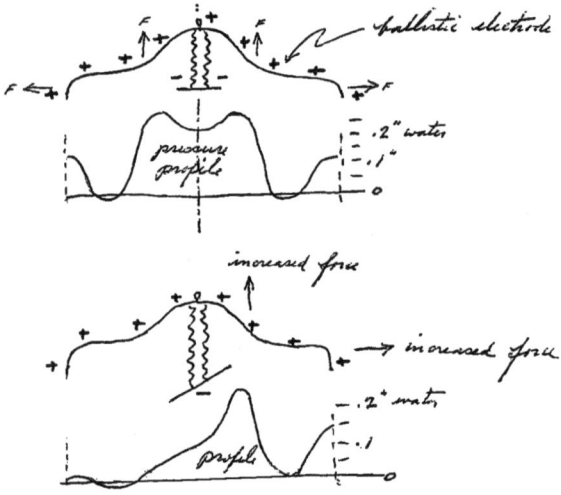

Figure 3.4. The pressure profile under Brown's vertical lift triarcuate ballistic electrode. (Drawing from a 1971 letter from T. T. Brown to T. Turman; see appendix A, letter 3)

It was found that, by canting the center electrode, the pressures could be unbalanced so that one side or the other could be lifted. This could provide horizontal stability in a large prototype. An alternate way of doing this is to provide three independent electrodes in triangular configuration instead of one center electrode. These electrodes can be differently charged in order to change the electric field configuration under the ballistic canopy and this did away with the necessity for a mechanical moving part. Horizontal stability could be maintained entirely electrically.[13]

Brown recognized that the forces involved may not be just electrogravitic, but may also involve a more conventional electrostatic force phenomenon such as electrophoresis (the force exerted on a charged particle in the presence of electric fields) or dielectrophoresis (the force exerted on dielectric materials in the presence of nonuniform electric fields). He referred to this new field of study generally as *electrohydrodynamics*—the study of high-intensity electric field phenomena and their influence on nonconducting (dielectric) media. He regarded this area of investigation to be the counterpart to the more widely known field of magnetohydrodynamics. Electrokinetics Corporation's "Electrohydrodynamics" report also describes vacuum chamber tests in which a 6-inch-diameter ballistic electrode saucer was made to levitate when energized at 150 kilovolts DC, the lift force reaching a detectable level above 10 kilovolts. The graph reproduced in figure 3.5 indicates how thrust and input electric current were found to vary as air pressure was decreased, with the supplied DC voltage being kept constant.[14] The report notes that when the pressure fell moderately below one atmosphere, current rose catastrophically and went off-scale when the pressure had decreased to about a hundredth of an atmosphere (10 mm of mercury). At these low pressures, the air spontaneously ionized, producing a glow discharge that shorted out the electrodes. As a result, the electrogravitic thrust plummeted, only to reappear when the pressure had dropped to the very low value of 4×10^{-6} atmospheres (0.003 mm of mercury). At and below this hard vacuum pressure, the glow discharge diminished, along with a precipitous drop in current to the device. Significantly, the graph demonstrates that *thrust remained constant at 17 grams (0.6 g/cm²), despite the major drop in supplied current.* The report states:

> A significant feature of the curves is that, except for this limitation [the glow discharge gap], thrust remains constant with the reduction

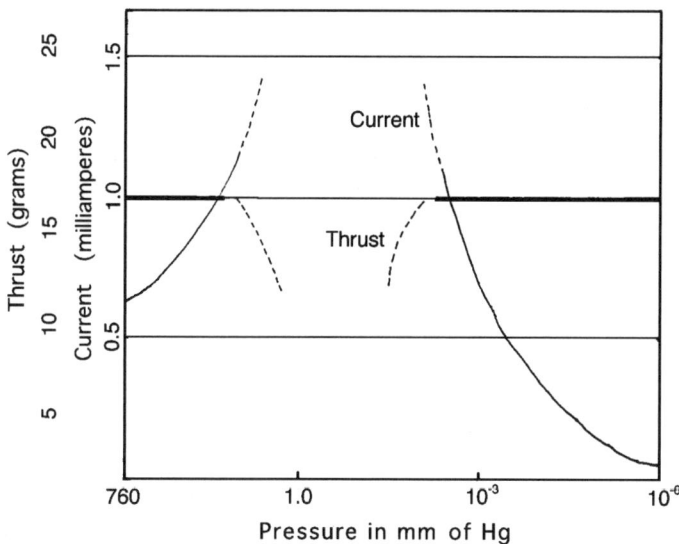

Figure 3.5. Typical plot showing how thrust and current vary with vacuum pressure for an electrohydrodynamic thrust device. (Sketch provided by T. T. Brown to A. Wagner)

in pressure to 10^{-6} mm of Hg, while current consumption falls off sharply—demonstrating the system's improved efficiency as a hard vacuum is approached.

For this reason the strong indication remains that thrust results primarily from electrostatic field stresses, rather than plasma flow. Thus electrohydrodynamics may prove more efficient in a hard vacuum (10^{-12} mm of Hg) than in air where the induced plasma actually seems to result in unnecessary power consumption.[15]

Because the rate of ion flux passing between the electrodes of the device correlates with current, the ion wind should accordingly have dropped precipitously with the drop in current. Hence, the data clearly demonstrate that the source of the observed thrust is due not to ion wind, as critics often charged, but rather to some other force. This raises a question in regard to the canopy pressure measurements described in chapter 2. In the vacuum chamber tests, there would have been no vortical ion movement below the positive electrode, nor any air-induced pressure differential of the sort proposed in explaining the operation of the aero-marine vertical takeoff vehicle. The suggestion in the "Electrohydrodynamics" report that the upward thrust results from "electrostatic field stresses" is explored later in this chapter.

Another test, presumably carried out with a larger electrode and energized at a higher voltage, achieved even greater thrusts:

Laboratory devices weighing 100 grams (approx. 3.5 ounces) less power source have produced a thrust of 110 grams, for an electrical power expenditure of 500 watts (250,000 volts @ 2.0 milliamperes). This experiment was performed in air (1 atmosphere). Supplementary research indicates much greater efficiency results (same thrust for less power input) when operated in a vacuum (10^{-4} mm Hg or better), when the current drops to about 2.0 microamperes.[16]

Here, the report makes the astounding disclosure that under hard vacuum conditions, a force of 110 grams (1.1 newtons) could be achieved for a power expenditure of just 0.5 watt (250,000 volts times 2×10^{-6} amps), or 2,200 newtons of thrust per kilowatt. This is about 150 times the thrust-to-power ratio of a jet engine! Also, it is 10,000 times greater than the thrust-to-power ratio of a space shuttle main engine. The report notes that under harder vacuum conditions of 10^{-12}mmHg, such as exist in space, Brown's electrokinetic propulsion device would be expected to achieve even higher efficiencies due to a further reduction in ion leakage power consumption. The report also compared the observed thrust with that of the National Aeronautics and Space Administration's (NASA) ion propulsion engine, which weighed ten times as much as Brown's saucer and produced only 28 grams of force for an input power of 1,200 watts, or just 0.23 newton per kilowatt. Hence, NASA's ion propulsion engine was 10,000 times less efficient than Brown's electrokinetic disc.

These measurements of thrust indicated that the force lifting Brown's electrified disc was almost 100 million times greater than what could reasonably be generated by an ion wind. For example, an upward thrust of 110 grams is equivalent to a force of about 10^5 dynes. By comparison, a 2-microampere ion wind resulting from electrons accelerated in a 250-kilovolt potential would yield a force of somewhat over 10^{-3} dynes, almost 100 million times less than the lift produced by Brown's apparatus.

Ion Wind Force

The upper-limit estimate for ion wind force may be arrived at through the following calculation: An electron current of 2×10^{-6} amps would comprise an ion wind flux of $\Phi = 1.25 \times 10^{13}$ electrons per second. A voltage drop of 250,000 volts would accelerate these particles to a velocity $v = 2.96 \times 10^{10}$ cm/sec, or 0.988c, which would yield a Lorentz factor of $\gamma = 6.6$. The ion wind force, or electron momentum flux, would equal $F = 9.1 \times 10^{-28}$ g/electron $\cdot \gamma \cdot \Phi \cdot v = 2.2 \times 10^{-3}$ dynes. If the ion wind was made up of aluminum ions instead of electrons, the resulting force would be only thirty-eight-fold greater. By comparison, the lifting force Brown observed amounted to 110 gm $\cdot$ 980 cm/sec^2 = 1.08×10^5 dynes.

This vertical-lift device performed much better than the tethered electrokinetic discs that Brown flew in his demonstration for Will Cady, the scientist from the ONR. It developed fourteen times greater thrust when electrified at a voltage that was more than five times higher. Also, it consumed thirty times less power when operated under vacuum conditions.

Moreover, Brown's apparatus would have yielded a far greater thrust if it had used barium titanate for its dielectric instead of Pyrex. Barium titanate ceramics can have a dielectric constant, K, of around 5,000 when charged with a DC potential, but when rapidly charged and discharged at frequencies of hundreds of megahertz, its K value may drop to about 50. The lower value may be the more relevant here because, as discussed in section 3.4 of this chapter, it appears that he was cyclically varying the voltage potential across the plates of his electrokinetic apparatus at frequencies of around 750 megahertz to maintain a high thrust condition. The "Electrohydrodynamics" report notes that thrust on the electrokinetic apparatus was observed to increase directly with the K value of the dielectric and according to the square (or in some cases the cube) of the applied voltage. Consequently, a dielectric having a K equal to 50, which is about twelve times greater than the dielectric constant for Pyrex, would produce a twelvefold-greater thrust. Also considering that barium titanate has a mass density 2.7 times that of Pyrex, it should develop a proportionately greater thrust when subject to a gravitational gradient. So, one might expect overall a thirty-two-fold improvement, allowing Brown's apparatus to generate a phenomenally high levitating force of about 3.5 kilograms! If this high-K dielectric device consumed only half a watt of power, its thrust-to-power ratio would have had the unusually high value of 70,000 newtons per kilowatt, almost five thousand times that of a jet engine.

The measurements reporting a power consumption of only half a watt may have been referring to just the DC power consumption. If substantially greater power was needed to establish an AC field across the apparatus, then the thrust-to-power ratio value estimated above would have to be reduced accordingly. If tests were conducted with barium titanate, curiously, they are not mentioned in the report.

Brown probably did not openly discuss such test results because they were so phenomenal. If reports were written, it is possible that they are currently classified.

Consider the usefulness of this technology for spaceflight. Suppose that a high-K electrokinetic thruster was able to achieve a thrust-to-power ratio of 70,000 newtons per kilowatt. Energized by a 3-kilowatt power source, a bank of 6,000 asymmetrical capacitors could develop a thrust of 21 metric tons (210,000 newtons), enough to propel a small, 100-ton spacecraft to Mars in just over five days. Using chemical or nuclear rocket propulsion, the same trip would require anywhere from five to seven months and demand a far greater fuel load. If this 100-ton spacecraft were placed in space and left to accelerate under the influence of this 0.21-g force (21 tons/100 tons), its propulsion efficiency would increase linearly with time. By the end of one second, when it had attained a velocity of about 2 meters per second, its propulsion efficiency would already have reached 7,000 percent.* Moreover, its efficiency would continue to increase tenfold with every additional tenfold increase in flight time. Because kinetic energy far in excess of the inputted electric energy appears to be created out of nowhere, Brown's device, by its nature, violates the first law of thermodynamics. Such iconoclastic results become the new norm when one steps into the era of field propulsion.

3.3 ▪ NONLINEAR FIELDS

Unlike his tethered flying disc models, Brown's levitating saucer models had no bumper wires for generating positive and negative ions. An idea of how they achieved their high antigravity thrust was put forth in Brown's "electrokinetic apparatus" patent, filed in 1958 and awarded in 1965, which discussed the assembly shown in figure 3.6. The patent

---

*A 0.21-g accelerating force will accelerate a 100-ton craft at the rate of 206 cm/s². After one second, this craft will attain a velocity of about 2 m/sec and have a kinetic energy of $E_k = 2.1 \times 10^{12}$ ergs. During this second, the craft's 3-kilowatt electric power consumption, E_e, would have resulted in a total energy consumption of 3,000 joules, or $E_e = 3 \times 10^{10}$ ergs. Hence, the propulsion efficiency would equal $\varepsilon = E_k/E_e = 7,060$ percent.

attributes the thrust of the device to its ability to produce a nonlinear field gradient between its positive and negative electrodes, the gradient being steeper at the negative electrode than at the positive electrode. This was accomplished by curving the upper, positively charged surface into the form of a parabaloid and by reducing the size of the negative electrode. The negative electrode was pictured either as a small-diameter sphere or as a disc placed at the positive electrode's geometrical focus. Regarding the importance of having a nonlinear field gradient across the dielectric member, Brown wrote:

> I have discovered that if two electrodes are mounted on opposite ends of a dielectric member, and a field emanates from these electrodes which produces a linear gradient through the dielectric member as shown by dotted line 30 of FIGURE 3 [dashed line in figure 3.6], then no thrust is produced by the dielectric member. However, if the field is distorted to produce a nonlinear gradient such as graphically represented by line 32 in FIGURE 3 [solid line in figure 3.6], then a thrust will be produced which thrust will be related to the degree of nonlinearity of the field gradient. One way to produce a gradient which varies nonlinearly is to shape one of the electrodes in a form of an arcuate surface . . .[17]

Brown's patent suggests that the electric field gradient could also be made nonlinear by using a conical dielectric member that tapers toward the negative electrode or one whose dielectric constant K progressively changes along its length, that is, one that preferably decreases toward the negative electrode. It also points out that the force is directed from a

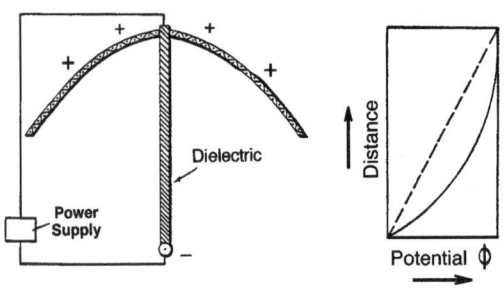

Figure 3.6. An electrogravitic thrust-producing device described in Brown's 1965 patent (left), electric potential, Φ, plotted versus distance along the length of the dielectric rod (right).

region where the electric flux density is high toward a region where the electric flux density is low. For example, it states:

> By attaching a pair of electrodes to opposite ends of a dielectric member and connecting a source of high electrostatic potential to these electrodes, a force is produced in the direction of one electrode provided that electrode is of such configuration to cause the lines-of-force to converge steeply upon the other electrode. The force, therefore, is in a direction from the region of high flux density toward the region of low flux density, generally in the direction through the axis of the electrodes. The thrust produced by such a device is present if the electrostatic field gradient between the two electrodes is nonlinear. This nonlinearity of gradient may result from a difference in the configuration of the electrodes, . . . from the shape of the dielectric member, from a gradient in the density, electric conductivity, electric permittivity and magnetic permeability of the dielectric member, or a combination of these factors.[18]

Brown often emphasized that nonlinear electric fields were central to the phenomenon. In a letter to Turman in 1968, he wrote that this 1965 patent held the key to understanding electrogravitics:*

> The Patent No. 3,187,206 contains the essential teaching in electrogravitics. A definition of the electrogravitic force might be "the ponderomotive force developed within a high-K dielectric under electrical strain." The patent teaches the use of nonlinear electric fields such as those internally developed in truncated cones of dielectric material. . . . The belief that the phenomenon is gravitic in nature is based almost entirely upon the appearance of the effects of mass (in the dielectric material) on the force exhibited.[19]

Brown did not elaborate as to why a nonlinear electric field would produce increased thrust. He arrived at his conclusions from careful observation but offered no theory to account for them. However, by extending the theory described in chapter 1 for the electrogravitic

---

*This and other letters that Brown wrote to Tom Turman are reproduced in appendix A.

effects of charge densities, it should be possible to gain an understanding of how nonlinear electric fields might boost a capacitor's electrogravitic thrust. This involves understanding how a nonlinear field would produce a *virtual-charge-density gradient* across a dielectric and how this gradient, in turn, would generate a gravity field gradient. A virtual charge is a charge source, like an electron or proton, but is one that is not associated with any particle. It instead arises from the ambient electric field continuum whenever that field varies with distance in a nonlinear manner.

This virtual-charge concept is most easily understood within the "ether physics" context presented in the next chapter. Thus we will defer discussing the virtual-charge electrogravitic thrust effects on Brown's electrokinetic apparatus until after the subquantum kinetics ether concept has been introduced. It should be noted here, however, that with sufficient field nonlinearity, the electrogravitic thrust effects produced by virtual charges could far exceed those produced by the real charges being applied to the capacitor plates.

Part of the thrust developed by Brown's electrokinetic apparatus would also have been produced by unbalanced *electrostatic* forces acting on the charges on the capacitor's plates. An unbalanced residual force would have been present because the capacitor's nonlinear electric field would have exerted more force on the smaller electrode than on the larger electrode. Such a residual force would be absent in a conventional capacitor having equal-size electrodes. Such capacitors establish a linear electric field across their dielectric when charged, the electric potential gradient being the same at their negative pole as at their positive pole. The electric field gradient being created across the charged capacitor, then, would electrostatically attract the capacitor's negatively charged electrode toward the field's positive pole and the capacitor's positively charged electrode toward the field's negative pole. Because both capacitor plates would carry the same surface charge density and be subject to the same field gradient, these two attractive forces would be equal to and opposite one another. Hence, the capacitor would experience a compressive force pushing the plates in toward the dielectric. However, since these forces would balance one another, the capacitor as a whole would experience no net translatory motion. In this force analysis, we assume that the capacitor's electric field is anchored neither to the capacitor's charges nor to the capacitor's plates, but rather resides in the capacitor's environs as an independent entity capable of exerting forces on the very same charges that created it.

Now, let us instead consider an asymmetrical capacitor of the kind that Brown used in his electrokinetic saucer experiments, one whose positive electrode is larger than its negative electrode (figure 3.7). The electric field across this type of capacitor varies with distance in a non-linear fashion, the electric flux density and field gradient being highest at the capacitor's negative pole and lowest at its positive pole. Suppose the capacitor is in a vacuum and, hence, has no ionic charges around it. The electric field established in the vicinity of the lower negative electrode would induce an attractive force on the negative charges gathered there, which would be directed upward toward the electric field's positive pole. Also, the electric field established in the vicinity of the upper positive electrode end of the dielectric would induce an attractive force on the positive charges gathered there, which would be directed downward toward the field's negative pole (figure 3.7). However, since the electric field lines converge toward the negative electrode, the field gradient would be stronger there as compared with the positively charged

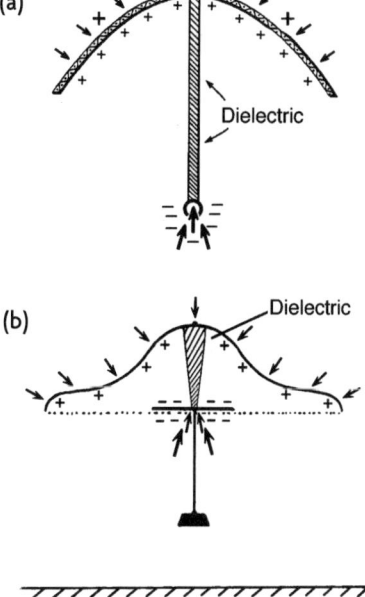

Figure 3.7. Unbalanced electrostatic repulsion forces induced on the dielectric members of two of Brown's saucer designs: (a) design illustrated in his electrokinetic apparatus patent and (b) apparatus used in test witnessed by Kitselman. (P. LaViolette, © 2007)

end of the dielectric. Consequently, the force pulling the lower end of the dielectric upward would be greater than the force pulling the upper end of the dielectric downward. As a result, the capacitor as a whole would experience a net force directed upward toward its larger electrode.

Moreover, the force vectors on the upper electrode would have only a fraction of their total force directed downward. The more peripheral regions of the positive electrode that make up most of the electrode's surface would have their force vectors angled inward, toward the dielectric axis, with a lesser vector component being directed downward (see figure 3.7). It then becomes clear why Brown fashioned his upper electrode into an umbrella-like arcuate shape, so that it curved downward to enclose all or most of the dielectric column. This pulled a greater number of positive charges toward the electrode's periphery and changed the direction of the attractive forces affecting those upper electrode charges, causing their vectors to be angled more horizontally, perpendicular to the dielectric axis. Thus the downward force on the upper electrode would be less than the upward force on the lower electrode both because the upper electrode force magnitudes would be less and because they would be vectored so that only a portion of their thrust would be aimed downward.

Such unbalanced electrostatic forces would produce a thrust toward the larger electrode even if the capacitor's polarity was reversed. This is because the direction of the residual electrostatic force is not linked to plate polarity, but to the direction in which the electric field diverges. The directional dependence for this electrostatic thrust differs from that for the electrogravitic thrust which, as explained in the previous section, is always directed toward the positive electrode. So, depending on the polarity of the DC field across the asymmetrical capacitor, the unbalanced electrostatic thrust would either reinforce or oppose the electrogravitic thrust.

In his patent, Brown mentions that his asymmetrical electrokinetic apparatus always produces a thrust toward its larger electrode, even when the electrode polarity is reversed, although he mentions that the thrust is greater when the larger electrode is positive rather than negative. This suggests that the electrogravitic force is being overpowered by the unbalanced electrostatic thrust that depends on field geometry rather than plate polarity. However, his statements here appear to be referring to the case where his thruster is energized with a DC potential. He does not comment on the correlation of thrust direction with plate polarity when an AC field is energizing the apparatus.

In responding to the unbalanced electrostatic force, the upward movement of Brown's electrokinetic apparatus would occur *with no recoil displacement of the electric field it was generating*. In fact, as the apparatus moves upward, the charges that generate this field would also move upward, so the field would move upward as well. Thus, by means of this unbalanced force effect, Brown's saucer, so to speak, "picks itself up by its own bootstraps."

Standard physics is somewhat split over the issue of whether a field might exist as an independent entity. For example, most would agree that electromagnetic waves propagate as entities independent of the displaced charges that first created them. However, the notion that an electrostatic field exists independently of the charges creating it poses a problem for theories that view electrostatic attraction as being mediated by entities such as virtual particles that are "mechanically" ejected and subsequently absorbed with a momentum recoil at the time of ejection and an equal and opposite momentum transfer at the time of absorption.

Subquantum kinetics, however, avoids this source-to-target momentum exchange requirement. According to subquantum kinetics, charges are able to create an electric field without suffering any momentum recoil from the effect that this field might produce on other charges. In other words, the source charges are blind to the consequences of the field they are producing. Also, when the field's voltage gradient accelerates these other charges, they respond with no recoil being transferred back to the field. In the familiar case of repulsion between two like-charged particles, each particle acquires its repulsive impulse by responding to the other particle's field with no recoil momentum being imparted to the field itself. This reactionless electrostatic thrust idea is generally consistent with a similar idea independently advanced by French inventor Jean-Claude Lafforgue (see discussion in chapter 4). The existence of electrostatic thrust in asymmetrical capacitors has been demonstrated in tests of devices developed by Lafforgue, which are reviewed in chapter 12.

As the nonlinearity and steepness of the applied voltage gradient increase, so too does the net thrust developed from the unbalanced electrostatic forces. The magnitude of the induced electrostatic force also depends on the amount of charge stored on the capacitor's plates. This explains why Brown emphasized using a high-K dielectric. For a given voltage differential, high-K dielectrics are able to store up more electric charge on their end electrodes, the amount of charge being directly proportional to their K factor. With a greater charge load, the plates are able to generate a proportionately greater unbalanced attractive force with respect to the ambient field. In agreement with this, the "Electrohydrodynamics" report states that under vacuum, the thrust on the triarcuate electrode was observed to increase in direct proportion to the K of the dielectric, that is, in proportion to the stored charge; recall the previous section.

Increasing the voltage across the dielectric would also cause the dielectric to store more charge because its ability to store charge is directly proportional to the applied voltage differential. At the same time, the higher voltage would increase the electric potential gradient across the capacitor and thereby augment the inward attractive force acting on each of its electrodes. Hence, for a given field geometry, a given increase in voltage should produce a far greater increase in thrust. Indeed, Brown found that the thrust on his saucer varied as the square or cube of voltage.

When Brown's electrokinetic apparatus was operated in an atmosphere, ionic forces also played a role, although in view of the results of the vacuum chamber tests described in the "Electrohydrodynamics" report, such forces could not have been very significant. Positive ions tended to be emitted on the underside of the canopy, the side facing the negative electrode, and they produced an upward-repelling force on the positive charges in the canopy. Also, negative ions emitted from the lower, negative electrode produced a negative space charge located somewhat above this electrode. This repelled the negative charges in that electrode, producing a force directed inward toward the electrode and angled downward. It is difficult to say whether the force produced by the upward-repelling positive ionic charges prevailed over the force produced by the downward-repelling negative ionic charges to produce a net upward thrust. Ionic forces appear to be more important in understanding the operation of the lifter devices that are described in chapter 12.

To summarize, electrostatic forces on the plates of a capacitor become unbalanced when the electric field intensity varies nonlinearly with distance between the plates. The net thrust increases as the field nonlinearity increases in accordance with the teachings of Brown's patent.

3.4 ■ AC FIELDS

Careful reading of Brown's 1965 patent indicates that he proposed applying an AC voltage across the high-K dielectric of his thrust-producing device. He may have gotten a clue to energizing his dielectric with a nonlinear AC field potential as a result of studying the results of his Paris vacuum-chamber experiments. Observing that the test rotor in those experiments developed a very high electrogravitic thrust during each of its spontaneous electron discharges, it would have been natural for him to steer his research in the direction of duplicating these high-thrust conditions by rapidly charging and discharging his vertical thrust electrokinetic apparatus with a high-frequency oscillating field.

Brown's 1965 patent suggests that a cyclically varying potential would repeatedly establish a nonlinear field gradient along the length of the dielectric member and increase the resulting thrust:

> In applying potentials to these various embodiments, it has been found that the rate at which the potential is applied often influences the thrust. This is especially true where dielectric members of high dielectric constant are used and the charging time is a factor. In such cases, the field gradient changes as the charge is built up. . . . One advantageous manner of applying potential is that of employing potentials which vary cyclically.[20]

In his patent, Brown proposed applying a high-voltage AC field in the megahertz radio frequency range to a hornlike device fitted with a conical dielectric (figure 3.8). The small disc (29) at the apex of the dielectric was identified as a "half-wave radiator," and the applied AC voltage was said to be of a frequency such that a half-wavelength spans the disc's diameter.

Note that the saucers that Brown tested, which succeeded in levitating 110 to 125 percent of their own weight, also used a disc electrode attached to the tip of a central dielectric column. Hence, these devices were essentially the same design as the microwave device pictured in figure 3.8, except that their positive electrode was curved rather than conical in shape and the disc antenna was somewhat larger. So the dramatic lift Brown obtained in these experiments may have been because he was applying radio frequency AC in addition to the high-voltage DC bias potential. In describing this experiment to Kitselman, Schaffranke, and Turman, Brown never mentioned that he was also using AC. Perhaps this was the key to the practical application of his technology, and for that reason, he wished to keep that aspect proprietary.

In the case of the device shown in figure 3.2, whose canopy was 15 inches in diameter, the negative disc electrode would have been about 4 inches (i.e., 10 cm) in diameter. In the absence of an attached dielectric, it would have been most efficiently excited at a frequency of around 1.5 gigahertz to radiate a 20-centimeter wavelength. However, in this case, in which the same size disc is cemented to the apex of a dielectric cone, the dielectric changes the disc's impedance so that the antenna would be driven more efficiently at a lower frequency, say

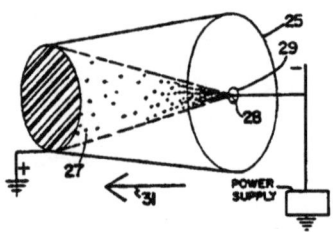

Figure 3.8. An electrokinetic apparatus proposed by Thomas Townsend Brown that used high-voltage AC to generate an electrogravitic thrust. Numbers indicate the following: positive electrode in the form of a frustrated metallic cone (25); frustrated dielectric cone (27) containing semiconducting particles near its tip (28); negative electrode in the form of a disc serving as a half-wave radiator (29); direction of electrogravitic thrust (31). (From Brown, U.S. patent 3,187,206, figure 4)

of 750 megahertz, which would be in the UHF range.* This would radiate a proportionately longer free-space wavelength of 40 centimeters. Provided that the negative disc electrode was spaced from the positively charged canopy by a quarter-wavelength distance (about 10 centimeters), the canopy (or horn) would act as a resonator cavity, allowing a 10-centimeter quarter-wavelength standing wave to build up across the two ends of its central dielectric.

Let us suppose that the DC potential bias applied across the capacitor dielectric was chosen to be -100 kilovolts and that the AC field amplitude was adjusted to have a comparable value of 95 kilovolts and was applied so that the negative electrode potential was left free to oscillate relative to the grounded positive electrode. The net potential across the capacitor, due to the summed AC and DC potentials, then would have varied between -5 kilovolts and -195 kilovolts. As explained in the next chapter, this repeating unipolar oscillation would have generated a virtual-charge gradient across the dielectric that would have produced an oscillating unidirectional electrogravitic thrust on the apparatus.

The electrogravitic thrust would have been stronger than that achieved with DC energization alone because with such a rapid charging cycle, the dielectric would have had insufficient time to appreciably polarize in response to each voltage onset. Hence, the dielectric's opposing electrogravitic dipole moment would have been unable to build up sufficiently to cancel out the imposed electrogravitic field, allowing a maximal thrust to be exerted throughout the dielectric. Unlike Brown's gravitator, described in chapter 1, whose forward thrust progressively diminished after being initially energized, the oscillating potential applied across his electrokinetic thruster would have caused it to receive a series of rapidly recurring forward thrusts.

Brown's patent suggests that the half cycle period of the AC voltage oscillation applied to the asymmetrical capacitor's negative electrode was

---

*For example, copper ferrite–barium titanate composites have a dielectric constant of about 50 in the frequency range of 10^8 to 10^9 hertz, the dielectric constant of such high-K dielectrics being much lower at high frequencies than at low frequencies. Knowing that the optimal driving frequency scales as $1/\sqrt{K}$ and that the dielectric covers only a small portion of the diameter of the radiating disc, we might surmise that the optimal driving frequency would be reduced by a factor of 2 rather than 7, making it about 750 megahertz.

comparable to the time taken for this voltage change to travel across the dielectric to the capacitor's positive electrode. Under such a circumstance, the applied oscillation would increase the nonlinearity of the field spanning the dielectric and thereby boost the thrust arising from both the electrogravitic effect and the unbalanced electrostatic force effect.

There is another aspect to this AC energizing that Brown did not discuss in his patent—namely, that a phase-locked stationary wave pattern would have formed beneath his disc and stored up the energy of each AC cycle. This would have created an electric and gravity potential gradient in the space around the disc that would have progressively increased over time, eventually becoming far greater than the gradient applied during any individual cycle. Tesla observed this effect in his experiments with high-voltage, high-frequency shock discharges. This important effect is well known to "black-project" engineers, who term it "field-induced soliton phenomenon." It is explored further in chapter 8, in the discussion of phase-conjugate resonance.

Rapid recurrent charging of the dielectric should also improve the thrust arising from unbalanced electrostatic forces. As the capacitor dielectric polarizes, the charges on the capacitor plates become partially neutralized by charges of opposite polarity supplied by the adjacent dielectric, so the residual thrust on the capacitor would tend to decline. By charging the plates quickly and repetitively, without any polarity reversal, the electrostatic thrust could be maintained at a maximal level.

3.5 ▪ ELECTROGRAVITICS GOES BLACK

When Brown began working on this AC electrogravitic resonator concept, he may have been getting too close to something that governments considered top secret. Around 1959, there was a substantial change in openness about antigravity research. Earlier, during the mid-1950s, aerospace companies did not hide from the public the fact that they were conducting electrogravitics R&D, although they generally kept the particulars of their own work confidential. For example, an article by A. V. Cleaver, from Rolls-Royce's Aero Engine Division, indicates that as of the beginning of 1957, secrecy had not been imposed but might be imminent. The article states, "The fact that there appears to be no very high security rating attached to it in itself suggests that definite results have not yet been achieved; if, and when, they are one would expect the usual 'clamp' to be tightened down."[21]

Openness continued, even into the early part of 1958, with the subject inspiring heated discussion at a January aeronautical science meeting in New York. Just prior to the event, *Product Engineering* magazine carried the following news brief:

ELECTROGRAVITICS: SCIENCE OR DAYDREAM?

A few weeks from now, at a special session of the Institute of the Aeronautical Sciences (New York City, Jan. 27–31), a group of dedicated men will discuss what some people label pure science-fiction, but others believe is an attainable goal. The subject: electrogravitics—the science of controlling gravity.

After exploring various notions of gravity, the article finally concludes:

Perhaps British aeronautical engineer A. V. Cleaver is right in insisting that if any anti-gravity device is to be developed the first thing needed is a new principle in fundamental physics—not just a new invention or application of known principles. Nevertheless, the Air Force is encouraging research in electrogravitics, and many companies and individuals are working on the problem.[22]

After the meeting, *Business Week* magazine reported the following:

If anyone had predicted 10 years ago that a cross-section of the nation's top physicists, aeronautical engineers, and mathematicians would be fighting for standing room to hear the chaste theory of gravity seriously challenged, he would have been labeled sun-stroked, senile, or worse.

. . . At an opening day meeting of the Institute of Aeronautical Sciences in New York last week, however, the impossible became possible. In record numbers—in a rush that stacked up scientists 20 deep at every entrance to the Sheraton-Astor's North Ballroom—the elite of research came to hear what it is that has reawakened scientific interest in the possibility of doing something about gravitation.

What has happened, they wanted to know, that has caused major aircraft companies as well as the government and various universities, to start serious inquiries into the possibility of controlling gravity? How do the recent discoveries in antiproton research fit into the picture? And even more importantly, how accurate are the reports (circulated by Tass) that Russian scientists hope to turn up some sort of machinery to cancel or modify the force of gravity sometime during 1958?[23]

Business Week went on to list an impressive array of companies and institutions backing gravity research, companies such as the Glenn Martin Company, Grumman Aircraft Engineering Corporation, Lockheed Aircraft Corporation, Sperry-Rand Corporation, the Army Transportation Research and Development Command, Princeton University, the University of North Carolina, and the University of California. Hughes Aircraft should also be added to the list of organizations that by 1958 had become involved in antigravity research.[24,25]

Yet this climate of openness began to change very soon after, as companies became increasingly silent about their involvement in gravity research. In the July 1959 issue of *Canadian Aviation,* Charles Carew wrote, "The author has not been able to determine whether the Glenn L. Martin Corp. has discontinued its antigravity program or made a significant discovery which has elevated it to super-top-secret category, since no information about the project has recently been available."[26]

This indicated that Glenn Martin had made a decided turnabout from its unusually outspoken support of electrogravitics, evident in its vice president's statements to the press in 1955. Most probably, antigravity research had begun to be funded by the military and as a result had continued under a cloak of secrecy. This could explain the difficulty Brown had been encountering in promoting his ideas. During this period, he had been slowly and patiently giving demonstrations for the Pentagon and key aerospace companies in the United States, hoping to generate some interest in his work, but success continually eluded him. In *The Philadelphia Experiment,* Moore wrote that "such interest as he was able to generate seemed to melt away almost as fast as it developed—almost as if someone (or perhaps something?) was working against him."[27]

As mentioned earlier, beginning in 1957 Brown was invited to work on a consulting basis with Agnew Bahnson Jr. to investigate electrogravitic propulsion. Together with Dr. Frank King, they had explored methods of applying AC fields to electrified discs. An examination of Bahnson's laboratory notebook shows that on January 5, 1958, he had suggested to Brown a variation of the AC electric field concept.[28] Bahnson's idea was to place a parabolic metal grid between a positively charged parabolic lift canopy and a negatively charged sphere, as shown in figure 3.9. Then he applied an oscillating high-voltage field between the lift canopy and the negative sphere. He chose a megahertz frequency for this oscillating field that would establish a resonant stationary wave between the two electrodes. He hypothesized that this electrostatic stationary wave would somehow store the energy of the applied AC field in an "ether-like" energy reservoir residing in the space immediately around the test device. He felt that this resonant condition might allow the latent energy in this stationary field to be used with a minimum of power consumption. So here we see Bahnson hitting upon this same key idea of a field-induced soliton phenomenon.

Bahnson also described the use of alternating current fields in a U.S. patent that he filed in September 1964 on an electrogravitic levitation device (see figure 3.10).[29] The data in his patent indicate that his test rig had developed a thrust of 100 grams at 150 kilovolts, with thrust increasing exponentially according to the 2.6 power of voltage. This performance was comparable to the levitating rigs that Brown had

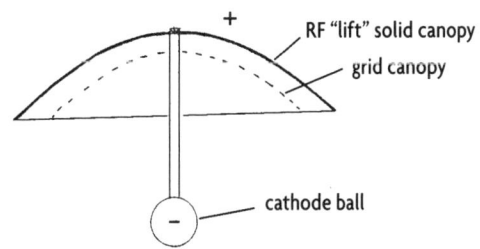

Figure 3.9. An electrogravitic lift device suggested by Bahnson. (After C. Yost, Electric Spacecraft Journal, *May/ June 1991, vol. 8)*

tested, which again leads us to suspect that Brown was using AC to get the impressive results that his friend Kitselman had witnessed.

Shortly after Bahnson filed his patent, tragedy struck. He was killed under somewhat unusual circumstances, when his private airplane reportedly struck a high-tension wire.[30] His patent, which was issued in December 1965, was assigned to his estate. His heirs, having no interest in further developing Bahnson's antigravity work, sold his patents to another company.

Brown's electrokinetic apparatus and his electric generator patents (3,187,206 and 3,196,296) were issued several months before Bahnson's, in June and July 1965. Brown's patents were assigned to the Electrokinetics Corporation, a company that Philadelphia businessman Martin Decker had formed in collaboration with Brown to develop Brown's electrokinetic devices. The company was located in the Philadelphia suburb of Bala Cynwyd, where Decker was operating an industrial compound. This was only eight miles from the General Electric Space Center in King of Prussia, where Brown had conducted vacuum chamber experiments in 1959. Brown had consulted for Electrokinetics since the early 1960s and had received a considerable

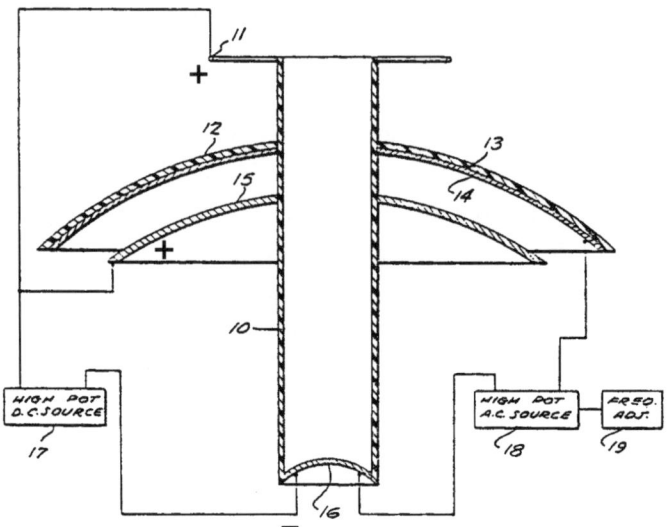

Figure 3.10. An electrogravitic thrust-producing device as illustrated in Bahnson's 1965 patent.

quantity of stock in the company in exchange for the assignment of his patent rights. In his biography of Brown, Schatzkin writes that in the summer of 1964, Decker had told Brown that his Electrokinetics stock had become worthless.[31] Nevertheless, the company must have continued to function, as patents of Brown's issued as late as 1967 are listed as being assigned to Electrokinetics. Some mystery appears to surround this company's circumstances.

Electrokinetics Corporation kept its electrogravitics work veiled in secrecy because in 1968, in a letter to Turman, Brown wrote, "The company to which I serve as consultant has not released some of the information you have requested and I am very much at a loss to know what to say."[32] This shroud of secrecy over Brown's work apparently continued into the 1980s because in February 1982, responding to an Illinois gentleman inquiring about the status of his work, he wrote, "I regret to advise you that electrogravitic research has been taken over in its entirety by a California corporation which has imposed secrecy—at least until their investigations are completed. No further publication or release of information is permitted, possibly until next year."[33] Was this California corporation Lockheed Martin, or was it Hughes Aircraft? Brown did not say.

After his brief consulting work for Electrokinetics in the early 1960s, Brown went into semiretirement, and by 1970 had dropped most of his applied electrogravitic work to busy himself with research on petroelectricity, or so it had seemed. Brown died of natural causes in 1985. Until the time of his death, details about the use of pulsating fields in his electrogravitics research were not forthcoming. Classification restrictions or concerns for trade secrecy presumably discouraged him from openly saying much about this aspect of his work.

Moore's biography of Brown paints a picture of someone who was ahead of his time yet not understood by most of his colleagues, of an inventor who was confronted by one discouragement after another in his attempts to secure government funding and who finally gave up electrogravitic propulsion research at the end of the 1960s. However, one source close to the Brown family indicates that this was not the case, that Brown was being kept in the loop of the secret aerospace research he had catalyzed and that the switch to petroelectricity research was primarily a cover. The public exposure he had received in the past due to media attention to his electric disc technology would have made him a potential security threat. Thus it is understandable that when the military began seriously funding his ideas, the work would have been contracted to a large aerospace company, with Brown being allowed to consult in private as long as he kept quiet about his involvement.

Although Brown was probably legitimately pursuing petroelectricity research, which also included the work he was doing at Stanford Research Institute, this openly acknowledged work must not have occupied him full time. He apparently had obtained high-level clearance and was also quietly consulting on a secret military project that was implementing his electrogravitic propulsion ideas. As described in chapter 5, Brown's electrokinetic technology was eventually incorporated into the B-2 Advanced Technology bomber, serving as its primary means of propulsion. In effect, the B-2 is the realization of the concept Brown first proposed in Project Winterhaven. Its electrogravitic technology would probably have remained a secret were it not for information leaked by a group of engineers who were part of the inner circle working on such super-secret projects.

It seems that the optimistic projections that electrogravitic vehicles would be commonly used for commercial flight have not come to pass as of the present date. Carew's investigations into the unpublicized gravity control research, being conducted all over the globe in 1959, at that time led him to believe that the technique for effectively controlling gravity would be mastered within the lifetime of his readers. The February 1956 *Aviation Studies* report was even more optimistic. It estimated that development of a prototype antigravity combat disc was only ten years away. It predicted that the twentieth century would be divided in half. Whereas air transport during the first half had used aerodynamic principles, heat engines, and flapping controls, it predicted that the second half would arise as a radical offshoot with no ties to past aviation science and that in this new era electrical energy would serve as the catalyst to motion. Gravity, the bitter foe in the first half of the century, would become the great provider in the second half. However, almost half a century later, we still find commercial aviation using the "sledgehammer" approach, employing jet and rocket propulsion technologies. Still, these early predictions were partially correct: gravity control did become practically applied, but not for commercial use. As described in subsequent chapters, antigravity vehicles have been developed for the military and are being flown in large numbers, but knowledge of their existence is being kept a closely guarded secret.

4

AN ETHERIC EXPLANATION

4.1 ■ THE NEW "CLASSIFIED" PHYSICS

As we discussed in chapter 1, the Biefeld-Brown effect proved to be puzzling to scientists right from the start, because of its departure from prevailing theories of gravitation held by classical field theory and general relativity. Einstein's space-warping equations, for example, failed to predict a connection between electrostatics and gravitation. The following passage from *Aviation Report* illustrates this confusion:

> Meanwhile Glenn Martin now feels ready to say in public that they are examining the unified field theory to see what can be done. It would probably be truer to say that Martin and other companies are now looking for men who can make some kind of sense out of Einstein's equations. There's nobody in the air industry at present with the faintest idea of what it [electrogravitics] is all about.
>
> *Aviation Report,* November 19, 1954[1]

Noting that modern physics did not shed much light on the electrogravitics phenomenon, the *Aviation Studies* February 1956 report speculates that an answer might be forthcoming from discoveries providing new insights into the physics of subatomic particles. It suggests that atom-smasher experiments and abstruse field theory calculations might turn up useful leads.

The scientific establishment provided little help in carrying out needed basic research into electrogravitics because its members refused to believe that such an effect could exist. The complacency of the conventional scientific world pertaining to this line of investigation is typified by the response of scientists at the U.S. National Bureau of Standards Laboratory in Boulder, Colorado. Of any government scientific laboratory, this one should have made it its business to be doing basic research into electrogravitic phenomena. Yet in 1985 I asked their expert on gravity measurement, Dr. James Faller, whether he knew of anyone who had done experimental research investigating a possible coupling between charge and gravitational mass. He replied that he knew of no such research. When I asked him why no one had carried out such a study, he answered, "Because there has been no interest."

Nevertheless, since 1956, when the Aviation Studies "Electrogravitics Systems" report was written, there have been vast improvements both in understanding the theory behind the electrogravitics phenomenon and in developing hardware, but most of this work has been carried out in Air Force black projects. In 1992, I had an interesting telephone conversation with a man who is one of the group of informants mentioned in chapter 5 whose stunning revelations about the B-2 bomber were published in *Aviation Week & Space Technology*. Although he gave me his full name, I will identify him as Ray for reasons of confidentiality. Ray claimed to have worked on a number of black R&D projects and to have been in contact with certain other black-world researchers.[2] He told me that the physics theories that academics and most laboratory physicists currently understand, teach, and write about are grossly in error. A very advanced and much more accurate theoretical framework has been developed by scientists of the black-programs community, but its fundamentals presently remain classified. From the standpoint of this new physics, modern physics concepts used in the conventional world, such as relativity theory, quantum electrodynamics, and quantum mechanics are referred to as "classical concepts," that is, they are regarded as terribly outdated.

According to Ray, unlike today's "classical" physics, the new physics does not begin with physical observables in developing its treatment of physical phenomena. Rather, it postulates the existence of an underlying reality consisting of an inherently unobservable subtle substance called an *ether,* or alternatively *aether,* which fills all space. It then defines all of its fundamental quantities at that subphysical level. Physical observables then emerge as mathematical solutions to equations defined in terms of these more basic ether processes. This new physics regards time and space as absolutes and views Einstein's notion of relative time and space as fundamentally incorrect. Physically observable phenomena, such as length contraction and clock retardation, which relativists normally interpret as alterations of the space-time continuum, emerge as manifestations resulting from motion through the absolute ether. Thus, the ether concept, so long spurned by the academic establishment, turns out to be central to this highly classified new physics.

Ray said that this ether physics embraces Brown's electrogravitics phenomenon as well as key research that Brown conducted while he was with the Navy, documents of which have remained highly classified. Perhaps he was referring to work Brown did in connection with the Philadelphia Experiment. Ray stated that this physics also embraces phenomena discovered by Tesla. Among other things, Tesla is known for his work with resonant AC circuits and with techniques for producing unconventional shock discharge Coulomb waves, sometimes called longitudinal waves. As described in chapter 1, the electrogravitic waves that Brown was producing with his communication device were of this sort. How Tesla's work relates to antigravity propulsion is further discussed in chapter 6.

As mentioned earlier, conventional physics is at a total loss to account for the Biefeld-Brown effect. Nor does string theory, with its ten-plus dimensional spaces, offer any insights, and now, after its forty-year reign, many physicists have become disenchanted with it, leaving the search open once again for a unified field theory that will work.[3,4] As of this time there has been no public disclosure of the classified ether physics or of how it explains electrogravitics. However, there is one very promising theory that we can talk about and that does predict many aspects of the electrogravitic phenomenon. This is the ether physics of subquantum kinetics.[5-11] Unlike string theory, which never resulted in any testable prediction, subquantum kinetics has to date had twelve a priori predictions verified, outdoing most standard field theories.*

Let us take a moment to review something about this new approach and examine how it accounts for the mysterious gravitational thrust that Brown was observing. Subquantum kinetics is an approach to microphysics that is based upon discoveries made in recent years in the disciplines of general system theory, nonequilibrium thermodynamics, and nonlinear dynamics. It was inspired from research carried out in the late twentieth century on certain types of nonequilibrium reaction systems that have the ability to spontaneously self-organize wave patterns of precise wavelength. Problems such as wave-particle dualism, field-source dualism, infinite energy absurdity, naked singularities, the cosmological constant conundrum, the wave packet dispersion problem, and many others that plague conventional physics do not appear in subquantum kinetics because it represents quantum phenomena in a very different way.

Like the classified physics of the black-project world, subquantum kinetics begins with an ether as its point of departure. It conceives quantum structures, such as subatomic particles and energy waves, to be concentration patterns that emerge in a primordial reaction-diffusion ether, one whose constituents both diffuse through space and react among one another according to a specified set of nonequilibrium reaction processes. This subtle medium is postulated to extend throughout space and to be composed of subquantum units, called etherons, that come in various types. In a similar manner, conventional physics postulates subquantum structures called quarks that come in various sorts distinguished by their "colors" and "flavors." However, subquantum kinetics, in its current Model G formulation, uses far fewer types of etherons as compared with the number of quarks that physics postulates. Model G involves just seven types of etherons for its specification: A, B, G, X, Y, Z, and Ω. Unlike quarks, which are characteristically unreactive, these etherons are postulated to react with one another and

---

*For information about those subquantum kinetics predictions that were subsequently confirmed, see www.starburstfound.org/LaViolette/Predict2.html.

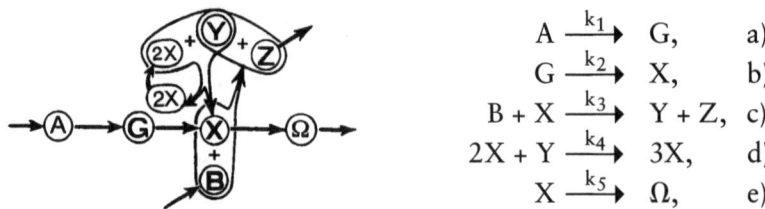

$$A \xrightarrow{k_1} G, \quad \text{a)}$$
$$G \xrightarrow{k_2} X, \quad \text{b)}$$
$$B + X \xrightarrow{k_3} Y + Z, \quad \text{c)}$$
$$2X + Y \xrightarrow{k_4} 3X, \quad \text{d)}$$
$$X \xrightarrow{k_5} \Omega, \quad \text{e)}$$

Figure 4.1. A schematic representation of Model G's reaction kinetic pathways (left), also displayed as a series of five separate kinetic equations (right). (P. LaViolette, © 1995)

transform from one etheron state to another according to a specified set of five reactions, which are collectively termed Model G (figure 4.1).

This reaction system is similar to the Brusselator, a two-variable reaction system developed at the Free University of Brussels, with the exception that it interposes a third variable G between the A and X reaction states, hence the name Model G. The k_i symbols in figure 4.1 are Kinetic constants that specify the rates at which each reaction proceeds forward. Together with the diffusion coefficients that describe the rate at which each etheron type diffuses through space, this set of reactions forms the essence of subquantum kinetics. The basic processes are extremely simple, yet from their interactions emerge physically realistic structures and a very rich array of behavior.

Subquantum kinetics identifies etheron concentration at any given point in space with the standard energy potential concept. In particular, an electric field characterized by a spatial variation in electric potential would correspond to a spatial variation in X and Y etheron concentration. A gravity field characterized by a spatial variation in gravity potential would correspond to a spatial variation in G etheron concentration.

Unlike traditional physics, which is founded on closed-system, mechanistic concepts, the continually reacting and transmuting reaction-diffusion ether of subquantum kinetics functions as an open system. Unlike closed systems, open systems allow the possibility for order to emerge from disorder. Under the proper conditions, the ether is able to spawn subatomic particles that have wavelike characteristics. They form spontaneously from energy fluctuations of sufficiently large magnitude that occasionally emerge from the ether's chaos. Thus, subquantum kinetics espouses a cosmology of continuous matter creation rather than a single big bang creation event.

According to subquantum kinetics, the etheron concentrations are in a state of continual fluctuation throughout space, manifesting as energy potential fluctuations. These are similar to the zero-point energy fluctuations proposed in conventional physics, except of far smaller magnitude, each being less than a quantum of action. Also, they do not necessarily arise as correlated matter–antimatter polarity fluctuations, but rather as individual unipolar pulses that can be of either positive or negative polarity.* This can manifest either as a positive polarity fluctuation—a region of high-Y and low-X ether concentration—or as a negative polarity fluctuation—a region of high-X and low-Y ether concentration.

On occasion, one such electric potential fluctuation "seed" will become large enough that over time it will grow in size and develop into a subatomic particle configured as a stationary electric potential wave pattern. The spontaneous growth of such an energy fluctuation would appear to violate the first law of thermodynamics, which holds that energy may be neither created nor destroyed. But such growth is permissible due to the open-system character of the ether, the action responsible for this growth coming from the ever-present reaction processes that underlie all particle and field phenomena. This matter creation process would occur so slowly that a well-equipped physicist would be unable to detect it in an Earth-based laboratory.

Subatomic particles would not emerge as mass points, as standard physics would conceive them, but as wave patterns configured of etheron concentrations whose magnitudes vary cyclically through space. Figure 4.2a illustrates the spherical, shell-like geometry of the X-Y wave pattern forming a proton, and figure 4.2b presents a stylized cross-sectional view of the variation of the X and Y ether concentrations in the core of a proton. This would chart the proton's electric potential field, a positive charge polarity corresponding to a high-Y/low-X core concentration. The negatively charged antiproton would have a low-Y/high-X

---

*This of course violates the conventional rule of charge conservation but is allowable in the subquantum kinetics framework when dealing with incipient zero-point fluctuations.

core concentration, hence X and Y magnitudes that were of reversed polarity.

The wavelength of this electric field pattern would equal the particle's Compton wavelength, which is numerically equal to the wavelength of the photon that would result if the particle's rest mass was converted entirely to energy. This Compton wavelength, λ, is mathematically quantified as: $\lambda = hc/E$, where E is the rest mass energy of the particle, h is Planck's constant, and c is the velocity of light. The proton's Compton wavelength, for example, would measure just 1.32×10^{-13} centimeters. The electron's wavelength would be about two thousand times longer. Subquantum kinetics calls this wavelike field the particle's Turing wave, in honor of the British cyberneticist Alan Turing. Turing was the first to demonstrate that reaction-diffusion systems could form such wave patterns through a process he termed *morphogenesis*.

One of the successes of subquantum kinetics is that its prediction that the nucleon's core electric field should be contoured with a Compton wavelength periodicity was confirmed almost thirty years later by nucleon scattering experiments.[12] The quark model, by comparison, failed to anticipate the nucleon's wave character.

The standard quantum mechanical view proposes modeling the location of the subatomic particle as a wave packet, a superposition of

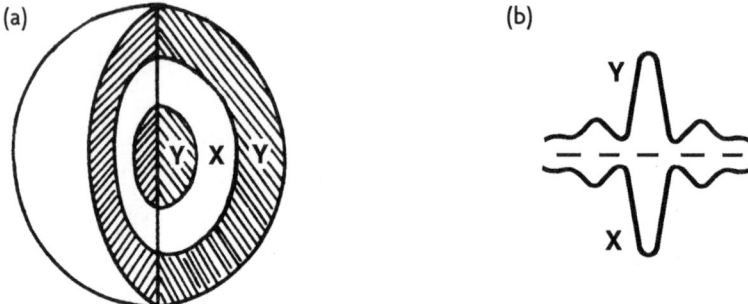

Figure 4.2. (a) A cross-section of the concentric, shell-like ether concentration pattern making up a proton showing alternating Y and X ether maxima. (b) Radial variation of the X and Y etheron concentrations forming the proton's stationary electric potential wave pattern. The peak-to-peak shell wavelength would equal the particle's characteristic Compton wavelength. (P. LaViolette, © 1995)

a series of linear electromagnetic waves. It has long been recognized, however, that the wave packet has the problem that over time it spontaneously broadens and eventually disperses, leaving the particle entirely unlocalized. The subatomic particles that Model G forms, though, do not disperse over time. Their periodic ether concentration structure is continually regenerated by order-creating etheric reaction-and-diffusion processes. The cyclical transformation of X into Y and Y into X depicted in figure 4.1 is what allows the model G reaction system to create and maintain the subatomic particle's stationary wave pattern.

An analysis of the Model G reaction system indicates that there is a stable steady state in which the core Turing wave of an initially neutrally charged particle becomes biased away from its zero-charge steady state. Upon making the transition to this state, the particle acquires a unit of electric charge that allows it to create its long-range electric field. This is not an assumption; it is a characteristic that emerges as a consequence of the postulated reaction system. In a positively charged particle, for example, the electric field pattern would become positively biased at its center, its high-Y core concentration rising and its low-X core concentration falling to adopt bias levels similar to those shown in figure 4.2b. In a negatively charged particle such as an electron, the electric field pattern would instead be negatively biased, the Y core concentration being depressed and the X core concentration being elevated. Recent particle scattering observations that elucidate the charge distribution in the core of the nucleon confirm this wave pattern bias prediction.[13]

Subquantum kinetics predicts that a particle's electric charge generates its gravitational mass and associated gravity potential field through the reverse, ether reaction G ← X, which converts X etherons (X-ons) into G etherons (G-ons). Although the ether reactions shown in figure 4.1 proceed predominantly in the forward direction (to the right in the figure), such reverse reactions also exist, although they proceed at an almost negligible rate. But even though the reverse X-to-G reaction produces a very small G flux, without it particles would be unable to generate their gravity fields. Through this reaction, an increase or decrease of X arising from either a negative or a positive charge polarity translates into a corresponding increase or decrease of G. Accordingly, subquantum kinetics predicts that there should be two gravitational mass polarities, each correlated with a corresponding electric charge polarity. That is, positively charged particles such as protons, which have low X-on concentrations in their cores, should produce a central gravity potential well, while negatively charged particles such as electrons, which have high X-on concentrations in their cores, should produce a central gravity potential hill.

Consider, for example, a positively charged particle such as a proton, which would maintain a high Y-on and low X-on concentration at its center. The low X-on concentration would correspondingly reduce the rate at which G-ons were created in the center of the proton via the X-to-G reaction. In other words, the particle's positive charge would generate a corresponding positive gravitational mass; see figure 4.3. The reduced X-on and G-on core concentrations resulting from the X-on and G-on production rate deficits would induce X-ons and G-ons to diffuse inward from surrounding regions. Also, the elevated Y-on core concentration arising from the Y-on production rate surplus would induce Y-ons to diffuse radially outward to the environment. These radial diffusive fluxes would generate the long-range electric and gravity potential fields that surround the proton's core.

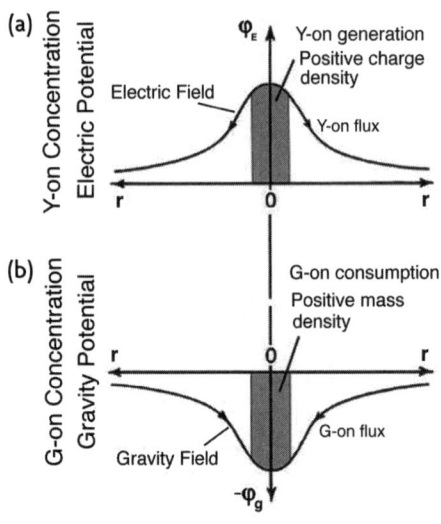

Figure 4.3. (a) The electric potential field of a proton (positive Y-on concentration) being created by a central positive charge density (Y-on production rate surplus). (b) The negative gravity potential field of a proton (negative G-on concentration) being created by a central positive mass density (G-on production rate deficit). Arrows indicate the directions of Y-on and G-on diffusion that create the respective fields. The X-on concentration profile, not shown, is the inverse of the Y-on profile. (P. LaViolette, © 2007)

The electric field would consist of a $1/r$ decline in the particle's X-on and Y-on etheron concentration biases, and the gravity field would consist of a $1/r$ decline in the particle's G-on concentration bias. Both fields mathematically conform to the requirements of classical electrostatics and gravitation. Their $1/r$ character is not assumed ad hoc, but rather emerges as a natural consequence of radial etheron diffusion. In his *Lectures on Physics,* the Nobel laureate physicist Richard Feynman proposed an etheron diffusion model not too different from this as a way of understanding how a subatomic particle generates its $1/r$ energy potential field.[14]

These concentration biases and etheron diffusive fluxes would be just the opposite for an electron. A negatively charged particle such as an electron would maintain high X-on and G-on and low Y-on concentrations in its core, which would constitute the negative charge and negative mass state. This X-on and G-on production rate surplus would induce X-ons and G-ons to flow outward to surrounding regions and the Y-on production rate deficit would induce Y-ons to diffuse inward from the environment. These fluxes would generate the electron's long-range electric and gravity potential fields. A proton's gravity well would have a matter-attracting effect on surrounding neutral matter, while an electron's gravity hill would have a matter-repelling effect on surrounding neutral matter.

Austrian astrophysicist Sir Hermann Bondi described this gravity-field-generating mass with the term "active gravitational mass," distinguishing it from passive gravitational mass, which characterizes the particle's tendency to respond to an external gravity field, and from inertial mass, which characterizes the particle's tendency to resist changes in velocity. Electrically neutral matter, containing equal numbers of protons and electrons, would have a net positive active gravitational mass, since the proton's positive gravitational mass is slightly greater than the electron's negative gravitational mass. That is, the proton's gravity well is only partially canceled out by the electron's gravity hill, and as a result, the electrically neutral atom is left with a residual matter-attracting gravity well. Particles producing gravity wells would attract neutral matter, whereas particles producing gravity hills would repel neutral matter.

It should be kept in mind that subquantum kinetics grew out of the intention to apply chemical kinetics concepts to microphysics and the realization that a reaction kinetic ether could provide a viable framework for describing the formation of subatomic particles. I did not devise it in a concerted attempt to explain the phenomenon of electrogravitics. In fact, I was initially somewhat concerned that the theory predicted two mass polarity states, each correlated with a corresponding electric charge polarity, but was later relieved upon discovering Brown's work.

Subquantum kinetics accounts for gravitational force in terms of the action of gravity potential gradients on material bodies and not by introducing ad hoc assumptions about dimensional warping. Gravity potential fields move a particle by affecting the ether reactions that generate the particle and form its Turing wave. This reaction-kinetic approach is something quite foreign to classical physics and quantum mechanics, which consider a proton or an electron as a relatively immutable structure. Not so in subquantum kinetics. The subatomic particle's etheron population is continuously transformed and renewed. A particle's Turing wave structure is re-created every instant, as etheron diffusion balances etheron creation or dissolution at each point in space. The subatomic particle, then, maintains its field structure in a dynamic steady state, or what the Hungarian systems theorist Ludwig von Bertalanffy would call a *Fliessgleichgewicht,* a "patterned flow equilibrium." Or, using a term coined by the Russian Nobel laureate Ilya Prigogine, we would refer to subatomic particles as "dissipative space structures."

The presence of an external field gradient will necessarily affect the position of a particle because it will disturb the equilibrium of the reaction processes that continually generate the particle's Turing wave pattern. That is, the field gradient will distort the particle's space structure by raising (or lowering) the etheron concentrations more on one side of the particle than on the other. Because the steady-state condition tends to create a symmetrical space structure for the particle, any such departure from symmetry will induce a stress, or a state of disequilibrium. This stress is identified with the electrostatic or gravitational force that this electric or gravity potential field exerts on the particle. The reaction system relieves this stress and momentarily gains greater symmetry by accelerating or moving the particle's regenerating wave pattern either up or down the field gradient, the direction of movement depending on whether the field's action is attractive or repulsive.

According to subquantum kinetics, force action is fundamentally a reaction-kinetic process rather than a mechanical process, one in which the particle's core field readjusts its concentration pattern in response to the disequilibrium caused by an imposed field gradient. Thus the field's induction of force and acceleration is accomplished with no recoil force being applied back on the field. No momentum transfer is involved. Similarly, the potential gradient generated by a charge or mass becomes established in the ether and is able to operate on charges and masses without any deference to the charges or masses that originally generated it. Thus, in subquantum kinetics it is perfectly acceptable for a field to accelerate the very same charges that generate it. This field autonomy is key to understanding how unbalanced electrostatic forces can induce motion in an asymmetrical capacitor, a subject we examined in the previous chapter.

French inventor Jean-Claude Lafforgue proposes a similar idea in his 1991 patent on an asymmetrical thrust capacitor. He suggests that when a capacitor is charged, the fields it generates have their seat in the local space-time continuum reference frame, allowing them to act on the capacitor without any reaction force being directed back to the capacitor itself. Thus, he suggests that the electric fields generated by a properly shaped asymmetrical capacitor can exert unbalanced electrostatic forces on the capacitor, whose residual is capable of displacing the capacitor relative to its initial rest frame. In his patent Lafforgue states:

> It is acceptable then to consider that F [electrostatic force] rests its support on E [electric force field intensity], that is to say, on the space-time continuum . . . It is the same for all the electrodes whatever their orientation and polarization . . . Whether the electrodes are at rest or in motion does not at all change the values of ρ, of σ, of q, of E, nor of F. So under the action of the "force of expansion," the isolated system moves and drags with itself σ and E and consequently F. Our isolated systems are therefore self-accelerated.[15]

Lafforgue's term "force of expansion" refers to electrostatic pressure, that is, the electrostatic force $F = \sigma E$ present on a portion of the capacitor plate divided by the surface area it acts through. Lafforgue does not refer to the notion of an ether. He uses instead the relativistic term "space-time continuum," and he adopts the convention of working with force field intensity, E, rather than the electric potential gradient $\nabla \varphi_E$. However, his conclusions about a capacitor's self-field action, which are based on experimental observation, are surprisingly identical to those predicted by subquantum kinetics. Hence, his work may be considered as independent support for the subquantum kinetics approach. Tests performed on one of Lafforgue's asymmetrical capacitors are discussed in chapter 12.

Classical physics assumes that subatomic particles produce only matter-attracting gravitational forces regardless of their charge polarity. Classical theory doesn't describe what a gravitational field is or how it exerts its force; it just identifies it and mathematically represents how gravitational force is related to the mass of a body and to an observer's distance from the body. As for antimatter, there are a variety of opinions among physicists. Some believe that antimatter particles should produce matter-attracting fields and others believe that they should produce matter-repelling fields.

General relativity, like classical physics, also asserts that all subatomic particles should produce attractive gravity fields regardless of their electric charge. Yet it also advances the additional proposition that these gravitational effects come about because the body's mass warps the surrounding metric of space and time and as a result induces an attractive motion in neighboring masses. Relativity does not state how matter manages to accomplish this warping feat; it simply states this as a given. Although fancy equations are presented to describe the warping assumption, they do not themselves provide any insight as to why matter, a physical quantity, might affect the geometry of space or the rate at which time passes. The initiate is asked to accept this on faith. Subquantum kinetics, on the other hand, does explain how a material body generates its electric and gravity potential fields, why these two fields are interlinked, and also how these fields induce forces on charges and masses. Furthermore, it describes how a material particle comes into being in space, another point on which conventional physics is a bit vague.

Most physicists are reluctant to consider ether-theory explanations of particles and fields because relativists have long drummed into them the misconception that the idea of an ether, with its preferred reference frame, has been disproved. Usually, they cite the Michelson-Morley experiment's failure to detect a directional variation in the velocity of light. However, many have argued that the experiment's null result was due to the fact that it was conducted underground, where the ether was stationary in the reference frame of the rotating Earth. Other experiments carried out by Georges Sagnac, Dayton Miller, and Ernest Silvertooth have since shown that the one-way velocity of light is not a constant.[16-20] Hence, the results favor the notion of an ether over relativity's relative-frame notion.

The vindication of the ether-frame concept even impacts us every day in a practical sense. To establish proper synchronization of the clocks in the global positioning system satellite array, computer software must make allowances for the change in radio signal velocity caused by the array's geosynchronous rotation relative to the local ether frame. Were this not done and the network was instead synchronized in accordance with the pronouncements of special relativity, a hiker would be unable to accurately establish his latitude and longitude coordinates. The military knows this quite well.

Moreover, general relativity has no exclusive claim over tests carried out to check its predictions, such as the gravitational bending of starlight by the Sun and the precession of Mercury's orbit. All of these have also been accounted for in terms of classical physics effects.[21] It might be added that these same gravitational effects are also predicted by subquantum kinetics. What general relativity refers to as a spatial warping of starlight photons by a celestial mass subquantum kinetics understands to be a refraction of photons (ether waves) by the body's gravity potential gradient (G-on concentration gradient). The gravitational time dilation phenomenon that all massive bodies experience and that relativity interprets again as an inexplicable warping of space is understood as a clock retardation effect that arises as a result of the reduction of the G-on concentration within the star or planet (as G decreases, so too does X, and in turn so too does the X-Y flux rate that is responsible for all physical action). The gravitational redshift of light emitted from a massive celestial body occurs for a similar reason. Subquantum kinetics also accounts for special relativistic effects. All of these effects emerge from the five basic equations diagrammed in figure 4.1 (see page 118). In summary, special and general relativistic effects emerge as corollaries of subquantum kinetics, but without requiring the magical warping of space-time.

So it is not surprising that highly classified black-budget programs embrace the concept of an ether in their attempts to understand the gravity-defying technologies that they have been developing. They have no obligation to please the academic physics establishment, which still teaches the rocket principle as the ultimate in space propulsion technology.

Norwegian researcher Dr. Björn Overbye points out that most physicists find it virtually impossible to visualize and understand relativity theory because it requires that one think in terms of four dimensions. He says, "Even experienced mathematicians and theoretical physicists who have worked with higher dimensional space for years admit that they cannot visualize them! Instead, they retreat into the world of mathematical equations."[22] According to the Nobel laureate Hannes Alfvén, "The people were told that only Einstein and a few geniuses that were able to think in four dimensions could understand the true nature of the physical world. Science was something to believe in, not something that should be understood."[23]

Subquantum kinetics, on the other hand, offers us a path back to visualization and understanding. It is based on a simple conceptual framework, but to follow it one must forget the misleading models that have been taught in the past, quantum mechanics and relativity theory being among them. Since the theory is not easily presented in the space of a few paragraphs, the reader is referred for more details to the journal articles cited at the beginning of this chapter and to my book *Subquantum Kinetics*, all of which deal more thoroughly with the subject.[24] The more philosophically inclined are referred to my book *Genesis of the Cosmos*, which presents a less technical introduction to the subject.[25]

4.2 ▪ VIRTUAL CHARGE ELECTROGRAVITIC EFFECTS

In earlier writings of how subquantum kinetics predicts an electrogravitic connection, I did not also consider that virtual-charge gradients might also produce electrogravitic effects. However, it is reasonable to expect that, like real charges, virtual charges also should induce gravity potentials. Moreover, gravitational thrusts on a capacitor dielectric arising from virtual charges may in some cases be far greater than those produced by the real charges creating the capacitor's electric field. So it is evident that the scope of the subquantum kinetics electrogravitics prediction should be expanded to include such virtual-charge effects. Let us proceed to derive this by considering a charged asymmetrical capacitor that establishes a nonlinear electric field across its dielectric.

Due to the math involved, the material presented in the next several pages may be a bit challenging for those who have not had a course in college physics. Some, then, may want to skip over it. But those wishing to acquire a nuts-and-bolts understanding of how antigravity technology might work are encouraged to read this section, if not now, at least at some later time, for this material is referenced frequently later in the book.

As was explained above, subquantum kinetics identifies an electric potential gradient with an X-on or Y-on ether concentration gradient. Consider a charged asymmetrical capacitor whose negative plate is smaller than its positive plate. With such a field geometry, the voltage gradient will vary nonlinearly with distance across the dielectric, the field gradient getting steeper toward the capacitor's smaller negative plate. Represented in terms of the X-on etheron concentration component, C_x, X-on concentration will rise nonlinearly, reaching a maximum at the negative plate. X-ons will continuously diffuse down this concentration gradient away from the negative plate, their flux per unit surface area being represented by the symbol Φ_x, called the X diffusive flux vector. It is related to the X-on concentration gradient, ∇C_x, by the formula $\Phi_x(r) = -D_x \nabla C_x(r)$, where D_x is the diffusion coefficient.

Suppose that we consider three adjacent volumes in the capacitor dielectric aligned along the axis of the capacitor and located near the capacitor's negative plate (see lower three boxes in figure 4.4). Because the field potential varies nonlinearly with distance across the dielectric,

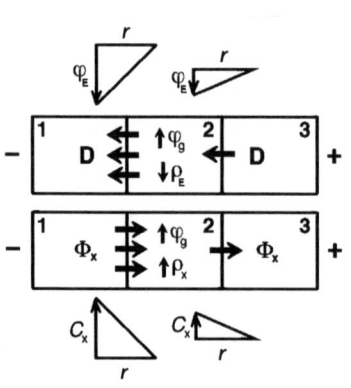

Figure 4.4. The flux vector divergence concept. The lower boxes show adjacent volumes of space along the polarization axis of a capacitor with their corresponding X diffusive flux vectors, Φ_x, arising from a prevailing nonlinear X-on concentration gradient. The upper boxes show, as in conventional theory, the corresponding electric flux density vectors, D, arising from a prevailing electric potential gradient. (P. LaViolette, © 2007)

the X-on concentration gradient, ∇C_x, between box 1 and box 2 will be steeper than that between box 2 and box 3. Therefore, more X-ons will be induced to flow into box 2 than will be induced to leave from box 2. As a result of this flux imbalance, the number of X-ons in that box will increase at the rate $\Phi_{x(in)} - \Phi_{x(out)}$. This has the effect of raising the X-on concentration in that box, which is equivalent to lowering its electric potential, negative potential being correlated with increased X-on concentration.

Because there is a net inflow of X-ons into box 2, the divergence of the X diffusive flux vector $\nabla \cdot \Phi_x$ for that box will be a negative quantity. This diffusive-flux-vector divergence by definition is calculated from the variation of X-on concentration $C_x(r)$ using the equation $\nabla \cdot \Phi_x(r) = -D_x \nabla^2 C_x(r)$. This negative Φ_x divergence, in turn, is equal to a quantity called the X production rate balance density, ρ_x, a scalar quantity that represents the rate, Q_x, at which X-ons enter a unit volume dV. This is mathematically expressed as $\rho_x(r) = Q_x(r)/dV = -\nabla \cdot \Phi_x(r) = -D_x \nabla^2 C_x(r)$.

The X production balance density, ρ_x, is the ether counterpart of the negative virtual electric charge density, $-\rho_E$.* Thus the excess influx of X per unit time into box 2 would act as though a negative electric charge density, $-\rho_E$, was present in that box. Also, the X diffusive flux vector, Φ_x, is the ether equivalent of the electric flux density vector, D, of con-

---

*The Y production rate balance density, ρ_y, would be the counterpart of the positive virtual electric charge density. We instead focus here on specifying the X-on component, since it is this component that is involved in generating the G-on gravity field.

ventional theory. Physicists also refer to **D** as the electric displacement, but we use here the flux density vector designation since it more closely approximates the terminology used for its ether counterpart. It should be mentioned that all of these conventional physics terms were originally developed in the context of 18th- and 19th-century ether models.

The upper set of boxes in figure 4.4 illustrates how these etheron fluxes would be expressed in conventional physics terms using **D** and ρ_E. A nonlinear electric potential gradient $\nabla\varphi_E$ is assumed to extend across the capacitor dielectric with the potential gradient steepening toward the negative electrode on the left. This field gradient induces a greater electric flux density to flow out from box 2 (to the left) than to flow into box 2 (from the right). Consequently, **D** has a positive divergence, which by definition creates a negative virtual-charge density in that box. In conventional terminology, this is expressed as $\nabla\cdot\mathbf{D}(r) = -\varepsilon\nabla^2\varphi_E(r) = -\rho_E$. Rearranging these terms we may write:

$$\rho_E(r) = -\nabla\cdot\mathbf{D}(r) = -\varepsilon\nabla\cdot\mathbf{E}(r) = \varepsilon\nabla^2\varphi_E(r) \qquad (6)$$

Here we also include in the expression the term $-\varepsilon\nabla\cdot\mathbf{E}(r)$, which expresses this equivalence in terms of the permittivity times the divergence of the electric field intensity, **E**, for those who are familiar with this alternate term.

However, in the case of real negative charges such as electrons, the **D** vectors would instead be inward directed making the divergence term $\nabla\cdot\mathbf{D}$ a negative quantity. If this subtle difference between real and virtual charges is not appreciated, one might mistakenly assign an incorrect polarity to virtual charges and calculate gravity thrust vectors pointing in the opposite direction. Taking the etheron model as an example, if real negative charges were present in box-2, X-ons would be entering the box through X-on creation attributable to the Model G reactions. The resulting X-on surplus would produce an outward diverging flux, making the divergence term $\nabla\cdot\Phi_x$ a positive quantity. In the case of a negative virtual-charge density, on the other hand, the X influx arises from the spatial redistribution of X-ons, which enter the unit volume as a result of the electric field's nonlinearity. Thus the diffusive flux vectors will instead point inward, making the divergence term $\nabla\cdot\Phi_x$ a negative quantity.

Based on equation 6 above, we conclude that virtual charge is formed wherever the field's electric flux density vector **D** acquires a nonzero divergence value, div **D** ≠ 0, or in other words, wherever the derivative of **D** differs from zero. Expressed in terms of the electric potential field, $\varphi_E(r)$, a virtual charge would arise wherever the second derivative of the field's electric potential becomes nonzero, that is, $\nabla^2\varphi_E \neq 0$. For the second derivative to be nonzero, the magnitude of the field's electric potential must vary *nonlinearly* with distance.

Now, going back to our ether flux model, let us consider how these virtual electric charge densities would produce a gravitational field across the capacitor. The positive X production rate balance density in box 2 (lower half of figure 4.4) results in the creation of a positive G production rate balance in that box due to the reverse reaction, G ← X, which in turn produces a local increase in G-on concentration, C_g. With this surplus G production rate, the volume acts as though it contains a negative virtual mass density that produces a local increase in gravity potential, φ_g. Consequently, a negative divergence of the X diffusive flux vector leads to an increase in gravity potential at that point.

Thus subquantum kinetics leads to a charge-mass equivalence similar to that stated in relation 2 of chapter 1, except here we broaden the definition of charge density so that we consider the electrogravitic effects of a virtual-charge-density gradient opposed to a real-charge-density gradient. Hence we may state that a virtual-charge density of magnitude ρ_E creates a proportional virtual mass density of magnitude ρ_m, that is, $\rho_m \propto \rho_E$. If the negative virtual-charge density varies with distance across the capacitor's dielectric, then there will be a corresponding variation in negative virtual mass density and a gravity potential gradient will form across the dielectric. Given that this virtual mass density creates a proportional negative gravity potential field, $\rho_m \propto -\varphi_g(r)$, we conclude that the gravity potential at a given point r should be proportional to the negative charge density at that location. By using equation 6 above, this may be mathematically expressed as:

$$\varphi_g(r) \propto -\rho_E(r) = \nabla \cdot D(r) = \varepsilon \nabla \cdot E(r) = -\varepsilon \nabla^2 \varphi_E(r) \qquad (7)$$

The gradient of this gravity potential field would create a gravitational force on matter that it spanned. As mentioned earlier, this force would arise because the G-on concentration gradient spatially distorts the etheric wave patterns of the subatomic particles it affects, perturbing them from their ideally spherical symmetric configuration. Because of their tendency to maintain a state of morphogenic homeostasis, the particles respond to this stress by moving down the gravity potential gradient toward the capacitor's positive pole.

In accordance with Newton's second law, the gravitational force acting on a body at point r would be proportional to the negative gradient of the induced gravity potential field at that location multiplied by the body's inertial mass, which is expressed as $F_g(r) = -Gm_o\nabla\varphi_g(r)$. Hence using relation 7 to substitute for gravity potential $\varphi_g(r)$, the gravitational force on a capacitor is expected to vary in proportion to the third derivative of the electric potential $\varphi_E(r)$, or as the derivative of the LaPlacian of the electric potential:

$$F_g(r) = k\ m_o\varepsilon\nabla(\nabla^2\varphi_E(r)) \qquad (8)$$

As before, the constant k in this equation is an experimentally determined electrogravitic proportionality constant that quantifies the virtual-charge-to-virtual-mass coupling relationship.

This thrust on the capacitor dielectric will persist as long as the applied electric field is not canceled out by the opposing electric dipole moment created by the polarization of the dielectric. By oscillating the electric field to repeatedly create virtual charge, the gravitational thrust may be maintained without complete cancellation.

One thing that becomes apparent from studying relation 8 is that the electrogravitic force should increase as the electric potential field across the capacitor becomes increasingly nonlinear; the more nonlinear the field, the greater the induced gravitational thrust. Force also increases in accordance with the dielectric's dielectric constant, ε, and its mass, m_o. Thus dielectrics with higher K and greater mass will deliver greater thrust. Brown stressed all of these points in his work.

Furthermore, like the electrogravitic force produced by a real-charge-density gradient, the electrogravitic force arising from a virtual-charge-density gradient will always be directed toward the positive pole. If the field polarity is reversed, the polarity of the virtual-charge density would also reverse, as would the direction of the gravitic thrust. Thus, if the smaller electrode in an asymmetrical capacitor was made positive instead of negative, the electrogravitic thrust would be directed once again toward the positive electrode, which in this case would be the smaller of the two electrodes.

Now let us consider a standard symmetrical parallel plate capacitor. Such a capacitor would develop no virtual-charge-density gradient when charged since its electric field potential would vary linearly across its dielectric. The only gravity field across its dielectric would be that arising from the charges on its plates. The negative charges on the negative plate would be producing X-ons and G-ons while the positive charges on the capacitor's positive plate would be consuming X-ons and G-ons. Consequently, the X-on and G-on concentrations would be highest at the capacitor's negative pole and would drop linearly with distance across the dielectric until they reached their lowest value near the capacitor's positive pole. These X-on and G-on concentration gradients would be accompanied by a diffusive flux of X-ons and G-ons flowing down the gradient in a uniform manner. Since any volume in the dielectric would experience the same etheron influx as eflux, the divergence of the X diffusive flux vector would be zero throughout the dielectric, as would be the virtual-charge density.

We may now attempt to calculate the gravity field developed across an asymmetrical AC electrokinetic capacitor having a design similar to that shown in figures 3.2 and 3.6. We may use electrogravitic relation 7 to determine the virtual-charge profile and gravity potential field that would be generated across the dielectric. Suppose that a 100-kilovolt DC bias potential were applied across the capacitor plates, with potential varying nonlinearly according to the inverse square of distance as shown by the dotted line in figure 4.5a. This would plot as the equation $V = -1/r^2$. Compare this with the field potential graph reproduced from Brown's patent (see figure 3.6). Note that this field is substantially more nonlinear than the $1/r$ potential field that would typically exist around a charged sphere. Let us also suppose that the capacitor's negative antenna electrode excites a quarter-wave sine wave oscillation across the dielectric with a node at the positive electrode. If it was a conventional symmetrical capacitor having equal area electrodes, the oscillating potential would vary with distance across the dielectric, as shown in figure 4.5b, in which the solid and dashed curves represent the potential distribution at voltage minimum and voltage maximum. However, since the capacitor is asymmetrical, with a field that varies with distance in a nonlinear manner, this sine wave amplitude will decrease sharply with distance toward the positive electrode.

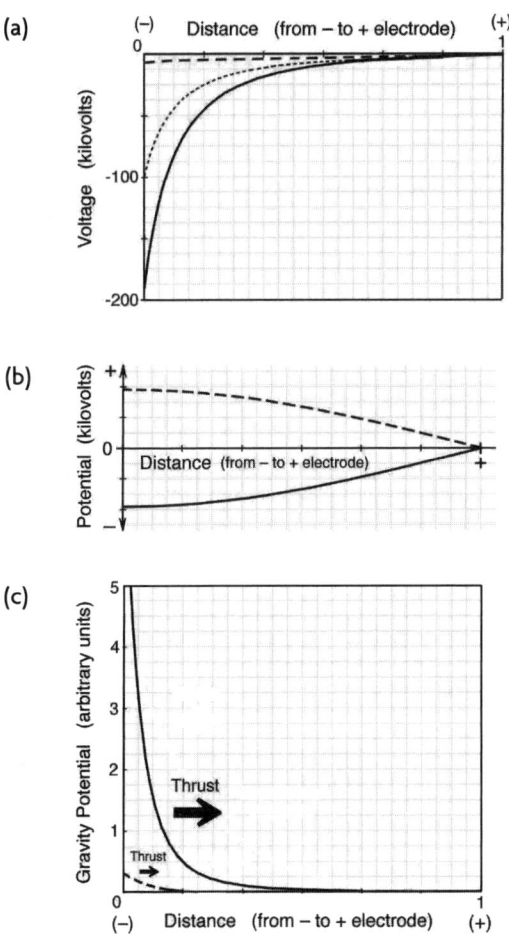

Figure 4.5. (a) The DC bias potential across the dielectric (dotted line), with the superimposed resonant oscillation (solid line represents maximum negative; dashed line represents maximum positive). (b) Sine wave voltage oscillation applied across the dielectric in Brown's electrokinetic apparatus. (c) The corresponding gravity potential profile. (P. LaViolette, © 2007)

If it were superimposed on the DC bias potential, the overall field would oscillate between the very negative potential profile prevailing when the sine wave oscillation was at its maximum negative voltage (shown as the solid line in figure 4.5a) and the low negative voltage field distribution prevailing when the sine wave oscillation was at its maximum positive voltage (shown as the dashed line in figure 4.5a). The amplitude of the sinusoidal oscillation at the negative electrode is adjusted to be about 95 percent of the bias voltage. Using the electrogravitic coupling expression that is presented as relation 7 and including a geometry correction to account for the capacitor's asymmetrical geometry, the resulting gravity potential distribution is computed to be that shown in figure 4.5c; see text box for details. The gravity gradient and resulting thrust vary from a minimum to a maximum as the voltage of the sine wave resonance at the negative electrode cycles between positive and negative maxima. Throughout the AC cycle, the thrust remains always directed toward the positive electrode, but oscillates in magnitude approximately 750 million times per second.

Gravity Potential Distribution in an Asymmetrical Capacitor with a DC Voltage Bias and Applied AC Voltage Oscillation

The gravity potential graphed in figure 4.5c was computed from the following equation:

$$\varphi_e(r) = \left(\frac{\partial}{\partial r} \left(\left(-2 \frac{\partial}{\partial r} \left(\frac{1}{r^2} \left(\frac{3}{\pi} \sin \pi (\pm 0.5 + (\tfrac{r}{12})) \right) - \frac{1}{r^2} \right) \right) 4\pi r^2 \right) \right) \frac{1}{4\pi r^2}$$

The central term in the brackets represents the electric potential and is given as:

$$\varphi_g(r) = -2 \left(\frac{1}{r^2} \left(\frac{3}{\pi} \sin \pi (\pm 0.5 + (\tfrac{r}{12})) \right) - \frac{1}{r^2} \right)$$

The sine term here represents the sine wave voltage oscillation at its positive and negative maxima. This is multiplied by $1/r^2$ to model the potential field's nonlinear variation with distance across the dielectric. The term $-1/r^2$, which has a similar nonlinear variation with distance r, is added to this to represent the DC field bias. The summed electric potential field has a minus sign, the voltage across the capacitor being negative as referenced from the grounded positive plate. This quantity is differentiated to derive the electric field gradient, $\nabla\varphi_E$, which by definition is proportional to the negative electric flux density vector, $-\mathbf{D}$. However, since we are dealing with an asymmetrical capacitor with a positive electrode surface area that is larger than the negative electrode surface area, in considering the total electric flux entering or leaving a given spherical shell increment, we must multiply $\mathbf{D}$ times $4\pi r^2$, which accounts for the increased surface area toward the positive electrode. Taking the derivative of this and dividing by $4\pi r^2$ gives the divergence of $\mathbf{D}$ in that volume increment. Then, multiplying by -1, we get the virtual-charge density in that increment, $\rho_E(r)$. The negative of this, in turn, gives the gravity potential, $\varphi_g(r)$, which is the quantity plotted in figure 4.5c.

Note that if the voltage potential across the dielectric instead was to vary only as $1/r$, as do fields radiating from a charged sphere, then in that case the divergence would be zero, since the flux per unit surface area would not change with distance. Hence, such a field, although non-linear, would not produce a virtual-charge-density gradient.

The gravity potential gradient is seen to be steepest at the negative electrode and to rapidly decrease in magnitude as the positive electrode is approached, and the gravitational thrust on the dielectric declines in a similar fashion. Consequently, the thrust developed by the device could be maximized by incorporating high-mass semiconducting particles in the dielectric near its negative pole, where the gravity gradient is highest.

The amplitude of the AC quarter-wave resonance should be kept below 100 percent of the negative DC bias voltage. Otherwise the net electric potential across the dielectric would become slightly positive during the small fraction of the oscillation cycle when the sine wave oscillation was at its positive maximum. This in turn would produce a slightly negative gravity potential having a small positive potential gradient that would generate a thrust directed toward the negative electrode. During the brief time it was produced, this small opposing thrust would subtract from the gravitic thrust developed during the majority of the oscillation cycle.

The results obtained in Brown's Paris vacuum chamber experiment may also be explained in terms of the virtual-charge electrogravitic concept. The sudden rotor thrusts observed after each spark discharge could be due to the creation of a virtual-charge-density gradient across the rotor's capacitor element. The fanning field geometry of the spark discharge would have created a nonlinear electric potential gradient between the capacitor plates that in turn would have created a virtual-charge-density gradient between the plates. This, then, would have momentarily generated a gravity potential gradient across the capacitor element. With the disappearance of the spark, these virtual charges would have also disappeared, the intervening field having become linear once again. While the virtual charges were present, however, they would have generated a substantial gravitational thrust on the rotor.

It is also possible that the rapid recharge of the negative electrode, which took a matter of milliseconds following each spark discharge, served to increase the nonlinearity of the field between the plates thereby increasing the created virtual charge and the accompanying gravitational thrust. Brown's observation that he obtained greater thrusts when he used a barium titanate dielectric between the capacitor plates may be explained on the basis of this virtual-charge concept. In accordance with equation 8 above, the electrogravitic force exerted on an intervening dielectric should vary in direct proportion to the dielectric's permittivity. A dielectric such as barium titanate has a K-value of about 10^4 when slowly charged, but when rapidly charged over a few milliseconds, its K-value will be much lower, perhaps around 1,500. So during the course of a voltage change of a millisecond duration, a 1,500-fold-greater virtual-charge density and 1,500-fold greater thrust would be created, compared to the case in which such a dielectric was absent. The high mass density of barium titanate would be another factor contributing to the production of greater thrust, barium titanate being about six times as dense as water. That is, for a given gravity potential gradient generated across the capacitor, dielectrics having greater mass would produce greater thrust.

4.3 ▪ TOWNSEND BROWN'S ETHER PHYSICS

To account for electrogravitic phenomena, Brown, too, came to theorize about the existence of an ether and to reject the idea of relative frames. His Vega laboratory notebook contains a section titled "Structure of Space," in which he qualitatively explores the subject of the existence of an ether and sets forth some of its more important properties. Although the notebook's cover page is labeled "Vega Aircraft Corp.," these notes on the ether are dated between January and March 1943, which suggests that Brown most likely wrote them when he was teaching at the Atlantic Fleet Radar School in Norfolk, Virginia. Interestingly, the ether theory that Brown explores has some similarities with subquantum kinetics.

In one passage, Brown explains his reason for considering the presence of an ether. He writes:

> For certain phenomena it is desirable and almost necessary to assume the existence of an aether in order to evolve a satisfactory explanation. An example is the force of gravitation, particularly the electrogravitational effects; The phenomenon of the movement of a dielectric is such an example . . . Much of the work [presented in these notes] is based on facts derived from actual experiments which cannot be satisfactorily explained without the existence of an aether possessing substantially these qualities.[26]

Brown proposes that the dielectric constant K and permeability μ, which are electromagnetic properties of free space, be identified with the ether. He then proposes that matter might induce a variation in the magnitude of K and μ, causing these quantities to attain greater values near a massive body. He associates this variation in K and μ with a gravitational potential field gradient and suggests that a mass acted upon by this field has a "tendency to migrate" toward regions of higher K and μ, that is, toward regions where the gravitational potential is more negative. He envisions a low K and μ region as manifesting a "high pressure" and a high K and μ region as a manifesting a "low pressure," and that a gravitating body would be migrating from a high-pressure region toward a low-pressure region.

As is done in subquantum kinetics, Brown proposes that potentials are the real existents and that a body's adjustive response to a field gradient is the essence of force. Brown's suggestion that a body migrates in an equilibrating response to the influence of a gravity gradient very much resembles concepts used in subquantum kinetics. However, he uses a mechanical analogy of a solid body's response to a pressure differential, whereas subquantum kinetics adopts a reaction-diffusion process analogy, which is fundamentally different. I believe the reaction-diffusion system concept is a better framework for application to microphysics because in addition to offering an understanding of how fields are generated and how they exert force, it predicts the autogenetic creation of subatomic particles having charge, mass, spin, and matter-wave properties.

Brown adopts a mechanical model when he suggests that the etheric field creates a pressure upon a material body. A similar concept has been expressed in many of the nineteenth-century ether theories. However, Brown's theory does not bring us any closer to understanding what force is. To say that the observed gravitational force arises from the summed collisional action of myriad energetic etheric particles merely begs the question; one is still left to wonder why these etheric particles should exert an accelerating force. Subquantum kinetics, on the other hand, addresses this question by providing an understanding of how a material body—an etheric reaction–diffusion wave pattern—migrates in response to the influence of an etheron concentration gradient (potential gradient). The wave pattern migrates because the etheron gradient alters the ongoing reaction and diffusion processes that are responsible for generating it and deploying it.

Like subquantum kinetics, Brown's theory makes a significant departure from the traditional general relativistic concept of assuming that masses warp space-time. However, does his theory explain how a mass might alter the K and μ values he ascribes to the ether? Based on the few quotes from his notes that his family has released to the public up to this point, there is no indication that it does. Brown indicates that his ideas about the ether are based on experimental results. Indeed, permittivity and permeability are observable quantities that are used to characterize the electrical properties not only of material media but also of a vacuum transmitting electromagnetic waves. However, it is a major leap of induction to assume that this aspect of the ether is the cause of gravitation. Beginning from observables, it is difficult to extrapolate the workings of an etheric realm, which are inherently inaccessible to direct observation. One risks making the error of the blind men and the elephant.

We know that the speed of light slows down in media having higher K and μ values, and we also know that the speed of a photon decreases while passing through the gravity well of a massive celestial body, which is responsible for the gravitational lensing effect. However, it does not necessarily follow that gravity mediates this effect by increasing the ether's K and μ values. Might not this speed decrease arise because a decrease in the gravity potential (etheron concentration) causes a gravitational clock retardation effect? Subquantum kinetics predicts the latter and proposes that the same retardation phenomenon that relativists term "time dilation" is also responsible for causing the gravitational redshift observed in the spectra of white dwarf stars.*

Subquantum kinetics also describes in detail how a mass locally decreases the G-on concentration to create a gravity potential well in its vicinity and also how a charged particle generates a corresponding decrease or increase in gravity potential, depending on its electric polarity. As such, it is the only unified field theory to predict the existence of electrogravitic coupling at low potential energies. Does Brown's ether theory correspondingly explain how electric charge might produce gravitational force effects by inducing changes in the ether's K and μ? With the small amount of information that has currently been made available, we are left only to wonder. Nevertheless, it is interesting to find that Brown was considering ether physics explanations at this early date in his electrogravitics research.

---

*Changing the K value of the ether, that is, the value of its electric permittivity, is equivalent in subquantum kinetics to changing the X and Y diffusion coefficients of the Model G ether reaction system. Changing these diffusion coefficients would change the X and Y etheric concentration magnitude (the electric potential). In subquantum kinetics, however, etheric concentration gradients (potential gradients) may be produced without altering the diffusion coefficients.

5

THE U.S. ANTIGRAVITY SQUADRON

5.1 ▪ ELECTROGRAVITIC SECRETS OF THE B-2 BOMBER

For many years, rumors circulated that the United States was secretly developing a highly advanced radar-evading aircraft. Rumor turned to reality in November 1988, when the U.S. Air Force unveiled the B-2 Advanced Technology Bomber (see figure 5.1). Although military spokesmen related some things about the craft's outward design and low radar and infrared profile, there was much they were silent about. However, several years later, some key secrets about the B-2 were leaked to the press. In its March 9, 1992, issue, *Aviation Week & Space Technology* magazine made the surprising disclosure that the B-2 electrostatically charges its exhaust stream and the leading edges of its winglike body.[1] Those familiar with Brown's work will quickly realize that this is tantamount to stating that the B-2 is able to function as an antigravity aircraft.

Aviation Week obtained its information about the B-2 from a small

Figure 5.1. The B-2 Advanced Technology Bomber in flight. (U.S. Air Force photo)

group of renegade West Coast scientists and engineers who were formerly associated with black research projects, which are defense projects so secret that even their very existence is classified. In making these disclosures, the scientists broke a code of silence that rivals the Mafia's. They took the risk because they felt that it was important for economic reasons that efforts be made to declassify certain black technologies for commercial use. Two of these individuals said that their civil rights had been blatantly abused (in the name of security), either to keep them quiet or to prevent them from leaving the tightly controlled black R&D community.

Several months after *Aviation Week* published the article, security personnel from the black world went into high gear. That sector of the black R&D community received very strong warnings, and as a result, the group of scientists subsequently broke off contact with the magazine. Clearly, the overseers of black R&D programs were substantially concerned about the information leaks that had come out in that article.

Northrop, the prime contractor for the B-2, had been experimenting for some time with the propulsive benefits of applying high-voltage charge to aircraft hulls. For example, at an aerospace sciences meeting held in New York in January 1968, scientists from Northrop's Norair Division reported that they were beginning wind tunnel studies on the aerodynamic effects of applying high-voltage charges to the leading edges of high-speed aircraft bodies.[2,3] They said they expected that the applied electric potential would produce a coronal glow that would propagate forward from the craft's leading edges to ionize and repel air molecules upwind of the aircraft. The resulting repulsive electric forces would condition the airstream so as to lower drag, reduce heating, and soften or eliminate the supersonic boom.* Their results showed that when high-voltage DC is applied to a wing-shaped structure subjected to a supersonic flow, seemingly new "electro-aerodynamic" qualities appear that result in significant air-drag reduction on the structure and the virtual elimination of friction-caused aerodynamic heating, as well as the elimination of shock wave and wave-drag phenomena.[4] Similar research was carried out in 1965 by the Grumman and Avco corporations. Interestingly, in 1994, Northrop bought out and merged with Grumman as part of its drive to place increased emphasis on defense electronics technologies.

Northrop and Grumman scientists apparently got the idea for investigating this sonic cushion effect either from Brown or from papers describing his work that had been previously circulated. For example, in his 1952 paper describing Brown's electrogravitic discs, Rose wrote, "The Townsend Brown experiments indicate that the positive field which is traveling in front of the saucer acts as a buffer wing which starts moving the air out of the way. This immaterial electrogravitational field acts as an entering wedge which softens the supersonic barrier, thus allowing the material leading edge of the saucer to enter into a softened pressure area."[5] This was accompanied by the diagram reproduced in figure 5.2a, which shows how the supersonic flow would be diverted around leading edge of a wing.

Brown also called attention to this effect in his 1960 electrokinetic apparatus patent, which describes using a flame-jet generator to place a high-voltage positive charge on a needlelike electrode at the front end of a rocket (see figure 5.2b). In one passage, he wrote, "By using such a nose form, which at present appears to be the best suited for flying

---

*Although the author of that article speculated that Northrop was negatively charging the aircraft's leading edge, the sonic barrier effects could also be accomplished with a positive charge, as Brown originally suggested.

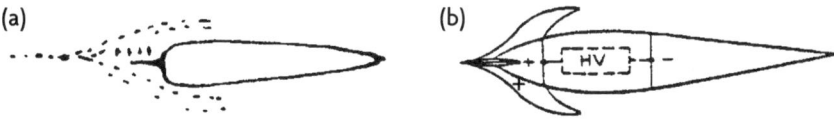

Figure 5.2. (a) Electrostatic deflection of the airstream around the electrified leading edge of a saucer-shaped aircraft. (From Rose, "The Flying Saucer," University for Social Research, April 8, 1952, vol. 7) (b) Brown's proposed use of a high-voltage needle electrode at the prow of a rocket. (From Brown, U.S. patent 3,022,430, figure 2)

speeds approaching or exceeding the speed of sound, I am able to produce an ionization of the atmosphere in the immediate region of this foremost portion of the mobile vehicle. I believe that this ionization facilitates piercing the sonic barrier and minimizes the abruptness with which the transition takes place in passing from subsonic velocities to supersonic velocities.[6]

Aerospace companies later put Brown's suggestion into use on rockets. A spike was placed at the nose of a rocket and caused to emit a high-voltage arc. Wind tunnel studies showed that the resulting electric field pushed the bow shock front away from the rocket nose so that it no longer contacted the main body of the missile and, hence, substantially reduced air drag. According to one Greek scientist working in affiliation with the U.S. Embassy in Greece, nose electrification is a standard technique used on U.S. rockets to stabilize them during take-off. Engineers are told to figure a 20 percent weight reduction during the first few kilometers' gain in altitude when determining the rocket's trajectory.

In the late 1970s, Russian scientists at the Ioffe Institute in St. Petersburg led by Anatoly Klimov carried out an interesting experiment that demonstrated how plasmas could reduce air drag. They fired a 3-centimeter steel sphere at a velocity of one kilometer per second through a tube filled with low-pressure argon gas. In one section of the tube, the argon gas was ionized to form a plasma. They found that when the sphere entered the plasma, its shock wave stood twice as far away from the sphere as it would in ordinary gas, and, more important, the sphere's aerodynamic drag was reduced by 30 percent.[7]

Interestingly, Northrop, which had past experience in leading-edge electrification, was contracted by the Pentagon in 1981 to work on the highly classified B-2. Northrop's expertise in this area must have been a key factor contributing to its winning of this contract, for *Aviation Week* reported that the B-2 uses "electrostatic field-generating techniques" in its wing leading edges to help it minimize aerodynamic turbulence and thereby reduce its radar cross-section.[8] The same article mentions that the B-2 also charges its jet engine exhaust stream, which has the effect of rapidly cooling its exhaust and thereby remarkably reducing its thermal signature.

Although these disclosures were framed in the context of enhancing the B-2's radar invisibility, in fact they are part of its field propulsion drive capability. With a positively charged wing leading edge and a negatively charged exhaust stream (figure 5.3), the B-2 would function essentially as an electrogravitic aircraft. Just as in Brown's model flying discs (see figure 2.1) and in his patented electrokinetic disc (see figure 2.8), the positive and negative ion clouds created ahead and behind the

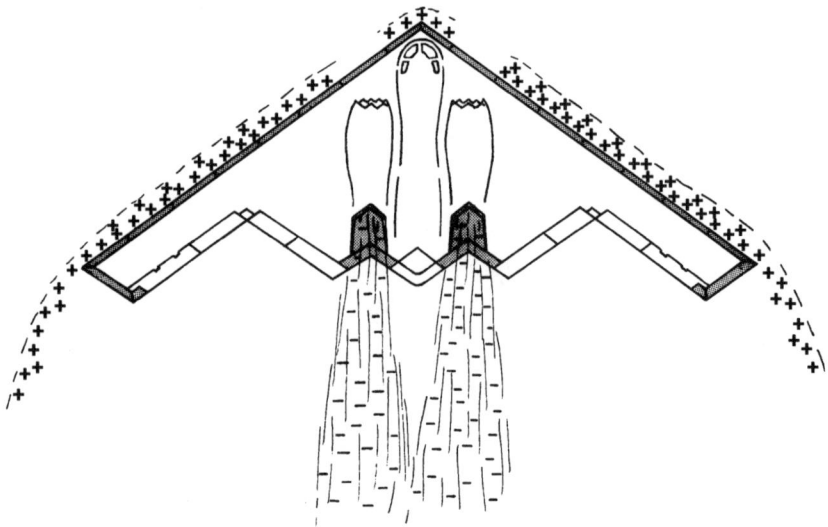

Figure 5.3. The profile of the B-2 as seen from above. The plane measures 69 feet from front to back and 172 feet from wing tip to wing tip. Cowlings on either side of the cockpit feed large amounts of intake air to the flame-jet high-voltage generators enclosed within its body. (P. LaViolette, © 1993)

B-2 would produce a locally altered gravity field that would cause it to feel a forward-directed gravitic force. In effect, the B-2 is a realization of the flying disc design Brown described in his electrokinetic generator patent as seen in chapter 2.

Rumors circulating among aviation industry personnel close to the project allege that the B-2 does use antigravity technology. A similar claim was made in the 1970s by Marion Williams, a former Central Intelligence Agency officer who had worked at the highly classified Area 51 facility, where the B-2 was test-flown.[9,10] Just before he died of cancer, Williams confided to his relative Andrew Basiago that design principles from crashed alien antigravity spacecraft were being utilized in the stealth bomber. Thus, our conjecture that the B-2 incorporates an electrogravitic drive may be substantially correct, although its design may actually have originated closer to home than Williams had been led to believe. The B-2, then, may be the first military antigravity vehicle to be openly displayed to the public! It may be the final realization of the kind of craft that Brown had proposed in Project Winterhaven and that the 1956 Aviation Studies report had disclosed was beginning to be developed by the military in late 1954. Consequently, the designation "B-2" might more appropriately stand for Biefeld-Brown effect.

The secrecy that has so tightly surrounded the B-2 most likely does not concern its radar-evading technology as much as it does its antigravity propulsion technology, although the two are probably closely intertwined. The use of such nonconventional propulsion technology would explain the B-2's high price tag, which averaged more than $2 billion per plane.

Although the black-world scientists mentioned nothing about electrogravitics in their *Aviation Week* disclosure about the B-2, they did admit to the existence of very "dramatic, classified technologies" applicable to "aircraft control and propulsion." They were especially hesitant to discuss these projects, noting that they are "very black." One of them commented, "Besides, it would take about 20 hours to explain the principles, and very few people would understand them anyway."[11] Apparently, what he meant is that this aircraft control and propulsion technology is based on physics principles that go beyond what is currently known and understood by the general public as well as most academic physicists. Indeed, by all normal standards, electrogravitics is an exotic propulsion science. Nevertheless, by beginning with an understandable theory, electrogravitics becomes a lot less mysterious. As mentioned earlier, subquantum kinetics provides one such viable theory.

The B-2's body design also raises suspicions that the aircraft is in fact an electrogravitic vehicle. A primary design criterion for an electrogravitic craft is that it have a large horizontally disposed surface area so as to permit the development of a sufficiently strong antigravity lift force. As Brown's experiments demonstrated, such an aircraft need not necessarily be disc shaped; triangular- and square-shaped forms also exhibit antigravity lift when electrified, although disc shapes give the best performance. The triangular planforms used in the B-2 and other advanced stealth aircraft may have been deemed better for reasons of their much lower radar cross-section.

Interestingly, one of the central features of the B-2's classified technology is the makeup of its hull's outer surface. Authorities tell us that the hull is composed of a highly classified radar-absorbing material. Ceramic dielectrics are a likely choice for the B-2. Unlike many lossy dielectrics that dissipate the energy of incident radio waves and therefore function as radar wave absorbers, ceramic dielectrics are lossless, energetically noninteractive, and, hence transparent to radar waves. More important, ceramic dielectrics also have the ability to store large amounts of high-voltage charge. By covering the hull with such an electric insulator, it would be possible for the B-2 to maintain a high-voltage differential between its positive leading edge and its negative ion exhaust stream. At sea level, the breakdown voltage is about 27,000 volts per centimeter, whereas at an altitude of fourteen kilometers, the breakdown voltage drops to about 10,000 volts per centimeter. So with its 69-foot (21-meter) front-to-back dimension, the B-2 at sea level in dry air should be able to maintain a voltage differential of up to 57 million volts before arcing over, whereas at fourteen kilometers, it should be able to maintain a differential of up to 20 million volts. Military spokesmen have said that the B-2 cannot fly in rainy weather, giving the reason that its coating of radar-absorbing material can be adversely affected. The real reason is that if the hull becomes wet, it can lose its insulating properties, and the leading edge electrode can short out to the rear exhaust duct.

Even after the hull's high-voltage electrification is shut off, the hull dielectric can retain a residual charge for some time because of the dielectric absorption effect mentioned in chapter 1. This could explain rumored incidents of ground crews having been zapped by touching a B-2 too soon after it landed.

The B-2's positively charged leading edge, another key component of its propulsion technology, was also a matter of special concern to Northrop designers. According to *Aviation Week,* the bomber's leading edges posed a particularly challenging production problem on the first aircraft. The leading edge ionizer is most probably a conductive strip or wire that runs along the B-2's sharp prow and is electrically charged to upwards of many millions of volts. As the craft moves forward, its electrified leading edge deflects the approaching airstream to either side, so that a large fraction of the generated positive ions are carried away from its body surface and are prevented from immediately contacting and neutralizing the negative ions in the B-2's exhaust stream. As a result, the B-2 is able to build up very large space charges ahead of and behind itself that would subject it to a large gravity potential gradient. This artificially produced gravity gradient should become steeper as the B-2 attains higher speeds and deflects its positive ions outward with increasing force. Hence the B-2's electrogravitic drive should operate more efficiently when the craft is moving at higher speeds.

Best results should be obtained when the B-2 is traveling at supersonic speeds. Positive ions from its leading edge should become entrained in the upwind sonic shock front and flow away from the craft through that sonic boundary layer, later to converge on the negatively charged exhaust stream. Military sources, however, claim that the B-2 is a subsonic vehicle. Its somewhat stubby cross-section and the angle of its wings might lead one to believe that this is so. Yet these design features should not pose a problem for supersonic flight, considering that the B-2 uses an electrostatic field to deflect the approaching airstream. Brown's saucer designs similarly had a stubby cross-section and yet were intended for supersonic travel. The Air Force probably avoided disclosing the B-2's supersonic capability to avoid raising curiosity about how the craft would generate the required thrust.

In both subsonic and supersonic flight, the deflected positive ions would form an ellipsoidal sheath as they circuit around the B-2 (figure 5.4). The B-2's forward positive ion sheath would act very much like an extended positively charged electrode whose surface has a parabolic shape. Thus, the electrogravitic force propelling the B-2 would arise not just from the leading-edge electrode, but also from the entire positively charged forward ion sheath. The positive- and negative-ion space charge distributions would very much resemble the charge configuration that Brown employed in some of his later electrogravitic experiments. Compare figure 5.4 with the parabolic electrogravitic devices shown in figure 3.7 that Brown had been testing. Brown noted that he obtained a greater electrogravitic thrust when the positive electrode was curved and made much larger than his negative electrode. At the time they exit the B-2's exhaust nozzles, the negative ions should be spatially much more concentrated than the positive ions emitted along the B-2's leading edge, so the field gradient from front to back would be very nonlinear.

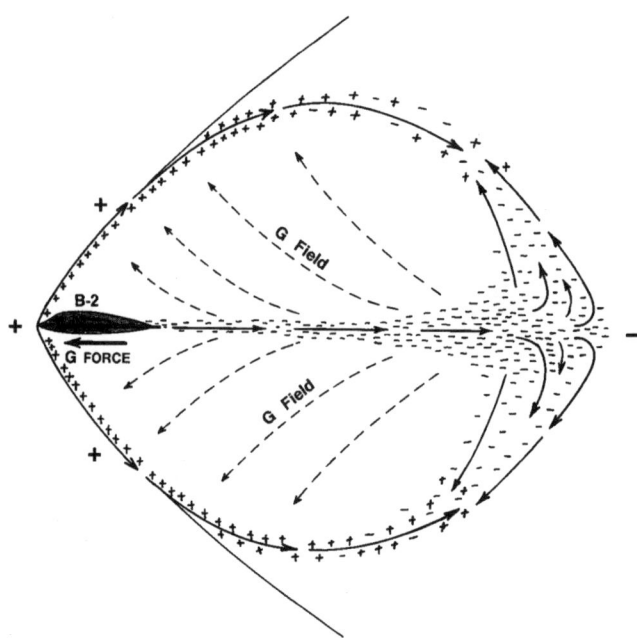

Figure 5.4. A side view of the B-2 showing the shape of its electrically charged Mach 2 supersonic shock and trailing exhaust stream. Solid-line arrows show the direction of ion flow; dashed-line arrows show the direction of the gravity gradient induced around the craft. (P. LaViolette, © 1993)

As mentioned in chapter 3, in describing the operation of his vertical-lift test rigs, Brown had voiced the necessity of establishing a nonlinear field gradient across the intervening dielectric to maximize thrust. As in these laboratory test rigs, the B-2 would have established a highly non-linear field from aft to fore while in flight. The field lines would have a very high flux density at the negatively charged exhaust stream exiting the rear of the craft and would have diverged out to a much lower field flux density at the greatly dispersed, positively charged ion sheath surrounding the front of the craft. This same asymmetry would characterize the polarization of the B-2's ceramic dielectric hull, the field lines being most concentrated toward the negatively charged exhaust ducts and most dispersed toward its positive leading-edge electrode.

The electrostatic field produced by the ions surrounding the B-2 would exert forces on the B-2's polarized dielectric body that would produce a net forward thrust, as shown in figure 5.5. The high concentration of negative charges at the rear end of the craft would repel its negatively charged tail forward. Electrostatic attraction forces would also assist the craft's forward thrust by pulling its negatively charged stern toward its positively charged bow shock. The electric field would fan out and therefore drop in intensity toward the B-2's bow, so opposing forces acting on the front of the craft would be weaker and would have force components vectored mainly crosswise to the craft's direction of travel. The rearward slant of the B-2's positively charged bow shock would also assist the craft's forward propulsion by producing forward vectored repulsive forces on the B-2's nose and wing leading edge. At faster velocities, the craft's bow shock would bend back to a steeper angle, thereby increasing the forward thrust delivered by these repulsive forces.

Although the charges are moving away from the aircraft at a very high velocity, they are continuously being generated and dispersed into the surrounding air. Consequently, their space charge distribution remains stationary relative to the craft. It follows the craft and continues to exert its propelling force. The electrostatic forces depicted in figure 5.5 are arrayed quite differently from the electrogravitic forces shown in figure 5.4, but both would assist the craft's forward propulsion. Not enough is known at this point to say which of these sets of forces would be more important in propelling the craft.

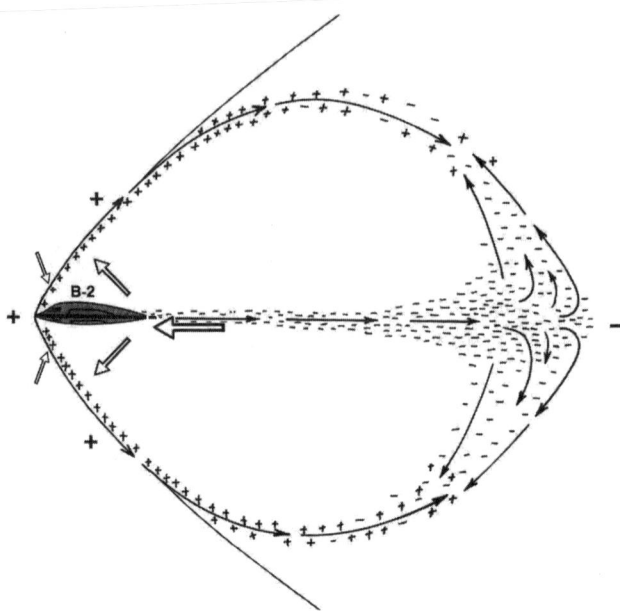

Figure 5.5. Side view of the B-2 showing the direction of the electrostatic repulsion forces (large white arrows) developed between the craft's charged body and the surrounding ion space charge. (P. LaViolette, © 2006)

As seen in figure 5.6, both of the B-2's leading edges are segmented into eight sections separated from one another by 10-centimeter-wide struts. Quite possibly, the struts electrically isolate the sections so that they may be individually electrified. In this way, through proper control of the applied voltage, it would be possible to gravitically steer the craft. Brown had suggested a similar idea as a way of steering his saucer craft.

The leading-edge sections positioned in front of the air scoops are, most likely, sparingly electrified so as to prevent positive ions from entering the engine ducts and neutralizing the negative ions being produced there. These two nonelectrified leading-edge sections would be ideal places to mount forward-looking radar antennae, since the ion plasma sheath produced by the other leading-edge sections would form a barrier that would interfere with radar signal transmission. In fact, the B-2's two Hughes Aircraft radar units are mounted precisely in these leading-edge locations, right in front of the air intakes. The ellip-

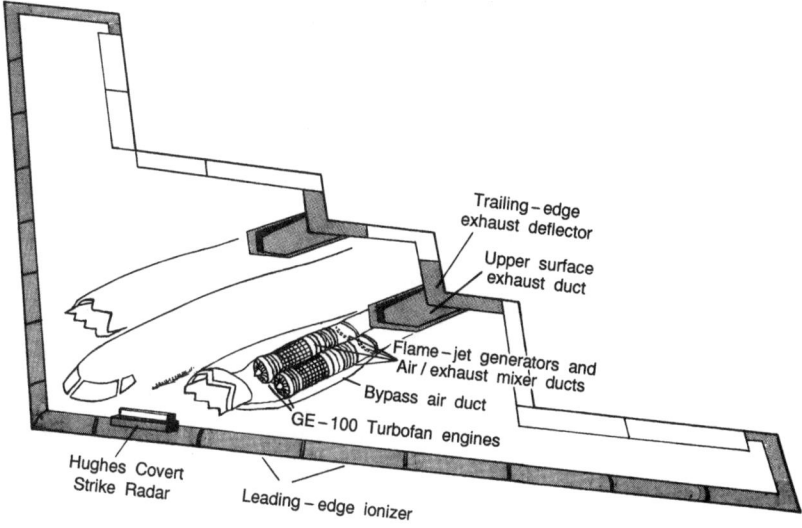

Trailing – edge
exhaust deflector

Upper surface
exhaust duct

Flame – jet generators and
Air / exhaust mixer ducts

Bypass air duct

GE – 100 Turbofan engines

Hughes Covert
Strike Radar

Leading – edge ionizer

Figure 5.6. A cutaway view showing the arrangement of the B-2's flame-jet generators. (P. LaViolette, © 1993)

soidal ion plasma sheath that envelops the B-2 would strongly attenuate incoming radar pulses as well as any signal reflected back by the craft, thereby substantially reducing the B-2's radar visibility. This ion sheath might actually attenuate radar signals better than the ceramic radar-absorbing material that composes the B-2's hull. In fact, the military continues to research ways of using plasmas to absorb radar signals in the hope that a plasma-enveloped plane would be radar invisible.[12]

The Hughes radar units may also be supplying microwave energy to the B-2's leading edge to assist the air-ionization process. Microwave frequencies emitted along the leading edge would readily ionize the approaching air and allow the B-2's high-voltage electric field to discharge a greater flux of positive ions. With increased ion currents, the B-2 would be able to generate a greater ion sheath space charge at a given velocity and thereby increase the electrogravitic and electrostatic thrust propelling the craft. *New Scientist* magazine reported that NASA's Langley Research Center in Hampton, Virginia, had conducted wind tunnel tests in which they used a microwave beam to create a plasma upwind of an aircraft wing in a Mach 6 airflow and found dramatic reductions in air drag.[13] Quite likely, the B-2 has been using the same technique, although a high-voltage radio frequency field might work just as well.

5.2 ■ THE B-2'S FLAME-JET GENERATORS

The excerpt from the October 1954 *Aviation Report* article quoted in chapter 2 suggests that there should be a division of responsibility in the program to develop a Mach 3 electrogravitic aircraft, that the "condenser assembly which is the core of the main structure" be developed by an airframe manufacturer and that the flame-jet generator that provides the electrostatic energy for the craft should be developed by companies specializing in jet engine technology. Consistent with that suggestion, we find that Northrop Grumman, a company experienced in aircraft electrostatics, was contracted to develop the B-2's airframe and that General Electric, a company experienced in the development of jet engines and superconducting electric generators, was contracted by the U.S. Air Force to develop the B-2's engines. Recall that the 1956 *Aviation Studies* report mentions General Electric as one of the companies involved in early electrogravitics work. Also, note that Brown had conducted vacuum chamber experiments at the General Electric Space Center and that the Electrokinetics Corporation, which had hired him as a consultant, was located just several miles away.

The Air Force states that the stealth bomber is powered by four General Electric F-118-GE-100 jet engines similar to those used in the F-16 fighter, but the B-2's engines quite likely have been modified to function as flame-jet high-voltage generators. The propulsive force lofting the craft, then, would come not only from the mechanical thrust of the jet exhaust, but also from the electrogravitic and electrostatic force fields set up around the craft that would be powered by the jet's generators. Such flame-jet generators also would account for the presence of ions, which *Aviation Week* says are present in the B-2's exhaust stream. As in Brown's saucer, the engine nozzle would acquire a high positive charge as it exhausted negative ions. Presumably, the engine is electrically insulated from the aircraft hull and surrounding ductwork and its positive charges are conducted forward to power the leading-edge ionizers.

The B-2's General Electric engines are reported to each be capable of putting out 19,000 pounds of thrust. Consequently, all four engines together should provide the B-2 with a total output of about 140,000 horsepower, which translates into an electric power output of about 25 megawatts, assuming a 30 percent conversion efficiency.* By comparison, the November 1954 *Aviation Report* concluded that a 35-foot-diameter electrogravitic combat disc would need to have access to about 50 megawatts of power in order to attain Mach 3 flight speeds. Thus it appears that the magnitude of the B-2's power output is in the right ballpark.

A total of about 50 kilowatts of power (50 kilovolts × 1 ampere) probably would be sufficient to get the engine ionizers started. This could easily be handled by electric generators mechanically driven by the jet turbines. Once the flame-jet generators were operable and power was being extracted out of the ionized exhaust stream, the power draw of the leading-edge and exhaust ionizers could be allowed to rise much higher, to tens of megawatts.

The B-2 may use superconducting generators for its more conventional means of generating power from its turbines. Such generators have the advantage of being nearly 100 percent efficient in converting shaft power to electricity and of being extremely lightweight, weighing less than one-tenth as much as conventional generators. The first superconducting generator was developed in the mid-1970s by scientists at the General Electric Research Laboratory working under an Air Force contract. Subsequently, the generators were being mass-produced for the Air Force.

When the B-2 was unveiled in 1988, one Air Force official commented that it uses a system of baffles to mix cool intake air with its hot exhaust gases so as to cool the gases and thereby make them less visible to infrared-guided missiles. Although infrared invisibility might be one side benefit, most likely the real purpose for diluting the exhaust is to greatly increase the flow volume and, hence, the ability of the exhaust stream to eject negative charges from the craft. Much of the air entering the B-2's intake scoops would bypass the inlet to the flame jet and be allowed to mix in with the jet's hot ionized exhaust (figure 5.6).

---

*This horsepower estimate is based on the assumption that the jets would be able to propel the craft to a velocity of about 600 miles per hour (Mach 0.8).

Actually, the jet's exhaust has an aspirator effect in that friction between the exhaust stream and the surrounding air creates a sheer layer that naturally entrains the bypassed air into the exhaust flow and thoroughly mixes the two. As a result, the temperature and velocity of the exhaust stream drop as its volume increases. At the same time, the sound that normally emanates from the exhaust's shear layer, which is the prime contributor to a jet's sonic boom, is substantially muffled, for all this occurs within the engine shroud. Aeronautical engineers call this air-mixing exhaust nozzle an ejector-type suppressor nozzle.

A series of electrified conical collars, similar to those described in Brown's patent 3,022,430 (see figure 2.10), located in the exhaust nozzle might inject additional negative ions into the mixed exhaust stream, thereby boosting its ion content. This augmented volume of ionized gases then discharges through the two rectangular exhaust ports positioned near the rear of the B-2's wing and contacts the titanium-coated overwing exhaust ducts, portrayed in figure 5.6. These open-duct sections may function as rear electric grids that collect million-volt electrons from the exhaust streams and recycle them to power the exhaust and wing air ionizers. This might be done in the same fashion as Brown had suggested in his patent (see figure 2.9). Additional high-voltage current could be recovered from the conical electrodes.

As the exhaust leaves the craft, it passes over trailing-edge exhaust deflectors, flaps that can be swiveled so as to direct the exhaust stream either up or down for flight control. This accomplishes more than just vectoring of the exhaust's thrust; it also changes the direction of the electrogravitic force vector. When the exhaust is deflected downward, negative charges are directed below the craft. As a result, the electrogravitic force on the craft becomes vectored upward as well as forward. When the exhaust stream is deflected upward, its negative ions are directed above the craft, resulting in an electrogravitic force that is directed downward as well as forward. Thus, by using these flaps, the B-2 is able to control its force field so as to induce either a gain or a loss of altitude.

Once the B-2 attained a sufficiently high flight speed, it would receive enough airflow through its scoops that it could maintain a relatively high flow rate of ionized exhaust, even with its engine combustion substantially reduced. Since hot exhaust is not essential to its operation, the high-voltage generator could just as well run on cool intake air with fuel combustion entirely shut off. As Brown pointed out in his electrokinetic generator patent, "It is to be understood that any other fluid stream source might be substituted for the combustion chamber and fuel supply."[14]

In such a "coasting mode," in which jet combustion is entirely shut off, the B-2 would be able to fly for an indefinitely long period of time with essentially zero fuel consumption, powering itself primarily with energy tapped from its self-generated gravity gradient. For example, during coasting, the kinetic energy of the scooped airstream would arise entirely from the craft's own forward motion, with this motion being due to the pull of the electrogravitic propulsion field. The kinetic energy of this ionized airstream is responsible for linearly accelerating negative ions down the B-2's exhaust ducts and, hence, for creating the multimegavolt potential difference relative to the positively charged engine body. The craft's high-voltage electron collector grids—the overwing exhaust ducts and other collector surfaces possibly hidden in the exhaust nozzle—recover a portion of this electric power to run the ionizers for the craft's flame-jet generator. Provided that this power drain is not excessive and that the plane's propulsive gravity field can be adequately maintained, the craft would be able to achieve a state of perpetual propulsion. As mentioned in chapter 1, such perpetual motion behavior is possible in devices having the capability to manipulate their own gravity field. Moreover, when the B-2 flies at a sufficiently high velocity, such that the flow rate of its scooped air exceeds many times the exhaust flow rate from its jet turbines, the electric power output of its mixed exhaust will be comparably larger, perhaps exceeding 100 megawatts.

When the B-2 was first put on public display, critics had suggested that it could not risk flying at high altitudes because it might create vapor trails that would be visible to an enemy. Edward Aldridge Jr., then secretary of the Air Force, was asked whether that problem had been solved. He replied, "Yes, but we're not going to disclose how." Clearly, to explain how the B-2 could travel at high altitude with its jet combustion essentially shut off and producing no vapor trail, he would have to disclose the vehicle's nonconventional mode of propulsion. Incidentally, in such a coasting mode, the B-2's waste heat output also would be greatly reduced, hence lessening its chance of being detected with infrared sensors.

The B-2's emergency power units (EPUs) probably play a key role in assisting such high-altitude flight. According to Bill Scott, author of the book *Inside the Stealth Bomber*,[15] each EPU consists of a small self-contained gas turbine powered by hydrazine, a liquid that rapidly decomposes into gases when activated by a catalyst. The expanding gases are made to drive a turbine that, in turn, drives an electric generator. Public disclosures state that the purpose of the EPUs is to supply electric power to the craft should the B-2's four jet engines happen to flame out or its four electric generators happen to simultaneously fail. More likely, they were designed to function as auxiliary generators capable of operating at high altitudes (or even in space), where the air would be too thin to sustain normal jet combustion. At high altitudes, the decomposed hydrazine gases would take the place of scooped air as the medium for transporting ions from the craft. That is, after passing through the EPUs, these gases would be electrified and expelled from the craft in the same fashion as would the jet exhaust. Brown noted that his electrogravitic propulsion system could run just as well using a compressed gas source such as carbon dioxide as the ion-carrying medium as it could using the exhaust from a jet engine.

When flying between an altitude of twenty-eight and eighty-three kilometers, the B-2 would have to shut off its hull electrification, since in this altitude range the air would become a very good conductor because of the glow discharge effect. By accelerating to an orbital velocity speed in the range of Mach 19 to 23 prior to reaching an altitude of twenty-five kilometers, the B-2 could coast through this forbidden region. Once in space, above an altitude of eighty-three kilometers, the vacuum would be good enough that the B-2's electrogravitic drive could once again be switched on. As mentioned earlier, it would rely on its hydrazine EPUs to power itself in spaceflight.

Figure 5.7 is a picture taken of a B-2 in transonic flight through humid coastal air. At transonic speeds, which range from just below to just above the speed of sound (Mach 0.8 to 1.3), some parts of the airflow over an aircraft become supersonic. In this speed regime, very-low-pressure areas form at various locations around an aircraft, and if the aircraft happens to be passing through humid air near the dewpoint, visible clouds can form in these low-pressure areas and remain with the aircraft as it travels. Figure 5.8 shows a cloud formed around an F/A-18 jet fighter flying at transonic speed.

Northrop Grumman has produced a movie clip showing the B-2 in various flight modes. It is available for public viewing at its website, www.is.northropgrumman.com/windows_media/b2_tx.wmv. One segment near the beginning of the clip, which lasts for one and a half seconds, shows the B-2 surrounded by transonic vapor condensation clouds as it flies through humid air. French astrophysicist Jean-Pierre Petit has posted this segment on his website and notes that the vapor cloud above the B-2's wing visibly luminesces as though it was being excited by a high-voltage field.[16] The reader is also referred to color stills from this video posted on Petit's website. Unfortunately, we were not able to secure permission from Northrop Grumman to reproduce the stills here.

The segments from this video show that the cloud itself has a yellow luminous hue, a color that differs from the white color that such clouds

Figure 5.7. A B-2 bomber flying through humid coastal air at transonic speeds with a vapor cloud condensing behind its bow compression wave. (Photo by Bobbi Garcia, courtesy of the U.S. Air Force Flight Test Center)

Figure 5.8. Vapor cloud around an F/A-18 jet fighter flying at transonic speed. (U.S. Navy photo by Ensign John Gay)

would normally exhibit in sunlight. Since at high voltages fog is more subject to electrical breakdown than is dry air, a high-voltage field could excite a glow discharge in an overwing vapor cloud to appear much like the luminescence seen in the video. An orange hue is also seen reflecting from the portion of the B-2's upper-wing surface that borders the vapor cloud. Interestingly, in the last two frames of the video clip segment, the vapor cloud almost entirely vanishes, yet this orange luminescence or glow reflection is still apparent on the B-2's wing, suggesting that the high-voltage field is still active. It is surprising that this cloud disappearance happens suddenly from one frame to the next, in less than a tenth of a second. It is not clear whether this change is due to a sudden change in air humidity or whether the B-2's electric field was being switched to a lower setting.

The B-2 is not quite as invisible to detection as is often claimed. For example, its flame-jet generator exhaust could generate a radio noise signal. If that's so, the random high-velocity movement of negative ions present in the turbulent exhaust stream would produce radio wave noise emission. This could explain the signal noise that one TV viewer reported at the time of one B-2 sighting. Also, although invisible to radar detection at microwave frequencies, at lower frequencies such as are used in television broadcasting, the B-2 produces a distinct reflection. Just like conventional low-flying airplanes, it causes a local distortion in the TV signals received by residential televisions. In fact, during the war in Yugoslavia, Serbs were monitoring TV disturbance patterns over populated areas as a method of alerting them to when a B-2 was in the area and to determine where one might be located at any given time. In retaliation, the Americans bombed their television transmitting tower.

5.3 ▪ AC ELECTRIFICATION?

It is possible that the B-2 superimposes an AC signal on its DC bias potential. The Aviation Studies "Electrogravitic Systems" report mentions using high-K dielectrics energized with 50,000 kilovolt-amps of power as a means for propelling a supersonic combat vehicle of the sort proposed in Project Winterhaven. This clearly implicates the use of high-voltage AC. So, we might venture that, in addition to the DC bias potential, a high-frequency AC field is applied between the wing leading edge and the rear exhaust ports. If the excitation frequency was chosen to be 30 megahertz, then a quarter wavelength would have fit across the ten-meter distance from the exhaust ducts to the wing leading edge. This would have allowed the applied AC field to resonantly build up to a high voltage potential, similar to what Brown was achieving with his electrokinetic apparatus. This could be done with a high-voltage class C amplifier designed to automatically lock in on the wing's resonant frequency. By repeatedly charging and discharging the craft's dielectric, the AC field would also have kept the craft's dielectric from fully polarizing and building up an electric dipole moment that might cancel out most of the field propulsion thrust effects.

We might venture that the same AC energization technique may also be used to provide vertical thrust to the B-2, thereby allowing it to hover. The B-2 is said to have a weight of about 158,000 pounds (72 metric tons) when empty and about twice that when fully loaded. For a wing area of 460 square meters, this works out to about 16 grams per square centimeter empty or 32 grams per square centimeter when fully loaded. By comparison, Brown's 18-inch-diameter vertical electrokinetic thruster was generating an upward force of 125 grams when energized at 170 kilovolts. This amounts to a lift of about 0.08 gram per square centimeter. So, to generate a force sufficient to support the B-2, a thrust-per-unit area only four hundred times greater would be needed. This could easily be accomplished simply by using a high-K dielectric for the thruster's central insulator and energizing the device at a higher voltage. The "Electrohydrodynamics" report mentions that thrust increased exponentially with voltage, according to the square or cube of voltage. Moreover, the data Bahnson presented in his 1965 patent indicates that thrust on an AC-energized test rig increased according to the 2.6 power of voltage. Extrapolating this, we find that Brown's vertical thruster would deliver greater than a hundred times more thrust if it were energized at 1,000 rather than 170 kilovolts. Also, if Brown had replaced his Pyrex insulator with a material such as barium titanate, having a higher dielectric constant and higher mass density, this would have boosted the thrust by an additional thirty-two-fold. So instead of just 125 grams of force, Brown's thruster could have produced an amazing 400 kilograms of force. If 380 of these asymmetrical capacitors were distributed over the B-2's lower wing surface, they would collectively produce an upward thrust of 152 tons, sufficient to loft a fully loaded B-2. Brown is likely to have made similar thrust projections in proposing his electrogravitics idea to the military. We may be erring on the low side in making this estimate, since dielectrics are known to exist that have K values more than four times higher than the K value of barium titanate.*

To ensure that the thruster electrodes did not arc over at these high voltages, the interior space of the arcuate canopy (shown in figure 3.2) could be filled with a low-K insulator. The entire thruster together with its high-strength canopy, central high-K dielectric, surrounding low-K insulator, and high-voltage step-up transformer might weigh only 20 kilograms, which would amount to 2 percent of the thrust that the device would be producing.

The "Electrohydrodynamics" report states that under vacuum conditions, Brown's electrokinetic capacitor drew just 2 microamps of current at 250,000 volts. At the 1,000-kilovolt potential proposed for the B-2 thrusters, this leakage current would probably extrapolate to about 30 microamps, or about 30 watts of power. Adding in the power requirement for the AC microwave source used to excite the negative electrode, the total power consumption might come to about 100 watts per thruster, or about 38 kilowatts total. Given that each thruster would be yielding 400 kilograms of force, this amounts to a thrust-to-power ratio of about 40,000 newtons per kilowatt, or about 2,700 times that of a jet engine.

As an alternative to Brown's electrokinetic thrusters, the B-2 could

---

*The black-project scientists mentioned earlier in this chapter disclosed information about the development of low-radar-observability dielectric ceramics made from powdered, depleted uranium.[17] The material is said to have approximately 92 percent the bulk density of uranium, which would give it a specific gravity of about 17.5. Thus, this new material would have a mass density about three times that of barium titanate and so would develop a comparably greater electrogravitic pull. Asymmetrical thrusters possibly incorporated in the B-2's wing may use a high-density dielectric of this sort adjacent to their negative electrodes where the field strength and gravitic thrust would be highest.

be lofted by a series of Lafforgue field propulsion thrusters of the type discussed in chapter 12. Theoretical projections suggest that such a capacitor, measuring 38 centimeters high, 8 centimeters wide, and 1 meter long, made with a K = 4,000 barium titanate dielectric, would be capable of delivering a lift of 2 tons when charged to 100 kilovolts. Currently, there is no laboratory data available on barium titanate Lafforgue thrusters to back up this projection, but if it is correct, it would imply that seventy-five such thrusters would be sufficient to levitate a fully loaded B-2.

Earlier, we spoke of General Electric's Air Force–funded development of lightweight superconducting generators, with the Air Force being the prime purchaser. Such generators might not only be used to run the B-2's electrical equipment, but might also be the principle means by which the craft generates AC power for its vertical lofting. Power from these generators would be fed to a network of high-voltage step-up transformers attached to each thruster. High-voltage AC power could also be conveyed between the leading-edge electrode and the overwing exhaust ducts to enhance the B-2's forward thrust. Power applied at a radio frequency of some tens of megahertz would have helped ionize the airstream approaching the wing's leading edge to soften the shock front, having the same effect as a microwave ionizer.

By having a distributed array of vertical thrusters, the potential of each thruster could be made to "float" so that those located closer to the bow of the B-2 would operate at a more positive DC potential than those at the stern. Also, the B-2 could accomplish pitch stabilization by selectively powering these thrusters. Activating more thrusters on its left side, for example, would cause the craft to execute a clockwise roll to its right. Thus, its thrusters would take the place of mechanical flaps on conventional planes. This selective energization could be carried out by an onboard computer, which would automatically control the stability of the B-2 with the help of a fuzzy logic servo system.

After the B-2 bomber was unveiled, scientists at the British Aerospace Corporation (BAE Systems) were eager to reverse-engineer its propulsion system. In 1996, a member of their Advanced Concepts Office privately told one visitor that they were aware that the B-2 flies by means of some form of antigravity propulsion and that the craft has a very massive power supply. Indeed, if the B-2 had superconducting generators and numerous high-voltage transformers on board, its power supply would have been quite massive.

In 1997, a three-star general told retired Air Force colonel Donald Ware he knows that "the new Lockheed Martin space shuttle [National Space Plane] and the B-2 [stealth bomber] both have electrogravitic systems on board"; and that "this explains why our 21 Northrop B-2s cost about a billion dollars each. Thus, after taking off conventionally, the B-2 can switch to antigravity mode, and, I have heard, fly around the world without refueling."[18]

Ware made this comment four years after I had presented my paper on the B-2's electrogravitic propulsion system at the 1993 International Symposium on New Energy.[19] After presenting this paper, I sent a copy of it to Bill Scott, editor of *Aviation Week and Space Technology*, the same magazine that had made the original disclosure about the B-2 charging the leading edge of its wing with high voltage. Scott, who has formerly worked for the National Security Agency, has himself flown the B-2 bomber during test-flight operations. Some time after sending the paper, I telephoned him and asked him what he thought. His response was, "[V]ery interesting, very interesting." He would say no more.

That same year Ben Rich, the man who had led the development of the F-117 Stealth Fighter at Lockheed's secret research and development Skunk Works, gave an alumni speech at his UCLA alma mater in which he stated: "We already have the means to travel among the stars, but these technologies are locked up in black projects, and it would take an act of God to even get them out to benefit humanity . . . Anything you can imagine, we already know how to do." Rich was right about the difficulty of breaking the military code of secrecy. In October 2007 I heard from a reliable U.S. government source that Boeing recently completed a classified electrogravitics propulsion project for the military that had certain novel features. The technology worked so well that they felt it could be of fantastic benefit if used on their commercial jet airliners. They reportedly applied for declassification of their invention for commercial use, but were denied permission.

6

GRAVITY BEAM
PROPULSION

6.1 ▪ EXPLAINING THE ELECTROGRAVITIC IMPULSE EFFECT

A high-voltage shock discharge produces a momentary gravitational thrust that we may refer to as the electrogravitic impulse effect. One example of this is the train of shock discharges that were emitted from Tesla's high-voltage magnifying transmitter (see figure 6.1). The shocks created thrusts in their direction of travel with minimal reversal occurring during their intervening relaxation periods. Tesla frequently remarked on the force that such impulses would exert on distant objects. He noted that when he stood near the source of the discharges, he could feel them as a great force or sharp pressure striking the whole front of his body.[1] These effects were most apparent as a stinging of the face or hands, which persisted even when he situated himself behind glass and metal shields as far as 50 feet from the shock source. By properly adjusting the discharger on his transmitter, he was able to either project forces outward or direct forces inward.[2]

Tesla referred to these longitudinal force field rays as radiant energy, although the usual use of this term was to signify the radiation of transverse electromagnetic waves. He fashioned a series of long vacuum tubes to project the radiant energy waves he was producing. These "beam-ray tubes" employed a single concave negative electrode in one end and, in many cases, had a thin metallic window, usually aluminum or beryllium, at the opposite end. Despite the hard vacuum that they were initially provided with, these tubes often developed anomalously high pressures and often exploded. In *Secrets of Cold War Technology,* Vassilatos notes that Eric Dollard, who duplicated many of Tesla's beam-ray experiments in the 1980s, also observed the anomalous force that these tubes developed. Vassilatos wrote that "vacuum bulbs so activated actually ruptured in tiny holes, and yet continued to produce their 'vacuum' discharges! Mr. Dollard and the witnesses of these experiments reported hearing a hissing issuance which emerged from the glass rupture holes. Once the activating energy was removed, the globes simply imploded."[3]

Dollard has demonstrated both mass-repulsion and mass-attraction effects being produced by radiant energy impulses. Tesla conceived these discharges as being waves conducted in a rarefied ether. Vassilatos wrote:

> In his article, Tesla describes the shield-permeating shocks as "sound waves of electrified air." Nevertheless, he makes a remarkable statement concerning the sound, heat, light, pressure, and shock which he sensed passing directly through copper plates. Collectively, they "imply the presence of a medium of gaseous structure, that is, one consisting of independent carriers capable of free motion." Since air

Figure 6.1. Tesla's magnifying transmitter operating, with Tesla sitting in the background.

Figure 6.2. Nikola Tesla in 1894 at age thirty-eight.

was obviously not this "medium," to what then was he referring? Further in the article he clearly states that "besides the air, another medium is present."[4]

Tesla's reference to etheric sound waves implies an ether medium that is compressible and that transmits waves longitudinally, much as air transmits sound. The ether he visualized was very different from the elastic solid ether proposed by the nineteenth-century ether physicists that was supposed to transmit electromagnetic waves by means of transverse stresses in its lattice, creating forces perpendicular to the direction of wave propagation. Tesla adopted this different view because the forces produced by his shocks were directed longitudinally, not transversely. As such, his concept of the ether comes close to the transmuting ether idea suggested in subquantum kinetics, which views a local energy potential as a localized high or low etheron concentration and an energy potential wave as a propagating etheron concentration magnitude. The alternate increase and decrease of etheron concentration that would characterize a passing wave very much resembles Tesla's idea of alternate compression and rarefaction of an ether gas.

Tesla ascribed the longitudinal forces he observed to the action of ether currents propelled forward by the ether shocks he was generating. However, as is suggested below, the net force imparted by these impulses is more likely due to the action of the potential gradient (the etheron concentration gradient) rather than to any mechanical momentum-type action arising from an associated ether wind.

Experiments performed by Eugene Podkletnov and his coworkers at a laboratory in Russia provide yet another example of the existence of the electrogravitic impulse effect. Using his knowledge of crystals and ceramic materials, Podkletnov developed a unique superconducting ceramic material, yttrium-barium-copper-oxide ($YBa_2Cu_3O_{7-y}$), and conducted a series of experiments in which he emitted high-voltage discharges from an electrode that had been coated with this superconducting material. In his early experiments, he applied thin coatings of this superconductor to the surfaces of metal spheres having diameters ranging from 25 to 50 centimeters. He would cryogenically cool a sphere, charge it to 500 kilovolts with a Van de Graaff generator, and then allow it to discharge across a gap to a second metal sphere. Both were contained in a helium-filled chamber (figure 6.3).[5] He observed that a weak gravitational pulse was emitted that was able to move a newspaper taped to the wall in an adjoining room. The force did not appear to diminish with distance.

In later experiments, which used a modified version of this spark gap, Podkletnov determined that this force was gravitational in nature. He succeeded in confining the impulse to a narrow beam that was capable of imparting strong longitudinal forces to very distant test masses. For these tests, he elaborated on the technology by enclosing his discharge apparatus in a vacuum chamber. Also, instead of a sphere, he used a 10-centimeter-diameter, 0.8-centimeter-thick superconducting ceramic disc for his emitter (see figure 6.4).[6-8] The disc was cooled to 50 to 70K, and an inner electromagnet coil induced a "frozen-in" magnetic field oriented perpendicular to the face of the disk to assist in collimating the discharge. An outer coil that girdled the discharge chamber was used to generate an auxiliary field to further enhance the collimating effect of

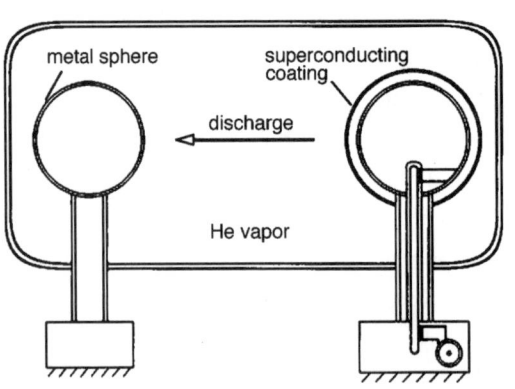

Figure 6.3. The initial setup of the Podkletnov impulse gravity generator. (After Podkletnov and Modanese, 2001)

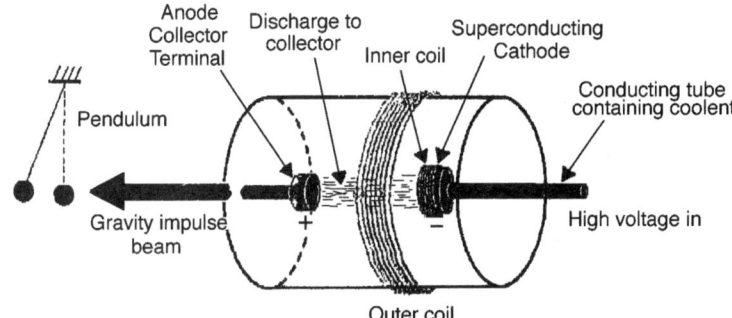

Figure 6.4. Gravity impulse beam generator developed by Podkletnov. (After Cook, Jane's Defense Weekly, *2002)*

the inner magnetic field. Podkletnov then used a Marx capacitor bank to generate a high-voltage electron pulse ranging from 0.5 to 2 megavolts, which he discharged through the disc and across the evacuated gap toward a 1.5-centimeter-thick copper anode of similar diameter.

When the capacitor bank was discharged, a coherent plane wave was emitted from the superconducting cathode as a flat, 10-centimeter-diameter glowing disc covering the entire electrode surface, which then propagated toward the anode. Using a laser beam as a sensor, Podkletnov and his associates were able to determine that the discharge had a rise time of less than 100 nanoseconds and a duration of the order of 10 to 100 microseconds. A gravity shock wave was apparently accompanying this electron discharge. While the electron discharge terminated at the beam generator's anode, a gravitational shock wave, apparently accompanying the discharge, would continue in the same direction, passing through the anode unstopped and emerging as a gravity impulse that was confined to a 10-centimeter-diameter beam matching the anode's cross-section.

When fired with a discharge voltage of 2 million volts, the emitted wave was found to produce a 14-centimeter deflection of an 18.5-gram pendulum bob suspended from an 80-centimeter-long thread and placed at a distance of 150 meters from the beam generator. The beam was able to exert this force after having first passed through a Faraday cage shield, an additional 2½ centimeters of steel, and a 30-centimeter-thick brick wall. This reminds us of Tesla's radiant energy shocks, which exerted forces even after having penetrated shields of copper and glass.

A quick calculation indicates that their pendulum bob experienced a momentary repulsive force of about 500,000 g with the passage of each 100-nanosecond shock front.*

Pendulum bobs of differing masses and made of various materials (e.g., rubber, glass, plastic, metal) were used, but all deflected by the same amount for a given discharge voltage. Since force on the pendulum scaled in direct proportion to the pendulum mass, Podkletnov and the physicist Giovanni Modanese concluded the effect they were seeing was gravitational in nature.[9] This mass effect rules out the possibility that momentum is being imparted to the pendulum by electromagnetic radiation pressure. Furthermore, the amount of electromagnetic energy produced by the discharge is far too small to explain the observed force effects. These pendulum results also rule out the possibility that this force might be due to a longitudinal "electrokinetic force," of the sort proposed by American physicist and professor Oleg Jefimenko, which would act only on free charges present in the target material.[10] If the force produced by the gravity impulse beam were due to such electrokinetic ion forces, differing force magnitudes should have been observed when differing pendulum bob materials were tested, and such was not seen. Figure 6.5 shows the amount of deflection that the pendulum experienced when the gravity beam generator was energized at various voltages.

Experiments conducted with smoke indicate that the air in the path of the gravity beam would briefly move forward and back with the passage of each emitted gravity impulse. Firing the gravity impulses through pressure-sensitive carbon paper at varying distances consistently produced a 10-centimeter-diameter black circle. This indicates that the beam was able to maintain tight coherence over large distances, with the force of the beam cutting off sharply outside of this circular boundary. In this fashion, this impulse beam is comparable to a laser

---

*In their 2003 paper, Podkletnov and Modanese reported a lower instantaneous acceleration of the order of approximately 500 g. Here they assume that the gravitational force is exerted during the entire duration of the pulse, which has a duration of 10^{-4} second. However, subquantum kinetics suggests that the gravitational force is delivered by the gradient at the forefront of their shock wave, which is of much shorter duration, having a rise time of less than 100 nanoseconds. This implies an instantaneous gravitational acceleration a thousand times larger than they calculate.

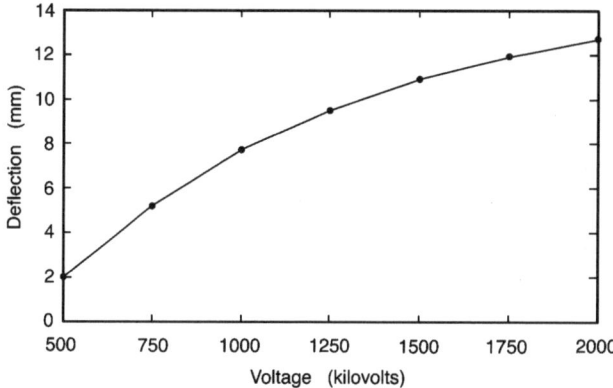

Figure 6.5. Graph of the pendulum deflection produced at various electron beam discharge voltages in the Podkletnov-Modanese gravity beam experiment. (After Podkletnov and Modanese, 2001)

beam, but achieves its coherence without the help of a resonator cavity. In an article in *Jane's Defense Weekly,* Nick Cook reported that a laboratory installation in Russia had demonstrated that, when fitted with a laser sight, this beam was able to knock over a set of books one kilometer away and that it exhibited negligible power loss even at a distance of two hundred kilometers.[11-13]

According to Cook, engineers at the Boeing Aerospace Corporation Phantom Works facility in Seattle were actively interested in investigating this beam technology with the aim of developing it into an R&D project. An internal company briefing document written entitled "Gravity Research for Advanced Space Propulsion" states, "If gravity modification is real, it will alter the entire aerospace business."[14] Other aerospace companies interested in Podkletnov's beam generator included BAE Systems and Lockheed Martin. Cook reported, however, that the Russian government had resisted allowing the gravity beam technology to be exported.

Subquantum kinetics predicts that Podkletnov's gravity impulse beam generator would produce no recoil when fired. That is, the back-directed impulse, which is delivered to the superconducting cathode at the time the electron pulse discharges, is canceled out by the equal and opposite forward-directed impulse delivered to the anode when the anode subsequently absorbs the electron discharge. However, the gravity field pulse, which continues its forward journey through and past the anode, would then produce additional forward thrusts on all masses through which it passed, in apparent violation of Newton's third law of motion. In this case, when these remote thrusts are included, it is no longer true that every action necessarily produces an equal and opposite reaction.

In 2003 I wrote to Dr. Podkletnov indicating to him my belief that his impulse generator should produce no recoil when it is operating and also that, based on subquantum kinetics, I expected that his pulses would propagate at superluminal speeds.[15] He wrote back that I was correct that, in fact, the device produces "no back mechanical reaction" when fired and also that his team had found that the pulses traveled superluminally. He said that they were able to determine that the pulses traveled at close to sixty-three to sixty-four times the speed of light, a result that they planned to check and recheck before submitting it for publication. He also wrote, "It is amazing that you could predict the effects that we have observed. We will be happy to learn more about your subquantum kinetics approach."[16] I subsequently sent him a copy of my book *Subquantum Kinetics,* and in 2004 he published a very favorable review about it in *Infinite Energy* magazine.[17]

In their 2003 paper, Podkletnov and Modanese acknowledge that conventional theories of gravity fail to explain the action of their gravity impulse beam. For example, general relativity predicts that gravity waves should induce quadrupolar forces in a target mass that are oriented transverse to the direction of wave travel. Instead, the gravity impulse beam is observed to produce repulsive longitudinally directed gravitational forces, hence, in line with the direction of wave propagation. This is just as Tesla had observed for his shock discharges. It also confirms a key prediction of subquantum kinetics that electrons would produce matter-repelling gravity potential hills (G-etheron hills) and that a change in the electric or gravity potential field should propagate forward as a longitudinal potential wave.[18,19]

Subquantum kinetics offers the following explanation of how these gravity waves might be generated: A shock discharge from a cathode to an anode would produce a wave having a sharp rise in electric field potential, followed by a more gradual relaxation. As described in chapter 4, section 4.1, subquantum kinetics predicts that an electron should generate a gravity potential hill. An electron discharge, then, would be accompanied by an in-phase gravity potential wave. This would appear similar to that shown in figure 6.6, in which the wave is shown traveling from right to left. The front of this wave would consist of a sharp rise in G-on concentration, that is, a rise in gravity potential. Its gradient would induce a gravitational force on encountered masses in its forward direction of travel, shown as from right to left in the figure. Hence, it would have a repulsive effect. The trailing part of the wave, which would have a declining gravity potential, would induce an opposing thrust that would create an attractive force on encountered masses. This drawing is highly idealized, since a pulse discharge typically produces an oscillating decline in voltage as it tails off.

A force (**F**) applied to an object over a period of time (*t*) yields a quantity called impulse, the product of force and time (**I** = **F**×*t*), which equals the resulting change in the object's momentum. So if a forward repulsive force exerted during the passage of the leading edge of the wave were to be ten times as great as the reverse attractive force exerted during the passage of its trailing edge but were to last only one-tenth as long as the force exerted during the passage of the trailing edge, the forward impulse would exactly equal the reverse impulse. So the wave's passage would have no net effect on the momentum of the target mass.

Consequently, to explain the findings of Podkletnov and Modanese, another important factor must be involved—virtual charge. The advancing

Figure 6.6. A propagating electrogravitic shock wave capable of producing a repulsive gravitational force. The wave would be traveling from right to left, and the net gravitational force and G-on flux would also be directed from right to left. (P. LaViolette, © 2007)

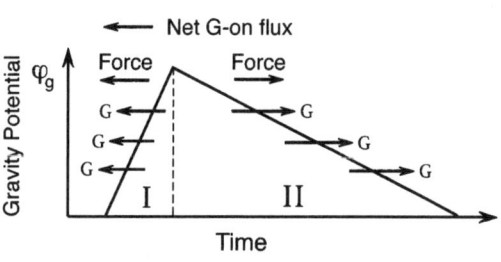

electron discharge would be accompanied by a negative electric potential wave, and not only would the front of this wave have a very steep drop in voltage with distance, but also its voltage would change with distance in a highly nonlinear manner.* So this shock front would be associated with the creation of a very high negative virtual charge density that would produce an enormous matter-repelling gravity field.

This sudden drop in voltage may be modeled with the exponential equation $V = -r^{10}$, which is plotted in figure 6.7a. Relation 7, from chapter 4, predicts that this should produce a gravity potential field varying as $\varphi_g(r) \propto \nabla^2 \varphi_E(r)$, hence as $\varphi_g \propto r^8$. This exceedingly steep gravity potential profile is plotted in figure 6.7b. The wave is plotted in the figure as traveling to the left, and its gravity gradient would be producing a force depicted as directed to the left, hence it would be repulsive. The damped sine wave oscillation that trails the shock front would produce a gravitational thrust that was many orders of magnitude smaller and oscillating from a reverse to a forward direction. However, the forward thrusts would always dominate, resulting in a net repulsive thrust.

When the electron discharge is absorbed in the impulse generator's anode, the electric field potential of the discharge goes to zero. Nevertheless, the gravity wave that was generated while the discharge was in flight continues to move forward. It passes through the anode and ultimately produces thrust effects on distant masses.

The steep gravity potential gradient at the shock's leading edge would induce a convective G-on flux in the direction of wave propagation, or, in other words, would create a G-on ether wind. The G-on fluxes described here would be accompanied by X-on fluxes traveling in the same direction (and by Y-on fluxes traveling in the opposite direction). In effect, with each firing of the gravity impulse beam, a puff of

---

*Relativistic length-contraction effects would also contribute to the steepness of this front. For example, at 500 kilovolts, the electron discharge would consist of 0.5 MeV electron with a mass twice as great as its rest mass. Electrons would be traveling at 87 percent of the velocity of light and would have a Lorentz factor of 2. Consequently, the voltage rise time would be compressed by a factor of 2, making the potential gradient twice as steep. At 2 MeV, the electrons would be traveling at 98 percent of the speed of light and would have a Lorentz factor of 5, which would make their potential gradient five times as steep.

Figure 6.7. (a) Voltage at the front of the electron shock discharge plotted as a function of time or distance. The wave travels to the left. (b) Corresponding gravity potential profile arising from the virtual charge density the wave would generate. (P. LaViolette, © 2007)

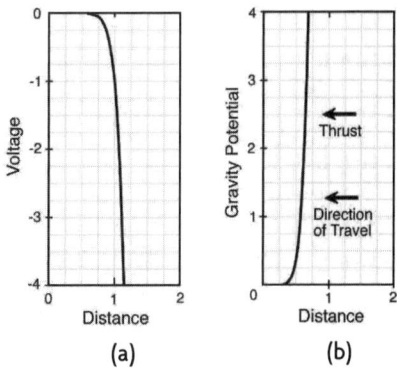

G-ons would travel rectilinearly away from the cathode. We may compare this to the firing of a cannon, in which the outward flight of the cannonball is accompanied by a forward-moving slug of air that creates a collimated ring vortex, or smoke ring. In a similar fashion, this G-on puff would be accompanied by an outward-moving G-on ring vortex that may help to collimate the impulse.

Podkletnov's gravity beam generator does not generate gravity impulses specifically because its cathode emitter is made of superconducting material. Rather, the gravitational repulsion effect of its shocks may be attributed to the electrogravitic coupling that exists between charge and gravity. The subquantum kinetics explanation given above for the production of the gravity impulse effect would apply equally well to the repulsion thrusts produced by Tesla's shock discharge pulses. The superconductivity of Podkletnov's cathode emitter more likely contributes to boosting the pulse's imparted force by sharpening and steepening its leading-edge field gradient. Also, it may help to cohere the gravity wave into a nondiverging beam.

We may surmise that this electrogravitic impulse effect manifests in essentially the same way as the Biefeld-Brown electrogravitic thrust effect. That is, it arises due to an inherent coupling between charge and gravitational mass. The impulse effect, though, exerts a much stronger instantaneous force than Brown's gravitators since its field gradient is much steeper. However, because this more intense thrust operates over a much briefer span of time, it must be cyclically repeated to produce a sustained propulsion effect.

In July 2003, Podkletnov had disclosed to me that at a higher discharge voltage, of around 10 million volts, the gravity wave pulse became so strong that it was able to substantially dent a 1-inch-thick steel plate and punch a 4-inch-diameter hole through a concrete block![20] Compared with the pendulum deflection produced by a 2-million-volt discharge, this kind of damage implies at least a thousandfold increase in the delivered force. Such a large impulse is not predicted by the trend line presented in figure 6.5, which shows pendulum deflection plateauing as pulse voltage increases. This trend projects a twofold increase in the impulse strength, not a thousandfold increase. Subquantum kinetics predicts that voltage gradients that are steeper and more nonlinear should deliver greater gravitic thrusts; recall equation 8 of chapter 4. So I theorized that for these more forceful gravity pulses, Podkletnov's research team must have powered their pulse generator with an improved Marx bank, one that was capable of delivering its charge much more rapidly to the beam generator's superconducting disc, allowing it to produce a gravity potential pulse having a steeper rise time.

To check whether my suspicions were correct, in 2007 I wrote to Dr. Podkletnov explaining my reasons for suspecting that he used an improved Marx bank to enable his pulse generator to generate these higher thrust pulses.[21] He wrote back confirming that this was indeed the case, that they had modified their Marx bank so that the pulse voltage on the superconducting emitter rose much more rapidly.[22] He stated that they observed that the faster the increase in voltage at the cathode emitter, the larger the generated impulse force. Since a faster voltage rise time would increase the nonlinearity of the pulse, their observations of a greater resulting thrust are consistent with the predictions of subquantum kinetics. Podkletnov also disclosed that this improved pulse generator exhibited increased thrust power even when energized with 5-million-volt pulses. Also, he noted that these powerful pulses would sometimes bend the generator's copper anode as well as damage the walls of the discharge chamber. It is perhaps because of these higher impulse results that the Russian government is resisting export of the technology. Indeed, technology with such capabilities could be misused as a weapon.

6.2 ▪ SUPERLUMINAL PULSES

Let us now examine some astounding evidence that shows that superluminal (i.e., faster than the speed of light) space travel is possible and at the same time refutes Einstein's outmoded special theory of relativity. One example of superluminal wave propagation is found in the gravity shock fronts produced by Podkletnov's beam generator. His research team was able to measure the speed of their gravity beam pulses by using an oscilloscope to mark the moments when the gravity pulse momentarily dimmed two laser beams directed across the beam's path. Knowing the distance between the laser beam cross-points and the times registered for each successive dimming, they were able to determine the speed of a gravity pulse. As mentioned earlier, Podletnov's team found that the pulses were traveling at sixty-four times the speed of light![23] They were only able to determine a lower-limit value since the speed of the pulses surpassed their oscilloscope's time resolution limit.

This controversial finding stands as a blatant disproof of the special theory of relativity, which maintains that nothing can go faster than the speed of light. However, the high speeds of these pulses becomes understandable when considered in the context of subquantum kinetics. According to subquantum kinetics, a light wave should have a speed of c, the velocity of light, relative to the local ether rest frame. Now, suppose that the field gradient of the advancing gravity potential wave accelerates a slug of ether to a high velocity relative to the surrounding laboratory ether reference frame. Let us say that it attains a velocity of $63\ c$. Theoretically, this should be possible since the ether is not bound by the same speed limit rules that apply to electromagnetic radiation. Now, if a light ray or shock front was moving within this ether wind slug in the same direction as the ether wind, we should find that, relative to the laboratory reference frame, this light ray would be traveling at sixty-four times the speed of light, $63\ c$ for the speed of the ether wind slug plus $1\ c$ for the light ray moving forward within it.

Podkletnov's team measured a far higher velocity for the concrete-smashing gravity impulses produced by their improved Marx bank pulse generator. Using a pair of synchronized atomic clocks to measure the arrival time of the impulses at separate locations, they were able to determine that the impulses were traveling at least several thousand times the speed of light, perhaps faster![24] Their faster speed may be attributed to their steeper field gradient, which would have propelled G-ons forward to a very high velocity.

In the 1980s, well before the experiments of Podkletnov and Modanese, American engineering physicist Guy Obolensky investigated the speed of electric field shocks to test Tesla's claims that his radiant energy shocks had traveled at superluminal speeds. In that work, Obolensky had shown that the sudden discharge of a 16-square-foot, high-voltage air-gap capacitor produced a surface wave that was able to travel along the length of a 7.07-meter-long transmission line at a speed of 1.23 c, hence 23 percent faster than light.[25]

In 2005 and 2006, I worked with Obolensky at his laboratory in upstate New York to investigate the superluminal speed of shock discharges. For this we used a high-voltage magnifying transmitter that Obolensky had built some years earlier and that incorporated many of Tesla's design features. Like Podkletnov's apparatus, Obolensky's magnifying transmitter is energized by the discharge of a Marx capacitor bank (figure 6.8). The electron shock discharge is conducted down the length of a horizontal, oil-filled tube called a Teslatron, which contains a lengthwise coil that helps to sharpen the shock

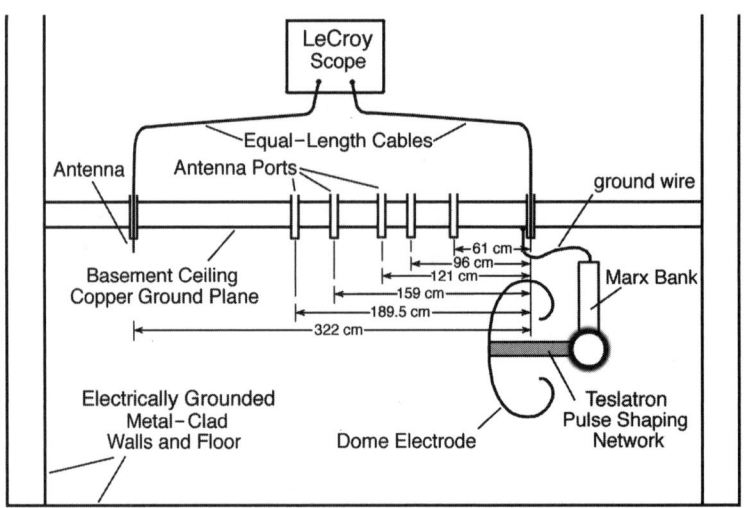

Figure 6.8. A test setup used to measure superluminal pulses radiated from a dome electrode. (P. LaViolette, © 2007)

front. Thus, it performs a function similar to that of Podkletnov's superconducting disc. The tube terminates inside a 1.2-meter-diameter, mushroom-shaped dome electrode that has a geometry similar to the dome on Tesla's Wardenclyffe tower. The electric potential of this dome "floats" at the shock's potential, so it functions much like the cathode in Podkletnov's beam generator, although it has no superconducting coating.

The shock discharge induces a damped sinusoid oscillation along the length of the Teslatron column, such that the initial negative swing in potential is followed by a positive swing, then a negative swing, and so on. This AC oscillation imprints itself on the advancing shock wave, with a typical AC pulse appearing, as shown in figure 6.9. Upon reaching the dome, the electron shock begins to fan out as it moves forward away from the electrode, forming an electric potential wave termed a Coulomb wave. This differs from a conventional electromagnetic wave in that the Coulomb wave exerts primarily longitudinal forces on charges it encounters, rather than transverse forces.

The negative swing in electric potential at the forefront of the Coulomb wave would carry a forward-moving negative virtual-charge density. The subquantum kinetics electrogravitic coupling relation

Figure 6.9. Voltage versus time oscillogram of a typical shock front pulse measured by Obolensky. Upper trace: pulse detected at 189.5 centimeters from the reference antenna; lower trace: positive current flow detected very close to the impulse generator's ground terminal. (Courtesy of A. G. Obolensky)

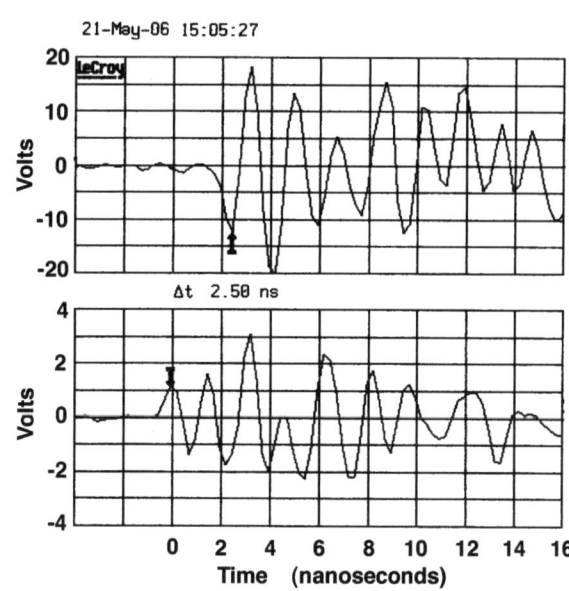

predicts that this would induce a gravity wave having a rising G field and a positive gravity potential gradient. Like Podkletnov's gravity impulse, this would exert a longitudinal repulsive force on masses it traversed. The positive swing in electric potential that immediately followed it would carry a forward-moving positive virtual-charge density that would induce a decreasing G field and an attractive force on masses it encountered. As the field continued to oscillate from negative to positive, the induced gravitational force would change between repulsion and attraction. Since the individual cycles in this wave train are sawtooth shaped, with differing rising and falling slopes, they should produce a net longitudinal gravitational force that presumably is repulsive. At a later date we hope to report measurements of the gravity impulse produced by this device.

In the case of the Podkletnov gravity beam, the beam's cross-section does not appreciably increase with distance from the beam generator. As a result, the pulse forefront should maintain its initial sharp gravity field gradient as it travels forward and should maintain its ability to accelerate G-ons in its path up to the same high speed. Hence, the beam's initial superluminal speed should not appreciably diminish with travel distance. However, subquantum kinetics predicts a different circumstance for impulses radiating outward from the dome electrode of Obolensky's magnifying transmitter. Unlike the collimated shock discharges emitted by Podkletnov's gravity impulse generator, those produced by Obolensky's magnifying transmitter fan out as they radiate away from the transmitter's dome electrode. In this case, because the impulse wavefront expands radially outward as it travels forward, the velocity of its generated ether wind would decline inversely with the impulse's distance from the dome (see box on page 181).

Since the speed of a superluminal wave would be the sum of the impulse's velocity (c) relative to the local ether wind frame plus the velocity (v) of the local ether wind relative to the laboratory frame, one would expect that the wave's net velocity would begin at a superluminal speed and decline toward c as the shock wave advances and the ether wind velocity tends toward zero.

The Decline of Ether Wind Velocity with Distance

In the case of an isotropic electrostatic or gravitational field, such as extends outward from the center of a particle, the field's potential gradient is observed to decrease as the *inverse square* of radial distance from the particle's center. However, in the case of an electric or gravitational shock wave, the gradient should decline according to the *inverse* of radial distance.[26] That is, provided that the width of the pulse does not change, the gradient should decline in accordance with the $1/r$ decline of the electric or gravity field potential. In these tests, the pulse width was found to remain relatively invariant, so one would expect a $1/r$ decline in field gradient.

In fact, tests that Obolensky and I performed showed that the velocity did decline with increasing distance as predicted. The data were best matched if ether wind velocity decreased according to the inverse of distance from the electrode grounding point.[27] This was the first experiment of its kind to determine whether a shock's superluminal speed might change with increasing distance from an emitting electrode. Obolensky's test arrangement was able to measure the shock wave's time-of-flight to six collinear antenna locations. These were situated at distances ranging from 61 to 322 centimeters, as measured from a reference point located where the current impulse from his Marx bank passed to the ceiling ground plane through a ceramic disc resistor (see figure 6.8). So this experiment was able to test the validity of the subquantum kinetics prediction that the speed of the shock wave should begin at an initial superluminal value and should subsequently decline in an asymptotic approach to the speed of light (c). It also simultaneously tested a specific claim made by Tesla that the impulses from his magnifying transmitter initially departed at a theoretically infinite velocity and subsequently slowed down, slowing rapidly at first and later at a lesser rate.

Obolensky's test setup used a 1-gigahertz-bandwidth LeCroy oscilloscope able to sample data at 250-picosecond intervals. It was much faster than the oscilloscope he had used in his earlier experiments. He used two monopole antennae to detect the electric field component of the ground current shock wave as it passed by. Each antenna was made from a single 12-centimeter-long wire attached to a 50-ohm coaxial cable terminator that led to the oscilloscope, both cables being of equal length and jacketed with ferrite surface wave suppressors. The oscilloscope, in turn, determined the time lapse between the two signal currents, and knowing the distance between the antennae, the pulse's propagation speed could be calculated.

Obolensky positioned one monopole antenna pickup immediately behind the ceramic disc grounding resistor that was close to but behind the dome's rim. This antenna sensed the positive impulse current that flowed into the laboratory ceiling ground plane with the shock wave's departure. He placed the other antenna pickup at one of the predetermined locations in front of the dome antenna. On successive test runs, he moved this second pickup to each of six antenna port locations to get pulse arrival time readings at these various distances from the ground-current-sensing reference antenna.

The lower trace in figure 6.9 depicts a typical shock current pulse detected by the reference antenna that is displayed as a positive voltage rise, its voltage maximum being indicated by an arrow. The upper trace in figure 6.9 shows the corresponding superluminal surface wave shock pulse detected by the second monopole antenna pickup positioned 189.5 centimeters from the ground-current-sensing reference antenna. Its first negative potential peak is also marked with an arrow, the surface wave's polarity being the inverse of the detected ground plane current impulse. The time interval between the two arrows indicates the time-of-flight of the superluminal surface wave. The timing of a given marker was accurate to approximately 125 picoseconds. The temporal width of the shock's lead wave cycle varied very little as it moved outward, the wave cycle having a duration of about 1.77 ± 0.09 nanoseconds.

Figure 6.10 shows the time-of-flight of the pulses as measured at various distances from the dome electrode's ground. The measurements are marked as black circles and a suggested model fit is represented by the small black diamonds. The 61-centimeter data point was given a zero time lapse since the oscilloscope measurements indicated that the pulse spanned this near-electrode distance virtually instantaneously.

Figure 6.11 shows the speed estimated for the shock at various

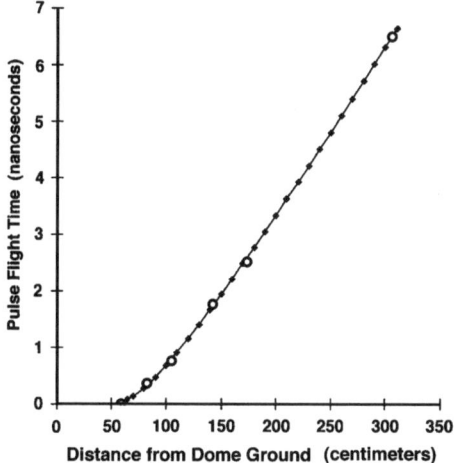

Figure 6.10. Graph showing shock front pulse time-of-flight as a function of distance from the emitting impulse generator ground. Black circles indicate actual data points. Small black diamonds plot the best fit to these data points based on the velocity-distance model plotted in figure 6.11. (Data taken by Obolensky and processed by LaViolette; © 2007, P. LaViolette)

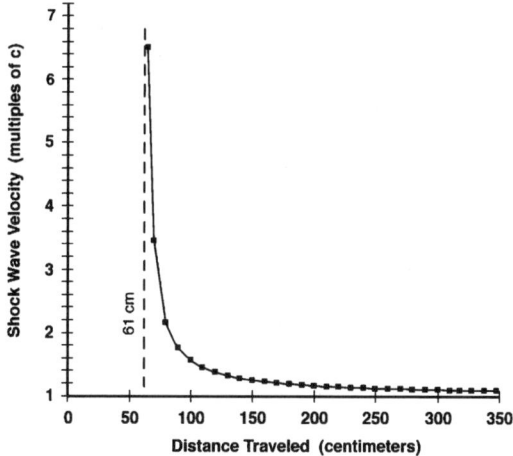

Figure 6.11. Graph showing superluminal shock front velocity plotted as a function of distance from the emitting dome's ground-current-sensing reference antenna. Gray squares indicate the model that makes the best fit to the six time-of-flight data points plotted in figure 6.10. With increasing travel distance, shock velocity declines toward c as subquantum kinetics predicts. (Data taken by Obolensky and processed by LaViolette; © 2007, P. LaViolette)

distances from the impulse generator's ground point based on the time-of-flight model fit performed in figure 6.10. This shows that at 80 centimeters from the ground-current-sensing reference antenna, the shock had a superluminal speed of approximately 2.32 times the speed of light, its speed progressively dropping toward the rest frame light speed (c) with increasing travel distance. The model shown assumes that ether velocity (v) varies with distance (d) as: $v = kc/(d - 61)^{1.1 \pm 0.1}$, in which k equals 33.6 centimeters and a distance 61 centimeters from the ground reference point is chosen as the model's zero point. Superluminal velocity (c') is then given as: $c' = c + v$.

The data strongly support the subquantum kinetics prediction that the superluminal speed of such a wave should decrease with increasing distance when radiated from a magnifying transmitter dome. Furthermore, they show that superluminal speed is a characteristic of shock discharges regardless of whether the discharges are emitted from a superconducting electrode of the sort used in the experiment by Podkletnov and Modanese. Finally, these results lend support to the unpublished findings of Podkletnov and Modanese that their gravity wave impulse had traveled at a high superluminal speed. In other words, taking the work of Tesla and that of Podkletnov and Modanese in context, we see that superluminal shock front propagation speeds are the norm rather than the exception.

Superluminal propagation speeds have also been observed in atomic bomb tests. Scientists working for the military have known since the early bomb tests in the late 1940s that the electromagnetic pulse shock wave from a nuclear explosion propagates outward at superluminal velocities when measured near the explosion epicenter. The enormous energy released in the explosion accelerates the fireball's free electrons radially outward at a relativistic velocity, generating a radially propagating shock pulse that, like a shock discharge from Obolensky's magnifying transmitter, moves outward at superluminal speeds. Subquantum kinetics attributes this breaking of the "light barrier" to the creation of a tremendous radial ether wind generated by the electric gradient of the advancing shock. For an isotropic explosion, the velocity of this ether wind would decline approximately as the inverse of the distance traveled, and similarly, the velocity of the electromagnetic pulse shock would be expected to follow this decline in an asymptotic approach to the velocity of light. At such a time when the data are declassified, it would be worth checking to see if the subquantum kinetics velocity decline prediction was born out.

Clearly, the results of Obolensky's magnifying transmitter experiment violate the fundamental premise of special relativity that energy cannot be transmitted faster than the speed of light. Beyond 1 meter from the dome's ground-current connection, the shock front travel distance exceeded the shock front's wavelength. Moreover, by the time the pulse had reached the 3-meter mark, its flight distance exceeded five pulse wavelengths. Hence, the superluminal speeds observed cannot be explained away as an allowable violation arising from quantum entanglement of the photon quantum, nor can they be ascribed to a change in shape of the shock wave profile. Thus, Einstein's theory is certainly disproved and the subquantum kinetics ether theory is vindicated.*

6.3 ■ INTERSTELLAR SPACE TRAVEL

By itself, the electrogravitic impulse effect could serve as an excellent drive for use in interstellar space travel. One could imagine a spacecraft outfitted with a large-aperture Podkletnov gravity beam projector that would be powered by a set of very high-power Marx capacitor banks operating at potentials of up to 2 million volts and conveying its discharge to a superconducting electrode seventy times larger in diameter than the superconducting disc Podkletnov fabricated for his beam generator, that is, 7 meters in diameter instead of 10 centimeters. The beam generator would be mounted at the rear of the ship and would direct

---

*Because of the revolutionary nature of these findings, we conducted a separate experiment that incorporated a double-check of our measurements. We measured the time the wave took to go from point A to point B, and the time was also measured for the wave to go from B to C. These points were collinear with the direction of shock wave propagation. The total of these two flight times was then compared with the measured time for the wave to go from A to C. The summation accurately corresponded to the observed A-to-C value. This indicated that the current pulses that the cables conveyed to the oscilloscope were reflecting the actual passage of the shock wave away from the dome electrode and that the wave's instantaneous position was being faithfully sensed by the pickup antennae.

its gravity pulses forward toward the ship's bow. Suppose that, like Podkletnov's laboratory apparatus, the ship's gravity beam generator were to produce impulse accelerations of greater than 500,000 g over the brief 100-nanosecond pulse interval. Also suppose that the ship's capacitor banks have sufficient time to recharge to deliver these pulses once per second, a rate sixty times greater than Podkletnov's apparatus was able to achieve. Averaging this gravity impulse over the 1-second interval between successive pulses, we find that this would be the same as if the ship experienced a continuous acceleration of about 0.05 g, or 5 percent of the Earth's accelerating force.

However, to run such a propulsion unit would require an enormous amount of power, 6,000 megawatts, which is 300,000 times more than the 20 kilowatts that Podkletnov and Modanese were using. Such power could be supplied by a set of large-sized nuclear power plants, similar to those operated by some electric utilities. Or it might be provided by an onboard "free energy" generator such as the gravity wave power generator described in the next section or by a Searl effect generator such as that described in chapter 10.

The beam would need to be designed to produce a very uniform acceleration across the surface of its electrode so as to minimize the development of gravitational sheer forces. If the craft's occupants were seated in the path of such a beam, they would feel no acceleration since every atom of their bodies, and the entire ship as well, would be uniformly accelerated by the gravitic pulses. The aperture of the beam would need to be made slightly larger than the craft so as to avoid the rapid drop-off in the accelerating force at the periphery of the beam. There would be no fear of meteor collisions because the beam's forward-directed gravitic pulses would clear out a path ahead of the ship that would be free of interstellar debris. By negatively charging the ship's bow, a repulsive gravitic field could be built up there that could deflect any meteor that happened to make a last-minute entry into the ship's flight path.

With an acceleration of 0.05 g, the journey to even the nearest star would take far too long to be practical for manned interstellar space flight. Such an endeavor would require engines capable of delivering gravitational accelerating forces on the order of 10 g. If the Podkletnov-Modanese beam generator was used, a pulse repetition rate two hundred times greater would be needed, or about one pulse every 5 milliseconds. However, the power demand would rise to the enormous figure of 1,200 gigawatts, which is three times the average rate of power consumption of the United States or six times the thrust power produced by the first stage of a Saturn V rocket.

Let us for the moment overlook the practicality of generating such a large amount of power on board a spaceship. Propelled by 10 g of acceleration, a ship would be able to accelerate to 2.8 percent of the speed of light in just one day. By one week, it would be up to 20 percent of the speed of light, and after one month it would be traveling at about 85 percent of the speed of light. Once up to this subthreshold light speed, having consumed three times the annual power consumption of the state of California, the crew could shut off the beam propulsion system and coast for the remainder of the journey. The time to reach Alpha Centauri, the nearest star system, lying 4.37 light-years away, would be just five years and two months. Upon nearing the Alpha Centauri system, the spacecraft would rotate itself 180 degrees and then would once again turn on its gravity propulsion beam to decelerate. Navigation could easily be done by using the "galactic GPS system," the network of pulsar beacons that is deployed throughout the galaxy. (See my book *Decoding the Message of the Pulsars* for more about the use of pulsars for interstellar flight navigation.)

Gravity beam technology, in its current state of development, is impractical from the standpoint of energy efficiency. A spacecraft having a mass of 700 tons, accelerated at 10 g with an energy consumption of 1,200 gigawatts, would have a thrust-to-power ratio of about 5×10^{-3} newtons per kilowatt, or about fifty times less than that of the NASA Lewis Research Center ion engine. Perhaps a hundredfold-higher efficiency might be secured if the craft's superconducting discs were powered by improved Marx capacitor banks of the sort Podkletnov used in generating the 10-million-volt concrete-smashing pulses. However, this would still not be much of an improvement over the efficiency of an ion engine.

If used alone, the gravity beam technology would be more practical if the beam generator and its power supply were to be located at a stationary spaceport facility with the beam being directed toward a specific destination to which one wished to travel. The spaceship would then simply need to be navigated to keep it positioned within the beam.

One advantage of the gravity beam technology would be its ability to propel the ether forward to a very high velocity within the confines of its beam, thereby creating an ether-frame dragging effect that would allow a ship to approach or even exceed the speed of light without harmful consequences to its passengers. As we have seen in the previous section, it is possible to transmit shock waves at superluminal speeds, presumably because of their ability to surf the forward-moving ether wind. So, we may conclude that it should also be possible to accelerate a ship to superluminal speeds. Unlike special relativity, subquantum kinetics allows the possibility that both matter and energy waves could be made to propagate at superluminal velocities. The gravity beam would not only propel the ship forward, but it would also push forward the ether within its superluminal tunnel. In standard physics terms, the gravity beam, in effect, would be accelerating the ship's local rest frame to superluminal velocities. As noted earlier, the Podkletnov gravity beam generator has been observed to produce gravity shocks that travel at enormous superluminal speeds. Based on this, we may conclude that it is technically possible to accelerate a "beam ship" to similar speeds, allowing it to travel toward its destination at hundreds or even thousands of times the speed of light.

Perhaps interstellar space travel could be made practical if Brown's electrokinetic thrusters, considered in chapter 3, were used in conjunction with the gravity impulse beam drive, thereby tremendously reducing the energy requirements. The power requirements for a trip to Alpha Centauri might then be brought down to around 50 megawatts, comparable to the output of a nuclear submarine reactor.

The initial phase of acceleration would be accomplished mainly through the operation of the electrokinetic thrusters. During this time, the gravity beam drive would be operated in a low power mode to conserve energy. Its main purpose would be to propel the ether forward so that the ship would not be exposed to an opposing ether wind. Then, after accelerating for several weeks up to a sublight speed of say 85 percent of the speed of light, the gravity impulse engine would be brought up to maximum impulse power, allowing the ship to accelerate through its ether wind tunnel to a superluminal speed of 200 *c*. The remaining journey to the Alpha Centauri environs would then take only one week. If there ever were a *Star Trek*–style impulse engine, this would be it, except instead of saying "Warp 5, Scotty," the captain might say "Grad 5, Scotty," consistent with the ether gradient concept of subquantum kinetics.* For longer interstellar journeys, a spaceship might accelerate to speeds exceeding 3,000 *c,* equivalent to Warp 11. Then a journey to our nearest satellite galaxy, the Magellanic Cloud, which lies 180,000 light-years away, would take only sixty years.

Is superluminal space travel possible? The answer is a resounding YES! It is no longer science fiction. It can be done using off-the-shelf technology coupled with a minimal amount of R&D. Commit $500 million and one hundred engineers and technicians to the project, and an interstellar drive unit could be built within, say, ten years' time.

---

*The warp factor scale developed for the *Star Trek* series calculates the attained super-luminal velocity from the warp factor number, using the formula: speed = *c* X (warp factor)$^{33.3}$. Hence, Warp 5 would allow a ship to attain a speed of 214 *c*.

7

PROJECT SKYVAULT

7.1 ▪ EARLY MICROWAVE RESEARCH

One evening in 1986, I went out for a beer with a friend of mine, a naturopathic physician by the name of Thomas Chavez. Like myself, Thomas had a keen interest in alternative, cutting-edge science. The topic of our conversation eventually turned to electrogravitics, and at this point my friend shared an interesting story. He told me that during the late 1950s, his father had worked as a physicist at the Rocketdyne Aerospace Corporation in Southern California and had been involved in some sort of super-secret antigravity research. At that time, Thomas had been just a young boy. He said his father normally told him nothing about what he did at work because of an oath of secrecy he had taken, but one evening, after returning home from work he had been unable to contain himself. Very exuberantly, he had exclaimed, "We got it to work, we got it to work!" When my friend inquired what it was that was made to work, his father drew him a picture showing a lens-shaped craft suspended in midair. He said, "We got it to lift off!" He would not say anything more about it, but that moment stuck in Thomas' mind and now he shared it with me. I knew him well enough to know that what he told me was entirely genuine.

Rocketdyne was first formed in the post–World War II era as a rocket engine R&D company. For most of its history, it was associated with North American Aviation. It was spun off from North American

Aviation as a separate division in 1955. Then in 1984, it remerged with its former company, which by then was named North American Rockwell as a result of the merger in 1967 of North American Aviation and Rockwell International. North American's aerospace and defense business had, among other things, developed the Apollo spacecraft and the space shuttle. At the time of the merger in 1984, Rocketdyne was producing most of the rocket engines used in the United States, but it appears it was developing much more than conventional rockets for its aerospace propulsion business. As we will discover below, its scientists were working on a next-generation propulsion system, a technology that goes far beyond the conventional rocket. At the end of 1996, Rockwell sold off its Rocketdyne division, along with most of its space and defense business, to Boeing Integrated Defense Systems. Then in 2005, Rocketdyne was resold to Pratt and Whitney, a business unit of United Technologies Corporation.

I frequently thought about my friend's story about this Rocketdyne project. It implied that the United States successfully demonstrated a field propulsion vehicle by the late 1950s, a time when Townsend Brown was still trying to interest the Pentagon and aerospace companies in his own electrogravitics research. The 1956 "Electrogravitics Systems" report did mention that North American was studying electrogravitic propulsion but that the company had not yet openly declared that it was working in this exotic field. No mention was made of its Rocketdyne division, which indicates that, at that early date, a very tight lid was already in place on Rocketdyne's antigravity project.

Some years later, in the summer of 1994, another piece of the puzzle dropped into place. At the time, I was attending a Tesla science symposium in Colorado Springs, where I was an invited speaker. I had just finished delivering my lecture on NASA's apparent suppression of electrogravitics technologies (discussed in chapter 13) and was surrounded by a small group of people asking various follow-up questions when someone handed me a quickly scribbled note, which I had a chance to read only much later. The note read:

> Sir, I've worked with the Biefeld-Brown effect for a number of years.
> I may be of help to you on verifying the effect. I believe I know your

mistake with the discs. I did correspond with T. Brown by mail and phone. Also associated with Project Winterhaven was a project with a slang name of "Sky Vaulting," a government funded project with North American Rockwell. If you are interested contact me.

P.S. NASA data is shared with the Department of Defense. Your key is with the Air Force. They are many years ahead of civilian research. NASA is a PR or a front to obscure Air Force research.

For purposes of confidentiality, I have chosen to withhold this person's name and refer to him only as Tom. The story he later told me about the Skyvault project was quite astounding. He said that he first heard about it in the fall of 1974, when working for an engineering firm in Texas. His supervisor, with whom he had come to be very good friends, one day told him about a top-secret government project that he had worked on between 1952 and 1957 while at North American Aviation, a company that was later renamed North American Rockwell. The project had been initiated by the Defense Department through North American's Rocketdyne division. Although Tom's boss had already passed away, Tom did not wish to reveal his name, so to facilitate the discussion, we will call him Murray. Well, Tom had heard from Murray that the purpose of this project was to develop an antigravity vehicle that used microwave beams as its means for propulsion. It is uncertain whether Skyvault was the official name of the project, but at least this is what the scientists at Rocketdyne used to call it.

Although Project Skyvault was initiated by the government in the early 1950s, investigations into this exotic microwave propulsion technique actually dated back to the late 1940s. Murray, who held a Ph.D., said that in those earlier days he had worked on projects that were associated with an initial phase of this research and that later he had continued this work at Rocketdyne, where he worked up until the 1960s. This microwave antigravity propulsion research project was still in progress in 1974, because Tom learned that a close friend of Murray's was then still working on the project at North American Rockwell, presumably in its Rocketdyne division. At that time, the whole matter was still very secret, because there was a lot that his boss couldn't tell him about the project.

Later, in 1975, Tom obtained what he felt was additional confirmation for the existence of Project Skyvault when the military sent his Texas-based engineering firm a bid request for building a vehicle launch gantry in New Mexico. From the blatant description of the shape of the gantry and the way it was to be built, he recognized that this was to be a launcher for a microwave beam antigravity craft. In this particular version, the power was generated on the ground and sent up to the craft as a microwave beam. The beam was emitted from upward-pointing microwave horns that were supported by the launch gantry. The craft was made of a special kind of material that was repelled by microwaves and, hence, was to be buoyed upward by the beam (see figure 7.1). A portion of the beam was returned to the ground to modulate the outgoing microwave beam. The craft was to be able to go straight up and down and could deviate only a small amount to either side of vertical.

In 1996, two years after my conversation with Tom, CBS-TV aired a weekly spy thriller called *Mr. and Mrs. Smith,* which starred Scott Bakula, an actor who also has had leading roles in various science-fiction series such as *Quantum Leap* and *Star Trek.* Interestingly, the "Space Flight Episode," number nine in the series, which aired on November 8, 1996, came very close to portraying Tom's story about the propulsion beam craft and launch gantry his firm was asked to bid on. The plot of this particular episode was based on the testing of an experimental disc-shaped vehicle called a "beam rider." The launching took place from a secret desert location. The test vehicle was lofted on a powerful microwave beam that was directed vertically upward toward the craft from a ground-based parabolic mirror. Since much of the early Rocketdyne research on Project Skyvault was done in the Los Angeles area, it is not surprising that this idea would one day find itself worked into a Hollywood script. However, even though there were four more episodes left to run, to the disappointment of many, *Mr. and Mrs. Smith* was canceled immediately after this episode had aired. As we shall see, the notion of using microwave beams for aerospace propulsion is not science fiction.

The discussion about Project Skyvault that is presented here and in the next chapter is based on notes I made of my conversations with Tom and on some material Tom had sent me. The latter includes copies of

Figure 7.1. Artist's conception of a Skyvault-type craft being launched on a ground-based microwave beam. (P. LaViolette, © 2007)

notes that he made of his 1974 discussions with Murray and a copy of a letter written by Murray's friend who was at the time still working on Project Skyvault (see appendix E).

According to Murray, the first indication that microwaves could be used for propulsion came about when it was discovered that microwave beams could move objects if the objects happened to be made from the right kind of material. The scientists believed that the microwave beam was somehow inducing a gravitational force on the object. The idea that microwaves could move objects was believable to Tom since he had heard of something remotely similar from a radar engineer friend of his who worked at Homestead Air Force Base in Florida. His friend had witnessed an experiment in which a low-power microwave beam from a klystron tube was aimed at pencils placed on a table and caused them to move around. Tom theorized that the microwaves must induce electric charge gradients in certain materials having nonlinear electrical properties and that the observed movement was actually due to the Biefeld-Brown effect imparting a thrust to the material.

The group that Murray had worked with had experimented with a whole lot of different kinds of samples to find out which ones worked best. Paper, silk, and some kinds of wood, for example, showed no movement. Brick and concrete also exhibited no movement, being essentially transparent to the microwaves. They found that some materials would move quite violently, whereas others would just vaporize. Aluminum foil would move but would disintegrate upon exposure. They carried out extensive tests, subjecting various kinds of materials to microwave waveforms of varying shapes, and accumulated data on the destruction and burning of the materials and on the effect of shock waves on those materials that responded. They found that the best propulsion effect occurred in materials that had a particular *magnetic* property. Tom attempted to find out more specifically what these types of materials were, but was told that the information was classified.

Murray said that their group had found that the effects were very frequency-sensitive, that is, that they were observed only within certain frequency bands that were characteristic of each material. If the frequency was off by a slight amount, the object could suddenly vaporize. He described an experience they had in their lab one time when they were experimenting with various frequencies—they had turned on their microwave generator and it had produced a bluish microwave beam that blew a hole through their laboratory wall and continued through an adjoining outside embankment as well. The beam was going into another building before they managed to shut it off. He said it "scared the living daylights out of them."

7.2 ■ELECTROMAGNETIC RESONANCE

Although Murray would not reveal what this unique class of materials was that could respond with a strong propulsive force, it is apparent that he was talking about materials that exhibit a strong resonance at a particular frequency. Such materials respond to incident microwaves in an unusual way. Take as an example a material that exhibits a resonant response to the electric component of an electromagnetic wave. Over most frequencies, the material's permittivity will have a positive value, and as a result, the applied electric field will induce a polarization in the same direction as its own field vector, as is commonly observed in most materials. Near a resonant frequency, however, the induced polarization will become very large, the material's large response being due to its accumulation of energy from the microwave beam over many wave cycles. The energy stored in the resonating medium can then greatly exceed that delivered by the incident-driving field. It can be so large that even changing the phase or sign of the incident wave would have little effect on the polarization oscillation.[1] As a result, when the frequency of the incident wave is increased slightly above this resonant frequency, the applied electric field will be out of phase with respect to the induced polarization oscillation, and as a result, the material will respond by exhibiting a *negative* permittivity, the induced polarization now being out of phase with the applied electric field. The electrons oscillating in the material will now resist the applied electric field, and as a result, the electromagnetic wave will exert a repulsive force on the material.

Physicists John Pendry and David Smith illustrated this repulsive force phenomenon by considering the example of a person pushing a swing. In an article in *Scientific American,* they wrote:

> Think of a swing: apply a slow, steady push, and the swing obediently moves in the direction of the push—although it does not swing very high. Once set in motion, the swing tends to oscillate back and forth at a particular rate, known technically as its resonant frequency. Push the swing periodically, in time with this swinging and it starts arcing higher. Now try to push at a faster rate, and the push goes out of phase with respect to the motion of the swing—at some point, your arms might be outstretched with the swing rushing back. If you have

been pushing for a while, the swing might have enough momentum to knock you over—it is then pushing back on you.[2]

In the same way, electrons in a material with a negative permittivity, ε, go out of phase and resist the "push" of the electromagnetic field. Such materials include silver, gold, and aluminum, whose resonances usually occur at optical frequencies.

The same repulsive force phenomenon occurs in materials that resonate with the magnetic component of an incoming electromagnetic wave. The magnetic permeability of the material, μ, which normally would be positive, becomes negative at frequencies slightly above the material's resonant frequency. The material's response then is to magnetically resist the magnetic field of the applied electromagnetic wave. Materials that naturally exhibit negative μ domains include ferromagnetic or antiferromagnetic materials that exhibit resonances. Such resonances usually occur at frequencies in the gigahertz range and tail off at higher frequencies in the terahertz-to-infrared range. For example, a group of Japanese scientists have reported negative permeability in a granular composite material consisting of 70 percent Permalloy when the material is exposed to microwave frequencies higher than 5 gigahertz.[3]

The special microwave propulsion materials that Murray said were being researched by the Project Skyvault engineers, which "had a particular magnetic property," were most likely materials of this sort exhibiting magnetic resonances in the gigahertz range. This would account for Murray's comment that the propulsion effects were very frequency-sensitive, that is, that each material had its own frequency band at which it would respond by developing a propulsive force. As mentioned earlier, negative μ domains in such materials are limited to a specific frequency range, with the greatest repulsive effects occurring when the incident wave has a frequency close to a material's magnetic resonant frequency. If the microwave beam was adjusted to have a frequency slightly lower, so that it matched the material's resonant frequency, then the material would absorb an enormous amount of energy from the beam and would store this energy in its resonant oscillation. In cases in which the material was being exposed to a very powerful microwave beam at this resonant frequency, it is possible that the energy that the material would capture would be great enough to vaporize it, just as Murray had said.

According to Pendry, the force that microwaves exert on a material at a given frequency depends on the strength of the material's interaction with that beam, which is proportional to the beam's scattering cross-section.[4] This force is always rather weak but can be significantly enhanced by tuning the beam to have a frequency close to one of the material's resonant frequencies. When the beam is at the material's resonant frequency, the material would present a high scattering cross-section and would strongly absorb the incident beam. At a slightly higher frequency, the scattering cross-section would continue to be high, but the ε or μ would now become negative and the material would begin to exert a repulsive force relative to the exciting beam.

A material would respond with an even stronger repulsive force if it were to exhibit electric and magnetic resonances in the same frequency range, allowing both ε and μ to become negative at a slightly higher frequency range. Such a material would have a negative index of refraction. The index of refraction (n) of a material is determined by the values of its permittivity and permeability; that is, $n = \sqrt{\varepsilon\mu/\varepsilon_0\mu_0}$, in which the constants ε_0 and μ_0 are the permittivity and permeability values in a vacuum. Most commonly occurring refractive materials such as plastic and glass have a positive index of refraction, with either one or both of their ε and μ parameters being positive. Materials with a negative index of refraction are not normally observed in nature, since electric resonances producing negative ε values and magnetic resonances producing negative μ values occur at differing regions of the electromagnetic spectrum. However, with proper engineering, it is possible to produce special materials, called "metamaterials," whose permittivity and permeability both are simultaneously negative over a specific frequency range, causing them to exhibit a negative index of refraction. Since negatively refracting materials are full of resonances, these resonances can be exploited to enhance the scattering cross-section and hence the propulsive force on the material.

The idea that it might be possible to produce a material with a negative index of refraction was first suggested in the open literature in 1967 by the Russian physicist Victor Veselago.[5] Beginning in the mid-1990s, researchers began experimenting to see if Veselago's prediction might be true. Finally, by 2001, Smith and his colleagues at the University of California San Diego successfully demonstrated the production of one such artificial metamaterial, which they made by constructing an array of straight wires and wire-loop split-ring resonators.[6-8] Using lithographic techniques, they fabricated a series of resonator elements into printed circuit boards having a straight wire on one side and C-shaped split-ring resonator patterns on the other side (figure 7.2).[9] These elements were then assembled in rows having a spacing of the order of 0.5 centimeter to compose a metamaterial matrix (figure 7.3). The array was found to exhibit both electric and magnetic resonances, causing the material's ε and μ values both to become negative over a frequency range of 10.3 to 11.5 gigahertz. They showed that a 10.5-gigahertz beam (2.8 cm wavelength)

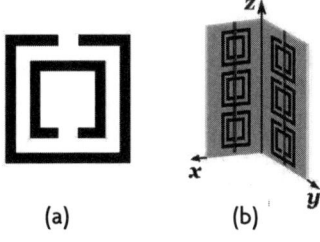

(a)   (b)

Figure 7.2. (a) A split-ring resonator. (b) Split-ring resonators combined on a circuit board with straight-wire segments to form an electric and magnetic resonator element. Many such elements together would be used to compose the metamaterial. The dimensions of the pattern are specially chosen to give the desired resonance effect. (After R. Shelby, et al., "Microwave Transmission through a Two-Dimensional Left-Handed Metamaterial," Applied Physics Letters, 78[4] [2001]: 489–91, fig. 1)

Figure 7.3. Resonator elements combined to form a metamaterial array. This metamaterial would exhibit a negative index of refraction over a specific microwave frequency range. (Photo courtesy of Richard A. Shelby)

refracted negatively as predicted. Soon after, Claudio Parazzoli, Kin Li, and coworkers at Boeing's Phantom Works Division constructed a three-dimensional wire lattice in the form of a 2.7-millimeter cube that negatively refracted a 10-gigahertz microwave beam.

Another group, at Bartol Research Institute at the University of Delaware, created a metamaterial in a very different manner by incorporating metallic magnetic nanoparticles into an appropriate insulating matrix.[10] This sounds similar to Brown's idea of embedding massive semiconductor particles, such as lead oxide, in the conical dielectric member of his electrokinetic apparatus. A diagram taken from his 1965 patent and included here in chapter 3 as figure 3.8 shows these particles as speckles concentrated near the tip of the dielectric cone. Brown disclosed that it would be advantageous to incorporate such particles to improve the thrust of his device when it was excited at microwave frequencies. His dielectric was made so that the particles became increasingly concentrated toward the tip, his intention being to progressively decrease the permittivity of the dielectric so that the voltage gradient at its tip would become increasingly high.

Although the term had not been invented at the time, Brown was in fact fabricating a metamaterial. Moreover, like the Bartol group, he may have been experimenting with embedding ferroelectrics in dielectric media to cause magnetic permeability to vary along the length of the dielectric. For example, in his patent he wrote:

> In applying potentials to these various embodiments, it has been found that the rate at which the potential is applied often influences the thrust. This is especially true where dielectric members of high dielectric constant are used and the charging time is a factor. In such cases, the field gradient changes as the charge is built up. In such cases where initial charging currents are also high, dielectric materials of high magnetic permeability like-wise exhibit varying thrust with time.[11]

In either case, by embedding such particles in his dielectric, Brown would have been producing domains having electric and/or magnetic resonances over a range of microwave frequencies, which, in turn, would have created regions where the dielectric's permittivity and/or permeability would be negative.

So, many decades before the University of California San Diego group demonstrated negative index of refraction in a metamaterial, Brown was experimenting with similar artificial materials but with the aim of enhancing the thrust in AC-excited dielectrics. He made no mention of resonances or negative ε and μ values, so maybe he was not entirely aware of all the reasons why these semiconductor particles were improving the lift of his dielectrics. He was conducting these investigations almost a decade after the Project Skyvault scientists had begun their highly classified early experimentation with similar materials, so it is not surprising that Admiral Rickover advised him to drop his electrogravitic investigations. Brown was apparently getting too close to work already in progress in Project Skyvault. Interestingly, in the early 1950s, Brown was conducting electric disc experiments at his Los Angeles laboratory, which was in the same metropolitan area where Project Skyvault was under way. One wonders if he had heard rumors of the Skyvault work.

Metamaterials have strange new properties not normally seen in nature. First of all, they refract electromagnetic waves more strongly than naturally occurring materials that have a positive index of refraction. The diagrams in figure 7.4 compare the trajectories for a beam passing: (*a*) through a medium having a positive index of refraction and (*b*) through a medium having a negative index of refraction. Regardless of their refractive index, materials with a positive index always refract incident rays into the right quadrant, which lies on the opposite side of the line that is normal to the refracting surface. Materials with a negative index (e.g., $n = -1$) always refract incident rays into the left quadrant, which lies on the same side of the normal line. For this reason, materials

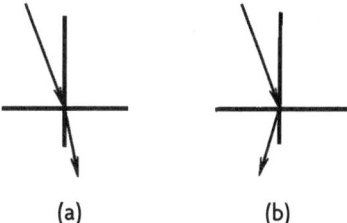

Figure 7.4. Refraction of a light ray in a material having (a) a positive index of refraction and (b) a negative index of refraction.

(a)   (b)

with a negative index of refraction are sometimes termed "left-handed materials."

Another unusual property of left-handed materials is that they can be used to make a lens that resolves far greater detail than one made out of a material with a positive index of refraction. Such a lens can depict details even smaller than the wavelength of light used to illuminate the object and can be made much lighter compared with the bulging convex lenses of conventional optical devices. Metamaterials refract light so strongly that a planar sheet serves as a convex lens, with the incident rays coming to a focus within the sheet. In order to allow the beam to come to a focus outside the lens, the metamaterial must be fabricated in a concave shape.

Also, metamaterials can be designed to be total absorbers of the incident radiation, making them ideal as radar-absorbing materials. In 2003, Yong Zhang and coworkers at the National Renewable Energy Laboratory demonstrated negative refraction in a common ferroelastic material made from a "twinned" alloy containing yttrium, vanadium, and oxygen.[12] They found that this crystal would negatively refract light of any frequency with no back reflection at the medium interface, which raises the possibility of making reflection-free optical lenses.

The U.S. Air Force Research Laboratories has supported a number of research projects with the recent upsurge of interest in metamaterials. Also, the Defense Sciences Office (DSO) of the Defense Advanced Research Projects Agency has an ongoing program for funding metamaterials research. In fact, as of July 2006, the description of its metamaterial program that DSO gave on its website frankly stated that one of the intentions of its research program was to develop magnetic metamaterials for "electric drive and propulsion":

> Metamaterials are engineered composites that exhibit superior properties not observed in the constituent materials or nature. The objective of the MetaMaterials Program is to develop, fabricate, and implement new bulk metamaterials that will fill the tremendous voids that exist in the design space for a number of applications of critical importance to the military Services. In particular, this program will develop (1) magnetic metamaterials for power electronics *and electric drive and*

propulsion, and (2) microwave and optical metamaterials for antenna, radar, and wireless communication applications.[13]

Thus, the notion that Rocketdyne was developing metamaterials in the early 1950s for an aerospace electric propulsion application seems plausible.

Interestingly, by early November 2007, DSO had removed this webpage and replaced it with a rewritten version that made no mention of its interest in using negative index materials for propulsion. It now mentions only the application of negative index materials to optics and to the development of "lightweight, compact RF structures," with an additional broad reference to "practical application" of the technology. Obviously, sometime between July 2006 and November 2007 Defense officials must have decided that the propulsion part of their metamaterial program was too sensitive to be mentioned publicly. One wonders whether this website modification might have been triggered by their becoming aware of the impending publication of this book. In early August 2007, a draft catalog copy announcing its forthcoming publication and noting its disclosure of Project Skyvault had been e-mailed to me for author review and by early October 2007 the finalized copy of the book announcement had been posted on the publisher's website.

7.3 ▪ SAWTOOTH WAVES

Murray said the Skyvault team found that the kind of response they got with a given material depended on the particular wave shape used. They achieved the best results with a sawtooth-shaped waveform consisting of asymmetrical triangular waves that have either a steep voltage rise or a steep voltage decline. As noted in chapter 6, the shock wave pulses produced by Podkletnov's gravity impulse beam generator have a sharp rise at their leading edge and are capable of generating strong repulsive forces. Tesla also observed longitudinal repulsive forces being produced by the energy wave shocks radiated from his magnifying transmitters.

Subquantum kinetics predicts that the magnitude and direction of the accelerating force depend critically on the shape of the sawtooth wave since different wave shapes generate different virtual-charge distributions, which in turn generate differing gravity potential gradients. Consider, for example, the wave shape shown in figure 7.5a. This plots the voltage potential profile, $\varphi_E(r)$, for a capacitor that is gradually charged through a resistance, R_1, and then rapidly discharged through another resistance, R_2, in which R_1 is greater than R_2. This produces a wave with asymmetrical rise-and-fall slopes (see figure 7.5 legend). When the x axis is chosen to indicate distance instead of time, the graph shows how the electric potential gradient varies with distance from the front to the back of the wave, in which the wave travels from right to left at the speed of light. The sawtooth profile in figure 7.5b (left) is a similar type of waveform, but one depicting voltage potential in a capacitor that charges rapidly and then discharges more gradually. In effect, the values for resistances R_1 and R_2 have been interchanged from those used in generating the curve shown in figure 7.5a. In this second example, R_1 is less than R_2.

Profiles 7.5c and 7.5d in figure 7.5 plot the corresponding gravity potential profiles, $\varphi_g(r)$, that would accompany each of these electric potential waves, based on the assumption that the virtual-charge densities that this wave creates generate corresponding virtual-mass densities and associated gravitational potentials. These are obtained by taking the negative second derivative of the voltage potential equation plotted in figure 7.5a or figure 7.5b. This is in accordance with the subquantum kinetics electrogravitic relation specified by equation 7 in chapter 4. The wave profile shown in figure 7.5a that charts a gradual voltage ascent followed by a rapid voltage decline is seen in the gravity potential plot, figure 7.5c, to initially produce a weak attractive gravitational force (small arrow) followed by a very strong repulsive force (large arrow). For the particular waveform plotted here, the repulsive impulse (force multiplied by time) is twenty-five times greater than the attractive impulse, even though the repulsive impulse lasts only about one-fifth as long. Thus, this wave would produce a net repulsive force on material bodies that it passes through. Changing the wave shape to that shown in figure 7.5b yields a gravity potential wave that has a very steep attractive gravity gradient at its leading edge and a gradual repulsive gradient at its trailing edge and produces a net attractive gravitational force (see figure 7.5d).

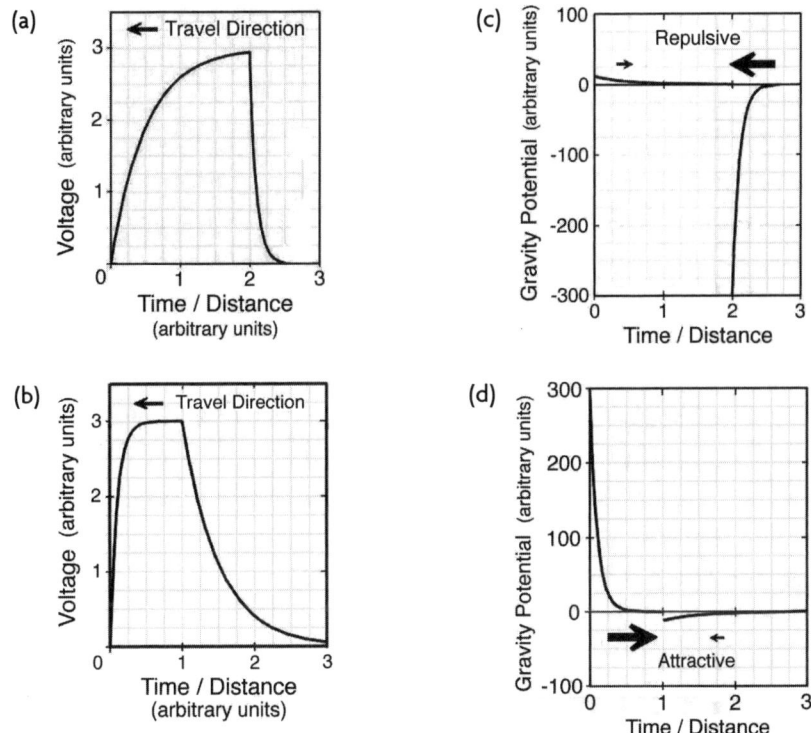

Figure 7.5. (a) left: Voltage profile for an asymmetrical RC-RC-type sawtooth wave having a gradual exponential voltage rise and rapid exponential voltage fall. (b) Voltage profile for an RC-RC wave adjusted to have a rapid exponential voltage rise and gradual voltage fall. (c) Gravity potential profile generated by the virtual-charge distribution of the RC-RC wave shown in 7.5a. (d) Gravity potential profile generated by the virtual-charge distribution of the RC-RC wave shown in 7.5b. Arrows in 7.5c and 7.5d indicate the magnitude and direction of the resulting electrogravitic thrust. (P. LaViolette, © 2007)

The voltage potential in 7.5a is represented by the following equations: $V = 3(1 - e^{-2x})$ for voltage ascent and $V = 3e^{-10x}$ for voltage descent, in which x represents time from left to right or distance from the front to the back of the wave. Thus, the wave has a descent rate that is five times its rate of ascent. The voltage in 7.5b is represented by the following equations: $V = 3(1 - e^{-10x})$ for the voltage ascent and $V = 3e^{-2x}$ for the voltage descent. Thus, the exponents in the equation for figure 7.5a have been interchanged in producing the equation for figure 7.5b. The corresponding gravity-potential profiles, 7.5c and 7.5d, were plotted by taking the negative second derivative of these voltage relations.

If the wave was made symmetrical so that its voltage rose at the same rate as it fell (R_1 = R_2), then the attractive gravitational force produced during passage of the wave's leading edge would just balance the repulsive force produced during passage of its trailing edge, and as a result, the net gravitational force would be zero.

If the polarity of the wave shown in figure 7.5a was changed so that its voltage was negative instead of positive, as shown in figure 7.6a, then the electrogravitic coupling relation predicts that the gravitational potentials would also reverse polarity, as shown in figure 7.6c. The gravity gradients would now change sign, making the net thrust attractive rather than repulsive. The same gravitational polarity reversal would occur if the profile shown in figure 7.5b was to go negative. Instead of producing a net attractive force, it would produce a net repulsive force.

Stavros Dimitriou, professor of electrical engineering at the Technical Education Institute in Athens, Greece, has investigated whether capacitors energized with sawtooth waves in the radio frequency range might exert gravitational forces on nearby masses.[14–16] In his master's thesis, he discussed an electric intensity waveform having shapes similar to those shown in figures 7.5 and 7.6, which he named an RC-RC waveform.[17] The RC acronym implies that the wave shape is determined by a capacitor of capacitance (C) being charged through a resistor of resistance (R). His electrogravitic wave research is discussed in chapter 11.

Another waveform shape that Dimitriou has investigated is shown in figure 7.6b. This one is similar to the RC-RC sawtooth wave shown in figure 7.5a, except that in this case voltage declines linearly, rather than exponentially.* The wave's voltage descent is created by discharging the charged capacitor through a constant-current (Norton) element; hence, Dimitriou refers to this type of waveform as an *RC-Norton* wave. The subquantum kinetics electrogravitic coupling relation predicts that such a wave would produce an attractive gravitational force during its voltage ascent and no gravitational force on voltage descent. That is, since voltage declines linearly, its second derivative would be zero. Hence, no virtual charge and no mass density would be produced during this decline phase.

---

*The wave in figure 7.6b is represented by the following equation: $V = 4(1 - e^{-x})$ for the voltage ascent and $V = 4(1 - 5x)$ for the voltage descent.

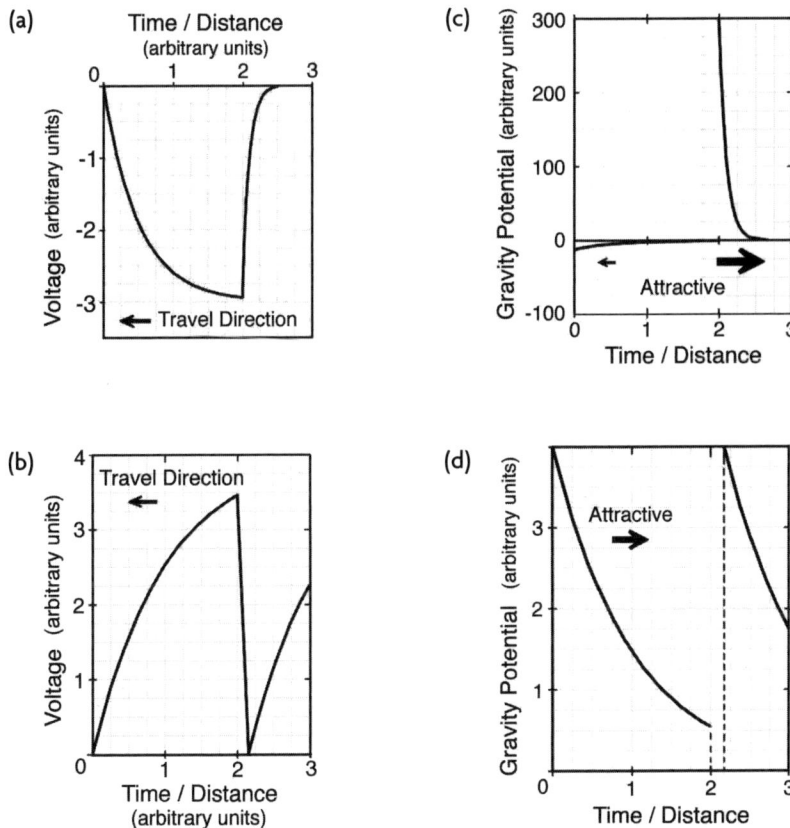

Figure 7.6. (a) Voltage of an asymmetrical RC-RC-type wave having a gradual exponential voltage drop and rapid exponential voltage rise. (b) Voltage of an RC-Norton sawtooth wave having a gradual exponential voltage rise and rapid linear voltage decline. (c) Gravity-potential profile generated by the RC-RC wave shown in 7.6a. (d) Gravity-potential profile generated by the RC-Norton wave shown in 7.6b. (P. LaViolette, © 2007)

If the gravity field is generated primarily by the electrogravitic effects of virtual charge, then a triangular sawtooth wave having a linear voltage rise and fall should produce no gravitational thrust, either on the voltage ascent phase or on the voltage descent phase. In another example, thrust will not be produced if the wave has a voltage profile that rises as the square of distance. Such a profile would have a concave parabolic leading edge similar to that of profile 2 in figure 7.7a. In this case, the negative second derivative of its r^2 voltage profile gives a negative gravity potential that remains constant over time (see profile 2 in figure 7.7b). Hence, its gravity gradient would be zero, resulting in no force exertion.

If we change the exponent characterizing the wave's profile so that the exponent is not equal to 1 or 2, but to a fraction or any other whole number, then the wave would be able to exert a gravitational force. In the case in which the profile varies as $r^{1.5}$, the gravity potential gradient creates a repulsive thrust in the same direction as the gravity wave's motion (see curve 1 in figure 7.7b). If the profile were to vary as $r^{2.5}$, the gravity potential would develop a slope of opposite sign that would produce an attractive force opposed to the direction of wave motion (see curve 3 in figure 7.7b). If the profile instead were to vary as r^3, the gravity potential would develop a steeper slope that would produce an even stronger attractive force (see curve 4 in figure 7.7b).

The shock discharge emitted from Podkletnov's superconducting cathode would similarly have been characterized by an exponential voltage rise at its leading edge, but one with a very large exponent. Hence, such discharges would have produced much steeper gravity gradients

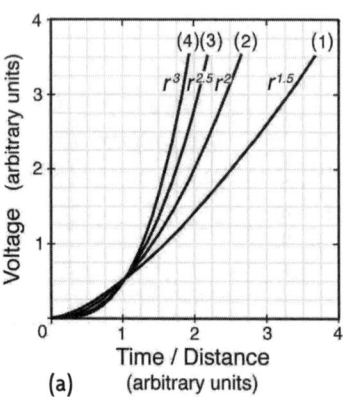 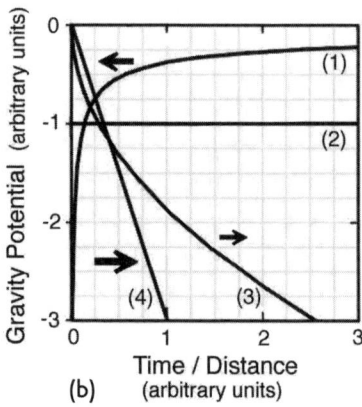

Figure 7.7. (a) Leading-edge profiles for waves whose voltage rises exponentially with varying degrees of nonlinearity, as: $r^{1.5}$ (curve 1), r^2 (curve 2), $r^{2.5}$ (curve 3), and r^3 (curve 4). (b) Corresponding gravity potential distribution generated by the resulting virtual-charge distribution. (P. LaViolette, © 2007)

and greater gravitational forces than those modeled here. Also, being a negative voltage rise, the gravitational force exerted by Podkletnov's beam is predicted by the electrogravitic coupling relation to be repulsive rather than attractive, as is observed.

The above analysis assumes that the gravitic effect of these saw-tooth microwaves arises primarily from the virtual charges that these waves produce, which themselves arise from the second derivative of the variation of voltage along the direction of wave travel. If, on the other hand, the gravity field of the emitted microwaves is produced mainly by the real electric charges that generate the wave, then the electrogravitic thrust would be proportional to the first derivative of the wave's voltage variation, and in this case linear sawtooth profiles and r^2 exponential profiles would produce gravitational thrusts. Further research is needed to know which of the two electrogravitic relations better characterizes gravity wave production at microwave frequencies, or whether thrust effects arise from a mix of both virtual- and real-charge electrogravitic effects.

One characteristic of the virtual-charge electrogravitic relation is that voltage profiles having a more nonlinear variation should produce greater gravitic thrusts. For example, a wave whose voltage rises as r^3 is predicted to produce a gravitational force 2.8 times greater than a wave whose voltage rises as $r^{2.5}$. Also, a wave whose voltage rises as r^4 is predicted to produce a gravitational force six times greater than a wave whose voltage rises as r^3. Those with exponents less than 2 would produce very weak forces. For example, a wave such as that in profile 1 in figure 7.7a whose voltage rises as $r^{1.5}$ is predicted to produce a gravitational force fifteen times weaker than that produced by a wave whose voltage rises as $r^{2.5}$. Corroborating this, Brown observed that the electrogravitic force developed by an electric field in fact increases as the nonlinearity of the field's voltage profile increases.

Dimitriou claims to have generated gravitational forces by energizing capacitors with sawtooth waves having an amplitude of around 15 volts (see chapter 11). Our attempts to duplicate his work in this low-voltage range, however, did not meet with success. Most likely the wave amplitude must exceed tens of kilovolts before this sawtooth wave thrust effect becomes large enough to be significant. This is consistent with the findings of Brown and Podkletnov, both of whom used waves in the range of 50 to 2,000 kilovolts to get their thrust effects. As we shall see, the Project Skyvault team was also using waves in the kilovolt range to get its propulsion effects.

In summary, the kind of propulsion results that Murray's Skyvault team would have been getting would have depended critically on the shape of the microwave waveform it was using. One question that should be examined is whether metamaterials develop a greater propulsive force (exhibit a larger interaction cross-section) when exposed to a microwave beam having an asymmetrical sawtooth wave shape as opposed to a symmetrical sine wave shape. If so, it is likely that the frequency-sensitive materials they were using in their research were in fact metamaterials.

7.4 ■ THE BEAM GENERATOR

According to Murray, during the early stages of their research, the Project Skyvault group used magnetron vacuum tubes to generate their microwave source beam. They worked with frequencies ranging from 7 gigahertz (7,000 megacycles) to upward of 1,000 gigahertz. By comparison, the magnetron tubes used in microwave ovens typically have frequencies of 2.54 gigahertz. The cavity magnetron has a central electron-emitting cathode surrounded by a positively charged copper plate, the anode (see figure 7.8). An axial magnetic field causes electrons emitted by the cathode to cycle in a circular orbit. They revolve at a frequency that depends on the applied voltage potential and the strength of the magnetic field. As they cycle, they induce microwave frequency oscillations in a series of cylindrical cavities spaced around the anode's inner circumference. Just as the length of an organ pipe tunes the pipe to a certain pitch, the diameter of these cavities can efficiently tune microwaves to a particular wavelength. These oscillations transfer to the cycling electron cloud and are then channeled out of the magnetron to form a microwave beam.

The microwave signal from the magnetron tubes used by the Skyvault group was sent into a wave amplifier cavity. This was essentially a metallic duct of rectangular cross-section whose long dimension

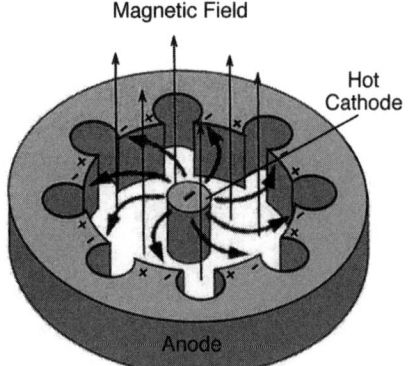

Magnetic Field

Hot Cathode

Anode

Figure 7.8. Cross-sectional view of a cavity magnetron.

was such as to fit a whole number of wavelengths of the microwave signal along its length. For example, if the magnetron emitted waves at a frequency of 100 gigahertz, the emitted wavelength would have been 3 millimeters. So if the cavity was made to have a length that was some multiple of 3 millimeters, then, as these waves reflected back and forth inside this cavity, they would develop a condition of resonance allowing them to build up a high-voltage amplitude.

By adding various types of microwave radar-absorbing materials to the resonator cavity, the inputted microwave signal could be changed from a sine wave into a sawtooth-shaped wave. For this, the Skyvault group may have used ceramic dielectrics such as barium titanate polarized with a high-voltage DC potential on the order of 10 kilovolts per centimeter. Once polarized, the high-K dielectric would have presented a highly nonlinear environment for the microwaves. The same wave transformation into a sawtooth shape would have occurred in Brown's AC-energized vertical-thrust apparatus described in chapter 3. The dielectric would have changed the shape of the input wave, causing it to have a more rapid rise of potential in the direction of the dielectric's polarization and a more gradual fall of potential during the other half of the cycle. The polarity of the sawtooth wave, whether it would rise sharply to a positive or to a negative potential, would depend on the polarity applied to the high-K dielectric. Microwave power from this amplifier would then have been conducted down a waveguide tube to a microwave horn, the horn's dimensions having been chosen so that its impedance would match that of the surrounding air to allow a microwave beam to efficiently radiate from the horn. Once polarized, the dielectric would have been able to retain its polarization without any outside input of DC power. In fact, the sawtooth waves would have acted to bias the dielectric's voltage potential.

The Skyvault team did not power their tubes continuously, but pulsed them about a thousand times per second using a mechanical pulser. This was a wheel in an evacuated chamber that spun at 60,000 to 100,000 revolutions per minute (1,000 to 2,000 hertz) and on each revolution actuated a set of platinum electrical contacts that briefly turned on high-voltage DC to power the magnetrons. The proper pulsing rate would depend on how much voltage and power one wished to extract from the tube. If the pulser was cycled at a faster rate or was in its on state for a longer fraction of the cycle period, more power would be radiated from the magnetron.

Murray said that they needed to make fine adjustments to the pulser's "square wave" signal envelope to get its pulse cycle amplitude and timing just right. In particular, the magnetron would have had to be turned on at just the right moment so that its waves would match the phase of the waves already reflecting back and forth in the microwave amplifier waveguide, thereby allowing its energy to properly add to the amplified signal. Magnetrons are very sensitive. If the pulse timing is wrong, the tube's energy potential can build up so high that the tube will burn out. Failure occurs when an arc jumps from the tube's cathode to its anode, burning off the cathode's electron-emitting thorium coating and rendering the tube useless.

Radar researchers later replaced this older mechanical pulser technology with thyrotron tubes, which were able to produce shock discharge pulses having a much sharper rise time. Thyrotrons had a fixed spark gap enclosed in sealed glass tube filled with hydrogen and used a third ignitor electrode to trigger the gap to discharge. These discharges would be much like Tesla's shock discharges, except that the magnetrons would convert these pulses into microwave frequency shocks. In *Secrets of Cold War Technology,* Vassilatos commented about the explosive forces that these radar bursts can produce, noting, "As these pulse methods were reaching their state of refinement, engineers found it possible to produce single DC impulses of extraordinary power. Components often ruptured when these explosive electrical applications were employed. Wires exploded. Gaskets and sealed electrodes ruptured. Magnetron tubes, high vacuum vessels, literally exploded. Here was that phenomenon of which Tesla spoke so highly."[18] The microwave bursts that the Skyvault engineers were experimenting with were most likely of this sort.

Murray said they were using the very best magnetron tubes they could find, which at that time were being used on military radar systems. To maximize the gravity wave propulsion effect, they had to operate these tubes well beyond their voltage specifications, powering them with up to 250 kilovolts. Murray did not say what the normal voltage range was for these special radar magnetrons, but for comparison, one unclassified research paper published in 1956 described the development of a 1.3-gigahertz magnetron that operated in the range of 50 to 75 kilovolts and delivered power outputs on the order of 10 megawatts during its ten-microsecond pulse period.[19] Magnetrons available in military black projects likely had achieved higher power outputs than this at a much earlier date.

In this "out-of-spec," high-voltage operating region, the tube's characteristics would have become highly nonlinear and prone to develop what is called the longitudinal sawtooth instability, which causes electrons circulating in the magnetron to begin to bunch up into clusters, transforming the tube's normal sine wave output into a series of sawtooth spikes. A similar effect has been reported to have been seen in the operation of the Synchrotron Ultraviolet Radiation Facility (SURF III).[20] When the sawtooth instability was present in SURF III, researchers observed bursts of coherent microwave radiation that were 10,000 times more intense than the normal synchrotron beam radiation and which consisted of sawtooth-shaped waves in the 10-gigahertz frequency range.

By operating the tubes beyond their specifications, the Skyvault team was apparently attempting to produce microwaves having a maximally abrupt rise time—hence, a very nonlinear negative potential onset curve. This in turn would have maximized the electrogravitic thrust that these waves were producing. As seen in our analysis of the gravity shocks produced by Podkletnov's gravity impulse beam, the subquantum kinetics electrogravitic relation indicates that such waveforms would have been repulsive.

Murray said that as a result of running the tubes beyond their specifications, the research team was blowing out magnetrons by the thousands. Members were willing to take this risk because they knew that this propulsion effect existed. Apparently, someone in the past fortuitously got the frequency and wave shape right and discovered the effect.

Initially, the equipment generating the Skyvault propulsion beam was quite bulky. The entire set up, which included high-voltage power generators, microwave generators, waveguide ducts, and wave-shaping resonators, required a building the size of a barn. Murray disclosed that in this early version, the conical test beam was projected upward and made to buoy a test vehicle that had a concave bottom wide enough to receive the beam. He disclosed that this concave portion was made from a ceramic similar to CorningWare.

Although CorningWare is optically opaque, it is partially transparent to microwaves. Thus, given the proper shape, it could be made to act as a microwave lens, which would look similar to an optical lens but would not necessarily be optically transparent. Such a lens could be made out of paraffin, ceramic, or glass. The important thing is that it be made of a material having the proper permittivity and permeability. So the Skyvault team could have used the craft's ceramic bottom as a lens to refract the microwaves that were being beamed up to it.

However, for a diverging microwave beam, one would expect that they would have used a converging lens to bring the waves to a focus inside the craft. One wonders whether this concave ceramic was actually a metamaterial that was engineered to have a negative index of refraction. One characteristic of left-handed (negative index) materials is that they have a concave shape in order to bring a microwave beam to a focus on the other side of the lens.

Although the beam generator for the Skyvault prototype craft was initially very bulky, with time the Skyvault team was able to make its equipment more compact. Murray said that eventually they got the apparatus small enough to put inside the craft. However, he didn't specify what kind of power supply was used. The craft were circular in shape and emitted a greenish blue microwave propulsion beam toward the earth. The beam was made to pass through an "iris type of convex lens" toward the ground, where it would reflect back up to buoy the craft upward. It is unclear what Murray meant by an "iris type of convex lens." An iris is a small opening at the end of a waveguide that allows microwaves to pass out.

Perhaps the microwaves were emitted through an iris at the end of the wave amplifier conduit and were then focused by a ceramic convex lens. The microwaves leaving the iris would have diverged and would have needed a convergent lens to refract them into a microwave beam. The diameter of the beam at the ground target region could have been adjusted by controlling the position of the lens relative to the iris.

This experimental version of the Skyvault craft, which was being developed in the 1960s, was apparently much more advanced than its forerunner, the version that Tom's engineering firm was asked to bid on in 1975. That is, by carrying its own onboard beam, it was far more mobile. Murray said that the craft was remotely controlled by signals relayed from a radio transmitter, probably situated on top of a mountain. The transmitter sent out encoded signals 6,400 times per second that controlled the craft's pitch, yaw, bank, and velocity. The vehicle had a range of nearly three hundred miles over the desert and could attain altitudes of 50,000 feet or more. Murray said that it could attain "extreme speeds." Initially, they did test flights of an unmanned craft. Later, they built and flew around a craft having a crew on board. Murray told Tom the vehicles he worked on had an estimated propulsion efficiency of 60 percent, and he imagined that by 1974 much higher propulsion efficiencies had been obtained. By comparison, a jet aircraft has a propulsion efficiency of only about 20 percent.

In the mid-1960s, after Murray had left the project, the Skyvault team began replacing their magnetrons with solid-state oscillators, called Gunn diodes, that were much more reliable. Murray had learned about this from a friend who had continued to work on the Skyvault project. Wanting to know more, Tom asked his boss if it would be possible for him to speak to Murray's friend. Murray contacted his friend, who told him that he would instead write Tom a letter, which he would send via Murray. The letter, which is written in a somewhat whimsical style, is reproduced in appendix E.

Murray would not divulge his friend's name, but for practical purposes let us call him Don. In his letter, Don said that Gunn diodes normally require less than a watt of power to operate, but that by working with the manufacturers they were able to engineer special high-power Gunn diodes suited to their project. These were able to produce up to 10,000 watts of microwave power, and the various diodes that they fabricated functioned over a frequency range of 1 to 500 gigahertz. Don did not specify whether this power rating referred to the average power of a single pulse or to the power that was put out when operating in a continuous mode.

The Gunn diode was first developed in 1963. Later, much higher-power, more-efficient devices called impact ionization avalanche transit-time diodes (IMPATT diodes) were developed that were capable of higher microwave power outputs. However, IMPATT diodes have the shortcoming that their signal has a much higher phase noise, meaning that their oscillation cycles may not be as precisely timed. (A description of how these devices operate is given in the accompanying text box.) Such diodes have the advantage of being simpler to use than magnetrons, and they are more reliable in that they do not burn out as easily. They are also able to produce much higher frequencies. In the case of gallium nitride Gunn diodes, frequencies as high as 3,000 gigahertz have been achieved. Commercially available Gunn diodes have an efficiency of only 2 to 5 percent, while IMPATT diodes have a somewhat higher efficiency of about 10 percent. This is low in comparison with magnetrons, which are able to achieve efficiencies of 60 percent.

Gunn and IMPATT Diodes

The Gunn diode is named after J. B. Gunn, the physicist who in 1963 discovered that a crystal of gallium arsenide would spontaneously oscillate at microwave frequencies when a sufficiently high DC voltage was applied to either side of it. This became known as the Gunn effect. Gunn found that gallium arsenide exhibits a negative resistance when subjected to an electric field of greater than 3,000 volts per centimeter. That is, below this critical threshold, the electric current passing through the crystal progressively increases with increasing voltage, as it does in most electrically resistive substances. In this low-voltage region, the crystal is said to exhibit "positive resistance." However, at a critical threshold, the current-voltage curve plateaus and begins to bend downward such that the current now decreases with increasing voltage, a phenomenon called negative resistance.

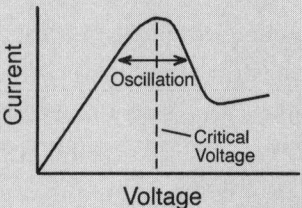

Due to this peculiar characteristic, when a gallium arsenide Gunn diode is biased above its critical threshold at about 5 kilovolts per centimeter, it will spontaneously oscillate at a specific gigahertz frequency. Some other substances found to exhibit this oscillatory effect are indium phosphide, cadmium telluride, zinc selenide, and wurtzite gallium nitride. This latter material oscillates when biased at a higher potential of about 150 kilovolts per centimeter.

An IMPATT diode is a silicon p-n junction diode that normally operates in a reverse-biased mode. Its principle of operation is different from a Gunn diode in that it involves impact ionization, which results in an electron avalanche electrical breakdown. It is similar to the Gunn diode in that it is a negative-resistance device that begins to spontaneously oscillate when its bias voltage is increased past a certain point.

Commercially available Gunn and IMPATT diodes most commonly have power outputs in the milliwatt range, although it is possible for civilians to purchase Gunn diodes that produce up to 30 watts of power. A thorough search for manufacturers of such solid-state oscillators carried out in the mid-1990s revealed that such diodes having power much higher than 30 watts are unavailable for public sale. The story is quite different in the case of oscillators being used for military applications. For example, in 2002 I learned through a personal contact that one defense aerospace avionics distributor had shipped out an experimental 1,040-gigahertz (300-micron) oscillator that reportedly had a rated output of 40 kilowatts! This company routinely shipped items marked as "microwave oscillators" to defense aerospace corporations such as Northrop Grumman, Boeing, Lockheed Martin, and BAE Systems as well as to defense R&D contractors such as Raytheon and SAIC (Science Applications International Corporation).

A survey of cutting-edge developments in the field of solid-state microwave devices indicates that the high powers quoted for the modified Gunn diodes used in Project Skyvault are not all that out of line. One literature review written in 1995 noted that gallium arsenide Gunn diodes were being combined to form units that could achieve the kilowatt level at frequencies above 30 gigahertz.[21] Also, in 2000 Purdue University researchers announced that they had developed a silicon carbide IMPATT diode that was able to achieve microwave power outputs four hundred times higher than silicon-based IMPATT diodes. Their simulations projected the possibility of achieving power outputs as high as 4.2 kilowatts at 10 gigahertz.[22] A lightweight microwave-emitting tube called the Pasotron (for plasma-assisted slow-wave oscillator), developed in the early 1990s at Hughes Research Laboratories and Hughes Missile Systems Co., was able to achieve even higher outputs.[23] This uses an electron gun that generates high-energy electrons that emit a beam of microwaves as they pass through a low-pressure glow discharge. The device produces microwave pulses lasting 100 microseconds with pulse voltages of 220 kilovolts, pulse power outputs of 1 to 5 megawatts, and effieciencies of between 20 and 50 percent. More recently, Pasotrons have been reported to produce 100-nanosecond pulses with microwave powers of 7 gigawatts. Although it is not a solid-state device, it has the advantages that it does not require a magnetic field for its operation, is much lighter in weight, and does not burn out easily. It is not known whether the Project Skyvault team tested Pasotrons at some point in its research.

7.5 ▪ THE BEAM AMPLIFIER

According to Murray's friend Don, the high-power Gunn diode used in the Skyvault vehicle, like the magnetron, was mounted in a waveguide box. This had an opening at one end and dimensions that matched the diode's oscillatory characteristics. In other words, the conduit's length was made to equal some multiple of the wavelength of the microwaves emitted from the Gunn diode so that the waves would resonantly reinforce one another as they reflected back and forth along the length of the waveguide. This resonance would increase the beam's voltage.

Although Don did not mention their voltage requirements in his letter, Tom told me that he had learned that these special Gunn diodes were designed to operate in the range of a few hundred thousand volts to a million volts. These voltages are unusually high in comparison with the voltages that commercially available Gunn diodes normally operate at, which is in the range of 5 to 100 volts DC. One is left to wonder whether this voltage might refer to the voltage rating of the diodes, that is, the voltage they were designed to withstand that could be generated in the amplifier cavity. The voltage of the amplified microwave beam, then, may have ranged up to several million volts.

As noted earlier, a simple waveguide cut to the proper dimensions would be able to increase the voltage of a microwave beam but not its total energy. But in his letter, Don seems to be talking about a different sort of amplifier, one capable of increasing the total energy of the beam. He said that this "amplifier" was needed to "extend the use" (i.e., the ability) of the Gunn diode so that it could "launch the . . . vehicle" (see his letter in appendix E). Although the modified Gunn diodes used in Project Skyvault had a power output far greater than those commercially available today, even a power output of 10 kilowatts would likely have fallen short of what was needed. The magnetrons that the project had been using in their earlier work must have had power outputs several orders of magnitude higher than this. So to match this, they would have had to boost the power of the Gunn diode beam in an "energy amplifier."

Most likely the Skyvault project was doing this with a *parametric amplifier,* a device commonly used by microwave engineers to boost signal strength. A parametric amplifier consists of a cavity containing

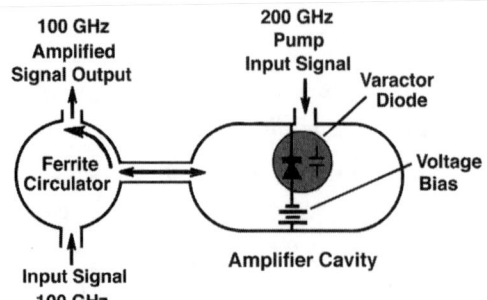

Figure 7.9. Diagram of a microwave parametric amplifier. (P. LaViolette, © 2007)

a nonlinear medium such as a varactor capacitor diode. The beam to be amplified is allowed to enter and exit the amplifier cavity through a port, and while in the cavity it passes through the diode where its energy is pumped up as the result of the action of a second microwave beam called the pump beam. The pump beam typically has a frequency twice that of the main oscillator beam and is oriented at 90 degrees to the main beam so as not to directly interact with it (figure 7.9).

The pump beam affects the main oscillator beam indirectly by varying the diode's parameter, its dielectric constant, at just the right time in the main beam's oscillation cycle.* For example, by decreasing the varactor's dielectric constant (K), the pump beam signal reduces the diode's electric permittivity (ε); this, in turn, decreases the diode's capacitance ($C \propto \varepsilon$) and increases its voltage ($V \propto 1/C$). By doing this at the phase of the cycle in which the main beam's voltage is approaching either a positive or a negative voltage maximum, the amplitude and power of the main beam may be boosted. In this way, the pump beam is able to progressively step up the power of the main beam. Often, the parametric amplifier is interfaced with a circulator that allows the main beam to circulate in a loop with some of its energy being diverted into the parametric amplifier for amplification. Parametric amplifiers are able to amplify a beam's energy anywhere from one hundred to one thousand times. Thus, a 10-kilowatt microwave oscillator signal could be boosted to create a 10-megawatt microwave beam.

One interesting thing about parametric amplifiers is that their

---

*Parametric amplifiers can also use a ferrite material as their nonlinear medium and work by instead varying the magnetic permeability μ of the ferrite. The outcome is essentially the same.

energy output can greatly exceed their energy input. The amount of energy that the pump beam requires to alter the permittivity or permeability of the amplifier's nonlinear medium can be much less than the amount of energy that the amplified beam gains through the parametric excitation process. The amount of this overunity output versus input depends on the type of nonlinear medium and its response in the frequency range used.

The magnetic resonance amplifier is an example of one such amplifier that operates in the audio frequency range rather than at microwave frequencies. It is based on the work and theories of the ninteenth-century American inventor John Ernst Worrell Keely and has been extensively researched by hobbyists. Circuit diagrams and research results on its operation are available on the Internet.[24–28] It uses a high-K dielectric such as a barium titanate capacitor hooked in series with a coil wound around a barium ferrite ceramic magnet core. By exciting it at a frequency of around 20 to 40 kilohertz, this nonlinear tank circuit is made to oscillate at its resonant frequency of around 8,000 to 11,000 hertz. Thus, the excitation frequency is chosen to be three times the resonance frequency, that is, three octaves (nine harmonics) above resonance. Power is drawn from the oscillating ferrite core through a secondary winding that is connected to a bridge rectifier. One such device built and tested by American researchers Joel McClain and Norman Wootan achieved a power output of 2.75 watts, for an input power of 0.7 watt or an overunity ratio of about 4.[29] At resonance, the voltage across the tank circuit ranged up to 1,000 volts when excited with a 30-volt AC pump signal.

Even higher outputs than this have been reported for parametric amplifiers in the audio range. For example, in 1949, Obolensky built a parametric amplifier that used Super Permalloy ferrite as its nonlinear medium and was able to achieve an overunity ratio of about a million to one when he pumped it at frequencies of 60 and 400 hertz.[30] Where does this excess energy come from? Physicists aren't really sure. Obolensky suggests that the energy is cohered from noise present at the atomic level in the amplifier's nonlinear medium and in the immediate space environment.

While it is possible to use a separate power source to generate the pump beam fed into the amplifier, it is also possible to draw off some of the energy surplus in the main beam that is being amplified and to recycle this to power the parametric excitation process. This could be done by connecting the pump beam waveguide tube to a fourth port on the circulator cavity (figure 7.10). The circulator would contain not only the fundamental frequency of the main beam but also its harmonics. So by making the length of this connecting waveguide equal to an odd number of half wavelengths of the main beam's second harmonic frequency, the fundamental frequency would be blocked and just the second harmonic (i.e., $2f_o$) would pass through to the parametric amplifier. As the fundamental frequency becomes more intense, so would its harmonics, and a greater amount of power would become available in the second harmonic for parametric excitation. As a result, the beam intensity would progressively increase.

Such a system, however, runs the risk of being unstable in that without proper regulation it could create an exponential buildup of energy that would ultimately result in an explosion. That is, energy could be created in the amplifier faster than it could be removed. Making such an amplifier work properly so that it is able to boost the wattage of the microwave beam without blowing the amplifier apart is quite tricky. It requires ingenious engineering—such as incorporating a fast-acting servo control that automatically changes the phase of the pump beam

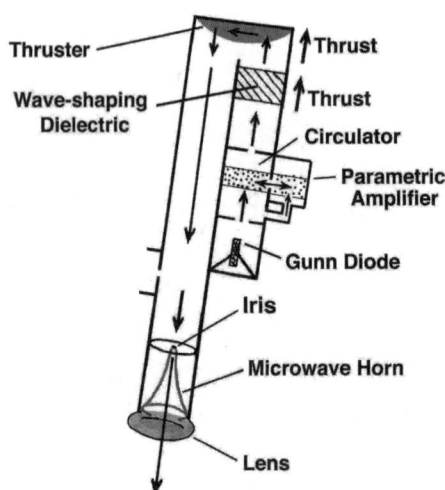

Figure 7.10. A possible arrangement of components making up the microwave beam generator mounted on board the Skyvault craft. (P. LaViolette, © 2007)

frequency when the amplifier's output power level gets too high. This would squelch the amplification process and halt the exponential rise in energy production.

Don did not mention the use of a sawtooth wave–shaping dielectric, so we do not know whether the wave-shaping dielectric was placed in the wave amplifier cavity or whether it was put in a separate wave-shaping resonator cavity. We will suppose the latter was the case. So once it had become amplified, the craft's Gunn diode beam would have exited the circulator and entered a waveguide containing a polarized high-K dielectric that would have shaped the wave into a sawtooth waveform. As in Brown's vertical-lift electrokinetic apparatus, the incident microwave beam would have exerted a substantial electrogravitic thrust along the length of the dielectric that would have helped loft the craft. Hence, this wave-shaping chamber was likely securely anchored so that its thrust would be transferred to the vehicle-support structure.

If the beam in the interior of the craft was directed upward through the wave-shaping dielectric, it may then have been made to pass into a convex slab of metamaterial having a negative index of refraction. As mentioned earlier, a microwave beam tuned close to the material's resonance would exert a strong repulsive force, so an upward-directed microwave beam would produce an upward propulsive force on the metamaterial slab as well. Metamaterials have also been found to efficiently refract microwave beams through tight turns. In fact, Pendry and Smith showed how a metamaterial slab having a convex lens shape refracts a beam through a 180-degree turn. In a similar manner, the metamaterial thruster at the same time could have been used to redirect the beam downward through an adjacent waveguide, from which it would ultimately exit the saucer via a focusing lens and proceed toward the ground (see figure 7.10).

Chapter 8 further examines Don's disclosure about Project Skyvault. We will find that, besides pushing upward against the craft, it was necessary that the microwave propulsion beam also project downward and scatter back to the craft from a ground reference point. In so doing, the beam could be made to resonantly store vast quantities of energy for supporting the craft, and the flight of the craft could be more precisely controlled.

8

MICROWAVE PHASE CONJUGATION

8.1 ▪ PHASE-CONJUGATE MICROWAVE PROPULSION

As mentioned in the previous chapter, my contact Tom had received a letter from a friend of his supervisor, a fellow we have called Don. In 1974, when Tom received this letter, Don was actively involved in Project Skyvault. The letter gave only a rough sketch of the project's microwave propulsion technology, since much of that work was then still classified. Nevertheless, Don gave enough information that, shortly after hearing Tom's story, I was able to come to an important conclusion about a key aspect of the craft's microwave beam propulsion technology. I was able to connect it to a field of optics research that in 1994 was just beginning to emerge in the open literature, but which apparently had been secretly under full development for aerospace applications back in the early 1950s. Before discussing this optical phenomenon, called *phase conjugation,* and how it might be applied to vehicle propulsion, let us summarize some additional details that Don provided about the propulsion system.

Don said that the microwave energy from the microwave amplifier cavity was directed into a horn-shaped waveguide that controlled or shaped the wave radiation pattern. Although he did not specifically mention it, this horn antenna was most likely situated between the iris and the ceramic convex lens that, as Murray had described, was used to control the diameter of the beam.

After leaving the lens, the microwave beam was allowed to pass through the air to a target area, which was presumably a ground surface location. Microwaves that reflected back up to the vehicle from this target region were then allowed to enter a cavity that contained a mixer diode (figure 8.1). There they were mixed with a portion of the outgoing microwave beam that had been locally diverted from the craft's beam generator. A mixer diode is a radar-absorbing material with nonlinear electromagnetic characteristics that is able to combine waves of slightly differing frequencies to produce a more complex wave having frequencies that are the sum and difference of the two frequencies. Interestingly, one material that has such nonlinear properties is barium titanate, the piezoelectric ceramic that Brown employed in his experiments. Don noted that a stable DC-voltage source was connected across the mixer diode to bias it and that the Gunn diode oscillator was biased in a similar manner.

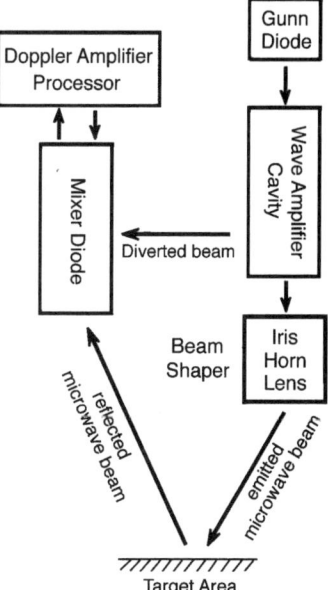

Figure 8.1. Murray's description of the microwave transmitter/mixer detector used in Project Skyvault. (P. LaViolette, © 2006)

Don described two types of antigravity microwave systems. The simpler version, called a homodyne, used the same conduits for both transmitting and receiving. He commented that this system "suffered from a lack of range" since its microwave beam typically targeted at no more than thirty miles. The second type of configuration, which he called the "Micro-X unit," used separate transmitting and receiving antennae and presumably could achieve a greater targeting range. It was probably named the Micro-X because it originally operated in the X microwave band, which extends from 8 to 12 gigahertz.

Up to this point, the apparatus that Don described sounds much like a basic radar unit. In the case of radar, the transmitter emits a powerful microwave pulse that then reflects back from the distant target and enters the radar receiver. This weak reflected wave is then combined in the radar's mixer chamber with a portion of its original outgoing signal. Suppose we let f_o represent the reference frequency of the outgoing signal and f represent the return signal that reflects from a target such as a distant aircraft. If the target is moving, the return signal frequency f will be slightly different from the original outgoing signal f_o, for an approaching target $f > f_o$ and for a receding target $f < f_o$. The two frequencies will differ by an amount Δf, equal to twice the Doppler frequency shift, that is $\Delta f = f - f_o = 2f_o v/c$, in which v is the velocity of the target relative to the radar transceiver and c is the velocity of light. A frequency φ (*phi*), equal to this frequency difference, $\varphi = \Delta f$, would come out of the radar's mixer chamber. Its magnitude and sign would indicate how fast the target is moving and whether it was approaching or receding. The greater the frequency difference, the greater would be the relative velocity inferred for the target.

However, the nonlinear media used in a radar mixer, as it turns out, can also act as a phase conjugate mirror. That is, when illuminated with a radar pulse echo coming in from the target, it produces an outgoing microwave pulse that exactly retraces the path of the incoming target echo signal. This is very different from normal mirror reflection.

Consider an example in which visible light reflects from a conventional mirror. When a laser shines its beam on a target object and some of that light scatters off from the object and strikes a conventional mirror at some angle to its surface, that light will then reflect away from the mirror's silvered surface at a similar angle. The reflected light will continue on a divergent path that takes it increasingly farther from the mirror and the light-scattering object (see figure 8.2a).

In the case of a phase conjugate mirror, on the other hand, the silvered mirror surface is replaced with a translucent piezoelectric crystal such as barium titanate, which has nonlinear optical properties. Again, consider an example in which a laser illuminates the target object and some of the light that scatters from the target enters this crystal, but in addition, part of the light coming directly from the laser is diverted and made to enter the crystal as beam P_1. This laser beam passes through the crystal and reflects back from a mirror on the other side to form a second beam, P_2, propagating in the opposite direction (see figure 8.2b). These counterpropagating beams are called the *pump beams*. The light that scatters from the target object toward the crystal is called the *probe beam*. The light waves forming the probe beam enter the interior of the crystal and interact there with the counterpropagating pump beams. The interaction of the probe beam with each of the pump beams produces an interference pattern that alters the index of refraction within the crystal to form a complex pattern of light-refracting surfaces collectively called *a holographic amplitude grating*. The holographic grating formed by pump beam P_2 then refracts the oppositely directed pump beam P_1 to form an outgoing *phase conjugate beam*. In contrast, the grating formed by beam P_1 refracts beam P_2 to again form an outgoing phase conjugate beam, with complementary characteristics. This combined phase conjugate beam is represented in figure 8.2b by the gray wave fronts. The phase conjugate beam has the special attribute that its light waves move outward from the crystal along trajectories that precisely retrace the trajectories that had earlier been followed by the probe beam's incoming waves. Hence, the phase conjugate beam converges back to the original light-scattering locations on the target object, and from there its waves rescatter back to the original laser beam light source.

The phase conjugate mirror, in effect, generates a new set of light waves that have the same angle, wave shape, and phase as the incoming light waves scattered from the target object but travel in a reverse direction, moving back toward the target. The effect appears as though time had been made to run backward, causing the incoming scattered

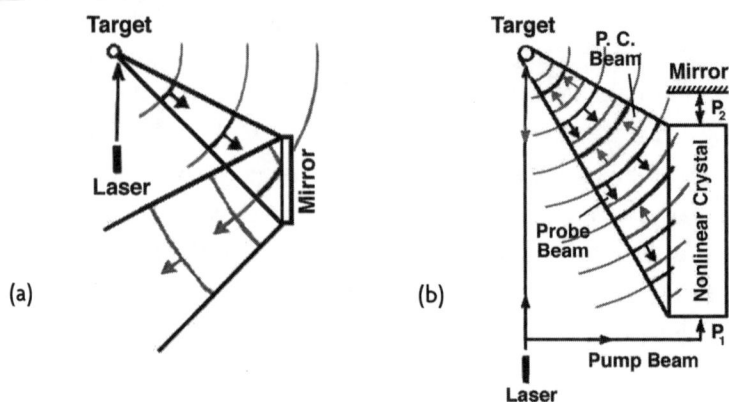

Figure 8.2. Comparison of an ordinary mirror and a phase conjugate mirror. (a) Light waves reflecting from an ordinary mirror. The black wave fronts represent the incident rays and the gray wave fronts represent the reflected rays. (b) Light waves reflecting from a phase conjugate mirror. The black wave fronts represent the incident rays and the gray wave fronts represent the phase conjugate rays that travel "in reverse," converging toward the upper left. The phase conjugate beam is generated from the interaction of the incoming probe beam with the two counterpropagating pump beams, P_1 and P_2.

light waves to reverse in their tracks. However, here the effect is accomplished with time running forward as it usually does and with light waves different from the incoming waves since the incoming waves are what are generating the grating pattern. Thus, unlike conventional mirror reflection, in which the reflected light continues to spread out into the surrounding environment, with phase conjugate reflection the light stays bottled up, confined to a beam that extends among the laser light source, its illuminated light-scattering object, and the nonlinear medium of the phase conjugate resonator. Since the light waves have no immediate exit, the energy trapped in this phase conjugate resonator can build up to very high intensities.

Physicists have come to call this arrangement a *four-wave mixer* since four beams intersect in the nonlinear medium: the incoming probe beam, the two counterpropagating pump beams, and the outgoing phase conjugate beam. Although the probe beam initially would be relatively weak, energy supplied by the two pump beams would emerge from this four-wave mixer as a powerful outgoing phase conjugate beam. Upon entering the laser resonator cavity, this phase conjugate beam would stimulate a more powerful outgoing laser beam that would be specifically directed toward points on the target that preferentially scatter light into the four-wave mixer. As a result, the probe beam entering the mixer crystal would progressively increase in intensity. Over time, a high-intensity light path would resonantly build up among the laser, the target object, and the four-wave mixer, and very little if any of the laser's light would be scattered into the environment.

The second law of thermodynamics tells us that in closed systems, order always tends toward disorder, that is, it predicts that light scattered from an object should disperse into space and eventually dissipate its energy. This is not so with light striking a phase conjugate mirror. The grating records the information carried by the incoming probe beam concerning where its light waves originated and in turn refracts (steers) the pump beam to create an outgoing phase conjugate beam whose waves precisely retrace the paths that had been followed by the incident rays of the probe beam. Thus, a state of initial disorder is made to tend toward a subsequent state of greater order. The scattered light is returned in a more concentrated state toward the point from which it emerged.

In most media, the electromagnetic wave interference pattern formed between the incoming probe beam and the pump beams would have no effect on light wave propagation, but in a polarized piezoelectric medium such as barium titanate, these field potentials are able to physically alter the index of refraction of the piezoelectric medium at a microscopic level. Through this multifarious patterning of its index of refraction, the crystal is able to complexly scatter light waves passing through it so as to reconstruct the "time-reverse" phase conjugate beam. If we were to think of the crystal as a computer hard drive, then the probe beam striking this crystal would, in effect, be writing data into this hard drive about all of the directions and phases of its various light rays. This stored information refracts the pump beam, directing it to form an outgoing phase conjugate beam.

The pump beams entering the mixer need not be formed from a laser beam split off from the illuminating laser. Rather, they may be spontaneously seeded from the probe beam, provided that mirrors are placed on either side of the mixer crystal at the proper angle. How such a passive phase conjugate resonator works is summarized in the text box on page 231.

The academic scientific community first became aware of optical phase conjugation through its members' work with lasers. Such laser experiments were conducted in 1972 at the Lebedev Physical Institute in Moscow. Subsequently, scientists in the United States and other countries began investigating the phenomenon.[1,2] This technology found a military application in the development of a Star Wars weapon that can track an enemy missile target by illuminating it with a beam of laser light and subsequently destroy the target by sending out a powerful laser pulse that converges onto the target, retracing the path of the light rays that had reflected from the target.

In 1994, when I first heard about Project Skyvault, the only reference to phase conjugation was in experiments that were being conducted at optical wavelengths with lasers, but Don's discourse on microwave beam mixing immediately led me to conclude that Project Skyvault was performing microwave beam phase conjugation. I conducted a literature search, but it turned up no references to the use of phase conjugation at microwave frequencies. Nevertheless, I concluded that if phase conjugation worked at optical wavelengths, it should work just as well at microwave wavelengths. Since that time, a significant amount of research on microwave phase conjugation has been published, indicating that my earlier conclusion was indeed justified.

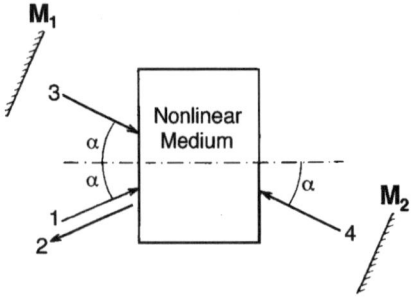

Figure 8.3. A self-pumped optical phase conjugate mirror showing an incoming probe beam (1) interacting with counterpropagating pump beams (3 and 4) between mirrors M_1 and M_2 to produce the outgoing phase conjugate beam (2).

Optical Phase Conjugate Resonance in Barium Titanate

When a probe laser beam, or "seed" beam of coherent laser light (see ray 1, figure 8.3), is directed to pass into a nonlinear dielectric medium, such as an electrically polarized crystal of barium titanate, and this dielectric is placed between two parallel mirrors, this probe beam will excite two counterpropagating "pump beams" (see rays 3 and 4, figure 8.3) to build up between the mirrors. These pump beams interact with the incoming probe beam to produce a stationary periodic electric field and refractive index pattern termed a *holographic amplitude grating*. Experiments have shown that the counterpropagating pump beams can self-excite to intensities sixty times that of the input signal beam without any additional energy input.[3,4] Higher amplification coefficients could be achieved by reducing the losses of the resonator cavity and by taking advantage of natural resonances in the nonlinear dielectric.

This passive resonator also functions as a phase conjugate mirror. The holographic amplitude grating produced by the interaction of the pump beam (ray 3 in figure 8.3) with the incoming probe beam (ray 1) refracts the counterpropagating pump beam (ray 4) to yield an outgoing time-reversed beam (ray 2), also called a phase conjugate beam. Similarly, the grating produced by the interaction of the beams (rays 4 and 1) refracts the counterpropagating pump beam (ray 3) to produce a similar outgoing phase conjugate beam (ray 2) in phase with that produced by the other pump beam (ray 4). The electromagnetic wave fronts in the outgoing beam (ray 2) are identical to the ordinary wave fronts in the probe beam (ray 1), except that they propagate backward instead of forward, precisely retracing the paths followed by the forward-moving wave fronts of the ordinary beam. Consequently, the barium titanate crystal functions as a self-pumped optical phase conjugate mirror. If the probe beam were to originate from a certain point, this phase conjugate mirror would reflect a beam that converges back to that point. This principle is used in Star Wars weapons designed to track and destroy missiles by using a laser beam.

However, if microwave phase conjugation was indeed being used in Project Skyvault, we get a very different picture of the historical development of this branch of science. It indicates that about two decades before academia discovered phase conjugation in laser experiments, this phenomenon was under intense investigation by aerospace black-project scientists studying its ability to amplify and bottle up microwave radiation into intense downward-directed beams from aircraft.

If made to function as a microwave phase conjugate resonator, a radar unit could be able to automatically lock on to and track its target and also to exponentially amplify the intensity of the incoming radar echo. Instead of being dispersed into the environment, most of the radar signal's energy would become confined to a nondispersing beam extending between the radar unit and the target. This has the advantage of allowing very intense microwave beams to be built up with very little input power, clearly an advantage if one needed to generate intense microwave beams for vehicle propulsion. Thus, Don's rather cryptic description of the Skyvault vehicle propulsion unit makes quite a bit of sense if he was describing a microwave device that operates similar to a phase conjugate resonator.

With some knowledge of how a phase conjugate resonator functions, we can understand Don's description of the Micro-X unit as follows. The Skyvault vehicle has a high-voltage DC power source on board that drives a high-power Gunn diode mounted within a resonator conduit mounted below the craft. Coherent microwave radiation emitted from the Gunn diode repeatedly reflects back and forth along the length of this conduit and becomes amplified in voltage. Dielectrics placed in the conduit transform the signal into a sawtooth-shaped waveform. The microwave radiation from this resonator passes through an iris and radiates downward through a microwave horn. A convex lens focuses the radiation into a beam that shines on the ground (see figure 8.4). A fraction of this radiation is absorbed in the ground and the remaining fraction is scattered upward, of which a small portion scatters back toward the craft. A second convex microwave lens mounted below the spacecraft intercepts a portion of this scattered radiation, which constitutes the incoming probe beam, and focuses it through another iris into another resonator cavity that contains the mixer diode. The mixer diode is essentially a block of material such as barium titanate or some other substance that has nonlinear electric properties and is electrically polarized with high-voltage DC. Since microwave wavelengths are about a thousand times larger than visible wavelengths, the mixer medium need not be a transparent or translucent crystal; an amorphous barium titanate ceramic works just as well.

Furthermore, a portion of the microwave radiation in the Gunn diode resonator cavity is diverted into the mixer diode cavity, where it enters the mixer diode. This pump beam passes through the mixer

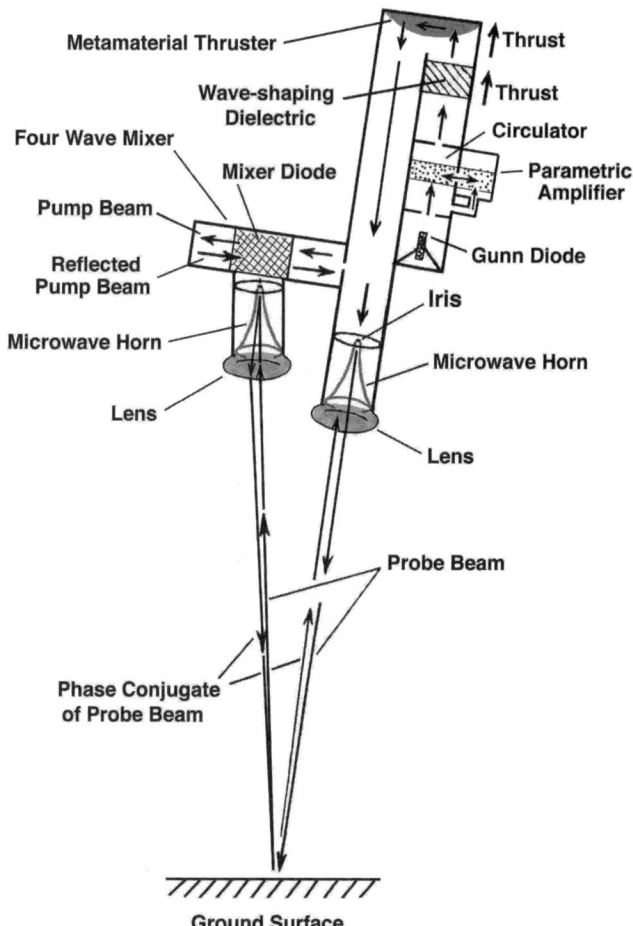

Figure 8.4. A rendition of the phase conjugate microwave resonator propulsion unit used in Project Skyvault. (P. LaViolette, © 2006)

diode and reflects from the far end of the resonator cavity. The cavity is designed to have a length that is a whole-number multiple of the Gunn diode wavelength to ensure that resonance is achieved. This establishes two counterpropagating phase-locked pump beams within the cavity. The mixer diode's electric polarization axis is oriented at the proper angle relative to these pump beams.

Within the mixer diode medium, the probe beam interacts with each of the two counterpropagating pump beams to produce a holographic electrostatic grating pattern. In turn, the grating pattern produced by a given pump beam interacts with the opposing pump beam to produce an outgoing microwave beam that is the phase conjugate of the probe microwave beam, or, in other words, is the phase conjugate of the radiation scattered back to the craft from the ground target site. As was the case with laser light phase conjugation, the waveform, angle, and phase of the phase conjugate microwave beam are configured just like those of the incoming probe beam, except that the waves propagate in a reverse direction. As such, these phase conjugate rays exactly retrace the paths followed by the ground-reflected rays and ultimately converge back to the probe beam's point of origin. That is, this time-reversed beam converges on the ground-reflection site and then continues to retrace the probe beam's path, ultimately entering the spacecraft's Gunn diode oscillator cavity.

This arrangement of two counterpropagating pump beams, an incoming probe beam, and outgoing phase conjugate beam, all interacting within the craft's nonlinear mixer diode, constitutes a four-wave mixer. Since the oscillation phases of the probe and phase conjugate beams are precisely matched, these ordinary and time-reversed waves reinforce one another to produce a resonantly amplified stationary wave pattern, or soliton, between the craft's mixer resonator cavity and the ground and between the ground and the craft's Gunn diode resonator cavity. As a result, the electromagnetic energy beamed to the ground from the Gunn diode amplifier cavity and also the energy beamed back to the ground as the outgoing phase conjugate beam become efficiently stored in this soliton beam. Based on the large body of knowledge that has been accumulated about the phase conjugation phenomenon, we know this energy storage soliton phenomenon to be fact.

This stationary wave in effect behaves as a giant energy capacitor. The soliton's intensity progressively builds up to a very high value, limited only by absorption energy losses when the microwave beam contacts surfaces that partially absorb the microwave radiation. The soliton beam develops specifically between high-reflectivity ground scattering surfaces. Ground scattering sites illuminated by the beam that are very absorbing to microwave radiation will return little radiation to the mixer, and hence, those ray paths will carry an insignificantly small fraction of the soliton beam's total energy flux.

Considering that passive phase conjugate resonators using barium titanate crystals have developed sixtyfold power gains over their seed (probe) laser beam intensity, we might speculate that Project Skyvault was able to achieve microwave beam intensity gains of at least two orders of magnitude and perhaps higher. If they used a Gunn diode that operated at a power of 10 kilowatts, then the soliton beam could have stored 1 to 10 megawatts of energy. If, on the other hand, the craft used a parametric amplifier to amplify the Gunn diode beam to a power of several million watts, then the soliton beam could have stored several gigawatts.

We can only speculate how much beam power was needed to lift the Skyvault craft. Don did not disclose this, nor did he mention the size of the Skyvault craft. Obviously, the power needed for a given propulsion beam would depend on the weight of the craft and on how many beams it used for levitation. With this method, it should be possible to lift a craft even the size of an aircraft carrier.

As mentioned earlier, sawtooth-shaped waves having a sharp rise in negative electric potential will produce repulsive forces on bodies they encounter. Also, artificial metamaterials having electric or magnetic resonances close to the microwave beam frequency are capable of responding with very large repulsive thrusts. As mentioned in chapter 7, the Skyvault craft very likely used such a material for its wave-shaping diode and in a beam refractor that reversed the path of its main beam to pass through a lens toward the ground. The mixer diode may also have been made out of such a metamaterial so that, in addition to producing an outgoing beam that would be the phase conjugate of the incoming probe beam, the diode would also be lofted by both the incoming probe beam and the downward-refracted pump beams.

Consequently, the sawtooth-shaped waves emitted from the Gunn diode resonator cavity will exert upward forces in the wave-shaping diode, in the beam reflector, and in the mixer diode that will together act to levitate the craft. The microwave beams that strike the ground will produce a downward force on the ground, but this force will not be as great as that lofting the craft's dielectrics, since it is unlikely that soil material typically encountered in the ground would have any resonant frequencies near the craft's microwave beam frequency.

Based on Don's testimony, we can conclude that this microwave-induced thrust would not be just an electrostatic effect, but also an electrogravitic effect in that it would repel the mixer diode through a mass effect. In such a case, this phase-locked soliton beam could induce a gravitational force on the craft that was opposed to the Earth's downward pull, thereby reducing the craft's weight and causing it to levitate. The idea that a microwave beam should have gravitational effects is not entirely unexpected. As discussed earlier, research by Brown and Podkletnov indicates that sawtooth electric potential waves produce gravity-like thrust effects. This electrogravitic coupling phenomenon is also a key prediction of subquantum kinetics.

A Skyvault spacecraft could maintain proper pitch stability by using three microwave beam generators spaced from one another so as to produce a tripodlike beam arrangement (see figure 8.5). A single mixer could be located at the center of the craft's lower hull to phase-conjugate the three beams. Alternatively, three mixers might be used, one near each of the craft's microwave beam generators.

In summary, by using phase conjugate technology, a craft would be able to build up a resonant energy beam between itself and the ground that would have a cumulative power far in excess of that which the craft would be feeding into it. Its advantage for propulsion may be readily seen when compared with a conventional microwave system that uses an onboard maser (*m*icrowave *a*mplification by *s*timulated *e*mission of *r*adiation) as a microwave transmitter. In the case of a conventional maser system, the emitted radiation would simply leave the craft, strike the ground, and scatter in various directions, with a minute fraction of the original radiation scattering back toward the craft. To get a

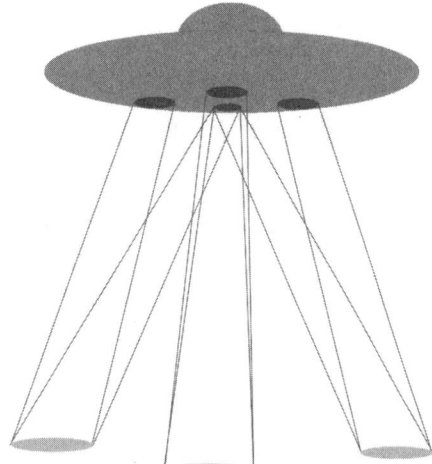

Figure 8.5. A possible arrangement of microwave phase conjugate resonators in a Skyvault spacecraft. (P. LaViolette, © 2006)

reflected beam by this method that would exert a measurable propulsive force, the craft's transmitter would have to be so powerful and its energy source of such a large size as to make this method of microwave propulsion totally impractical.

However, in the case of a craft employing phase conjugation technology, the initially minute amount of microwave radiation scattered from the ground and striking the craft would be phase conjugated and sent back to the ground as a time-reversed beam. This time-reversed beam would retrace the path of the scattered rays, targeting just those facets on the ground that were reflecting microwaves toward the craft and reflecting back from those facets to the craft's maser transmitter. So out of all the microwave rays being scattered from the ground, the phase conjugator would pick out just those that were striking the craft and send out its energy to trace in reverse the trajectory of those rays back to their transmission source. As the microwave energy was resonantly reflected back and forth among the craft's transmitter, the ground, and the craft's phase conjugator, a soliton beam would form specifically between the craft and the ground and begin to progressively increase in intensity. The waves would remain coherent despite repeated reflections, so losses would be minimal. Eventually, this beam would accumulate an intensity far in excess of the power being outputted from the craft's AC maser source and possibly would even draw in energy cohered from the surrounding environment. Given that this soliton beam exerts an upward repulsive force on the craft, with sufficient resonant amplification it would produce an upward force sufficient to support the craft. In effect, this microwave phase conjugate resonator would act as a gravity wave amplifier to generate an enormous electrogravitic thrust for supporting the craft. Ray, the black-project physicist I had spoken to, said that in the black R&D world, this resonant amplification effect is referred to as "field-induced soliton phenomenon."[5]

8.2 ▪ VEHICLE FLIGHT CONTROL

Another purpose of projecting beams out from the craft to the ground surface would be for flight control. As mentioned earlier, the ground-reflected probe beam enters the mixer diode, where it is mixed with the beam from the craft's main Gunn oscillator to produce sum and difference frequencies, the difference frequency being the Doppler output signal, whose frequency is determined by the speed of the vehicle's motion relative to the ground. Don said that in both the homodyne and Micro-X units this difference signal was fed into what he termed the "processing circuit," where it was amplified and processed "to produce the vehicle in motion." Unfortunately, he gave no additional information as to the nature of this processing or how it might result in moving the vehicle. He stated that the frequency of the Doppler output signal not only depends on the speed of the vehicle, but also "controls the speed of the vehicle" (see page 3 of Don's letter in appendix E).

In his letter, Don wrote that the Doppler frequency, φ, caused by the vehicle's movement in a 100-gigahertz microwave radiation field (f_o) was given by the equation:

$$\varphi = 2f_o\, v/c$$

in which φ is the frequency difference due to Doppler shifting of the outgoing signal, f_o is the frequency of the outgoing microwave signal generated by the vehicle, v is the speed of the vehicle in centimeters per second, and c is the speed of light (3×10^{10} centimeters per second). Unless he was just using an analogy here, Don seems to imply that they were using a 100-gigahertz oscillator for their propulsion beam—hence, one that operated in the W microwave band. The φ in the above equation represents the difference frequency, Δf, that would be produced by the mixer diode as a result of combining the incoming probe beam signal with the spacecraft's local oscillator reference signal, the pump beam.

If the vehicle was stationary relative to the targeted region, the waves scattered back to the craft would have the same frequency as the original outgoing microwave signal. Don noted that in this case the output in the mixer diode would be relatively simple; that is, no difference frequency would be produced. However, in a case in which the vehicle was moving relative to the ground, the reflected signal would be Doppler shifted, causing its frequency to differ slightly from the spacecraft's local oscillator frequency. In this case, he said the signal output from the mixer would be more complex, that is, there would be a difference signal output.

As an example, if the vehicle was beaming down a signal of $f_o = 100$ gigahertz and was receding upward from the ground at a speed of 1.5 meters per second, this 100-gigahertz signal would arrive at the ground redshifted by 500 hertz (10^{11} Hz $\times$ v/c), presenting a frequency $f_o - 500$ hertz. Upon being reflected from the ground up toward the receding vehicle, this redshifted microwave beam would appear to be redshifted by an additional 500 hertz relative to the vehicle. As a result, the incoming reflected signal would be redshifted by 1,000 hertz compared with the signal originally sent out by the craft's local oscillator. Consequently, when this redshifted frequency f is combined in the microwave mixer with the pump beam frequency f_o from the local oscillator, the mixer would produce a 1,000-hertz difference signal, which is the Doppler output frequency, φ. Harmonic multiples of this fundamental Doppler beat frequency would most likely also be present, although in a lesser amount.

It is quite likely that such a phase conjugate propulsion system would emit sound, since piezoelectric media physically move and oscillate when they are excited by electromagnetic waves. In fact, such materials are used in telephone speakers and sonic alarms. Similarly, the piezoelectric mixer medium in the Skyvault spacecraft, excited by this difference frequency, would emit a sonic vibration. If such a craft was to hover up and down at only 15 centimeters per second on its

100-gigahertz microwave beam, the fundamental harmonic of its beat frequency would have a very low value of around 100 hertz. Sonically, this would be at the lower end of the audible spectrum. As the vehicle proceeded to climb at an increasing speed, the sound pitch emitted from its mixer diode would accordingly increase, passing up through the audible range to ultrasonic frequencies. At an upward velocity of 30 meters per second (sixty-seven miles per hour), the craft would be generating a 20,000-hertz beat frequency (i.e., in the ultrasonic range).

Unconventional flying objects, however, have been sighted that emit somewhat lower microwave frequencies. For example, in one aerial chase of a UFO that took place near Meridian, Mississippi, in July 1957, electronic countermeasures equipment on board an RB-47 plane was able to pick up a 3-gigahertz microwave frequency emanating from the UFO. The signal occurred in 2-microsecond-long bursts that repeated six hundred times per second.[6] UFOs have generally been observed to radiate electromagnetic waves in the 0.3 to 3 gigahertz range, hence in the UHF band.[7] If the craft was to transmit a 1-gigahertz maser signal and travel upward at a velocity of 30 meters per second, its beat frequency would be felt in the low frequency audio range at 200 hertz.

Let us next attempt to interpret Don's laconic statements describing how the Skyvault craft's microwave receiver operates when the ship is in motion. As we noted above, when the craft is moving upward, the ground will be receding from the craft and, hence, upon reflection from the ground, the local oscillator frequency f_o will be Doppler redshifted to a slightly lower frequency by an amount $-\Delta f = -f_o(v/c)$. Also, as the ground-reflected probe beam enters the mixer diode, it will experience an additional Doppler redshift due to the craft's upward receding motion. Thus, upon entering the mixer, the probe beam will have been Doppler shifted relative to the local oscillator frequency by a total of $-2\Delta f$, which is equal to $-\varphi$.

However, the mixer diode will compensate for this frequency shift. That is, because the incoming redshifted probe beam and the pump beams differ in frequency by $2\Delta f$, they will generate a moving holographic grating pattern within the medium of the four-wave mixer diode. This is well known from experiments with optical phase conjugation. Upon interaction with this moving grating pattern, the pump beams will produce an outgoing phase conjugate beam that is frequency shifted relative to the local oscillator (pump beam) frequency by an amount equal to the frequency shift of the incoming probe beam, *but opposite in sign.*[8] Hence, the four-wave mixer will automatically produce a phase conjugate beam that is *blueshifted* in frequency by an amount $\varphi = +2\Delta f$. That is, the outgoing beam will be reverse Doppler shifted. The incoming redshifted probe beam and outgoing blueshifted phase conjugate beam then differ in frequency by $4\Delta f$ in the mixer diode reference frame. After reflecting from the ground and converging into the Gunn diode resonator cavity, the blueshifted phase conjugate beam will have been redshifted by an amount $\varphi = -2\Delta f$. So upon reaching the receding craft, its frequency will end up precisely matching the local oscillator frequency. Thus, a condition of resonance will be established with the outgoing local oscillator frequency.

In the Earth's reference frame, the frequencies of the ground-reflected probe beam and the phase conjugate beam will differ by an amount $2\Delta f$, numerically equaling the difference frequency, or beat frequency, that the ground-reflected probe beam generates in the four-wave mixer. As a result, these two counterpropagating beams will build up a phase-locked soliton beam between the craft and the ground that in the ground reference frame will have a beat frequency $\varphi = 2\Delta f$. This is equivalent to the frequency that Don's formula specifies to be the "frequency caused by movement of the unit" (or vehicle). Its value depends on the speed of the craft relative to the ground. This beat frequency will likely induce an audible sound in any material body on the ground that it happens to push against. This could explain UFO sightings in which witnesses have reported hearing humming sounds.

If we properly interpret Don's letter, he appears to say that the mixer's Doppler output signal may be used to control the speed of the vehicle. He talks of amplifying and processing this signal. Presumably, this amplified Doppler signal is fed back into the mixer. The amplified signal, in turn, would add its power to the outgoing phase conjugate beam and ultimately to the soliton beam. Thus, by controlling the amount by which the Doppler signal is amplified, a pilot would be able to control the soliton beam intensity and the amount of thrust that the beam would develop. In this way, he could control the speed of the craft. With more amplification, the craft would accelerate away from the Earth and with less amplification it would decelerate or even enter a descending flight mode. As the craft moved horizontally, its velocity relative to a beam's particular ground target location would continually change and as a result the frequency of the mixer diode's Doppler output signal would be changing accordingly. Whatever frequency the mixer happened to be outputting, the amplifier would be amplifying that signal at the amplification level that the pilot had set.

Thus, although the power level of the Gunn diode beam generator could also be changed, for finer adjustments the craft guidance system would be controlling the power level of the much-lower-frequency Doppler signal. This is reasonable since lower-frequency waves in the audio or radio frequency range are much easier to control than those in microwave frequencies. If the craft supported itself on three microwave beams, it would need some very sophisticated computer hardware to coordinate and properly control the Doppler signal power levels of all three beams.

It is conceivable that the same technique could be used to produce a tractor beam that would lock on to objects and draw them toward the craft. A microwave beam could be changed from a repulser beam into a tractor beam simply by inverting its sawtooth waveform to have a sharp increase of positive potential followed by a gradual decline. This could be done by reversing the polarity bias on the Gunn diode and on the barium titanate dielectric that would be used to shape the wave. If the craft had several phase conjugate resonators (i.e., more than one local oscillator and several mixer diodes), some might be used to create repulsive beams to support the craft, leaving another free to phase conjugate an attractive soliton beam that might be used to target a transportable object. By adjusting the power applied to its tractor soliton beam, the craft could control the movement of the targeted object as it made its approach. Similar technology could explain UFO sightings in which cars or people have been picked up by a force field and drawn toward a hovering craft.

In summary, the development of microwave field propulsion technology combines three areas of research: (a) research into the production of high-power microwave beams, (b) research into metamaterials that exhibit a negative index of refraction or strong electric or magnetic resonances at microwave frequencies, and (c) research into microwave phase conjugation. With the proper engineering development, it should be possible to produce a vehicle capable of free levitation.

8.3 ■ AEROSPACE INDUSTRY INVOLVEMENT

Is there evidence that companies have been doing work in the field of microwave phase conjugation in more recent years? Indeed, a survey of unclassified literature indicates that Rocketdyne has been relatively active in this field. For example, in 1990, scientists associated with Rocketdyne and with Rockwell International coauthored a paper titled "Microwave Phase Conjugation in a Liquid Suspension of Elongated Microparticles."[9] The nonlinear electric properties of the particle suspension described in the article would allow it to serve as an ideal medium in which four-wave mixing could take place. Also, it is not a secret that Rocketdyne has been interested in the development of high-powered microwave beams. For example, in 1993, scientists affiliated with the Rocketdyne division of Rockwell International and with the Titan-Beta Corporation reported tests of a high-power maser system capable of delivering a 2.86-gigahertz pulsed microwave beam having a peak power of 65 megawatts![10] The system used an SLAC 5045 linear accelerator klystron tube that functioned as a free-electron laser and was powered by a modulator unit developed by Titan-Beta. The unit delivered 3.5-microsecond pulses at a rate of 180 pulses per second. The paper does not mention what the beam was to be used for, but its power would have greatly surpassed that of the magnetrons used in the early days of Project Skyvault.

Hughes Aircraft is another company that was active in the field of microwave phase conjugation. Recall the story told by my friend Thomas whose father had worked at Rocketdyne, presumably on Project Skyvault, and had drawn him a picture of a lenticular levitating vehicle that had been successfully tested. His father had later moved on to work at Hughes Aircraft, also in deeply classified projects. When Thomas later asked his father if Hughes was doing research in electrogravitics, his father's terse answer was, "They are the world leaders." Later in 1992, I had the occasion to ask one high-ranking Hughes manager whether Hughes Aircraft was still pursuing its electrogravitics R&D. He answered, "Yes, but they keep their work very quiet."* The same could probably be said for the other aerospace companies that today continue to work in this field.

It is also known through papers published in the open literature that give author affiliations that a considerable amount of research was going on at Hughes in optical phase conjugation. One military application of such technology, mentioned in chapter 7, is the targeting and destruction of missiles by means of a pulse from a high-powered laser. Thus, considering that Hughes was also doing cutting-edge research in electrogravitics, it stands to reason that it was also applying its phase conjugation knowledge to microwave phase-conjugating systems on projects involved in developing vehicle propulsion systems. In fact, Hughes has had a long involvement in military radar systems. The forward-looking radar used on the B-2 bomber, for example, was developed by Hughes. We may conjecture, then, that Hughes Aircraft was heavily involved in Project Skyvault's research.

An indication that Hughes had been conducting research on microwave phase conjugation came in 1993 with the granting of its patent for radar cross-section enhancement using phase-conjugated impulse signals (U.S. patent 5,223,838). Researchers were applying the principle to radar as a way of locking on to a distant target. By receiving a radar echo that normally would be too weak to properly detect,

---

*Prior to its dissolution through corporate merger, Hughes had a long history of being involved in the development of very advanced, leading-edge technology. About 90 percent of its work was defense related, most of which is highly classified R&D. During the early 1980s, when the U.S. Government Accounting Office was cracking down on defense contract fraud, evidence surfaced that an astounding two-thirds of Hughes's contract costs could not be accounted for, far more than for any other corporation surveyed. Whereas other contractors received stiff fines for their unaccounted costs, Hughes managed to emerge virtually unscathed. Could federal authorities have understood that these excess cash flows were not cost overruns, but rather funds whose specific black-programs destinations could not be revealed or even admitted to?

mixing it in a mixer diode with a pump maser beam to create a holographic grating, and pumping the grating with a radar pulse to create an outgoing phase conjugate beam, they were able to create a soliton beam between the radar transceiver and target that would resonantly amplify the original radar echo to a detectable level. The patent makes no mention that similar technology could be used to create a soliton beam beneath an aircraft for the purpose of levitation. Any patent disclosing such an aerospace application would likely have run the risk of being classified, so there is no way of knowing whether such a patent in fact exists.

When all the above evidence is considered as a whole, a pattern emerges that points to Hughes having made a major effort in developing microwave phase conjugation field propulsion technology. Hughes has since been split up and sold off to a number of companies. The Hughes research facility in Malibu was a former hotbed of research on antigravity propulsion according to the testimony of a "Dr. B." in Steven Greer's book *Disclosure* (2001, p. 262). The laboratory, which currently goes by the name HRL Laboratories, is today jointly owned by Boeing and General Motors.

The technology of microwave phase conjugation is also being applied to aerospace communications. Ideas along these lines were discussed by University of Michigan researchers Leo DiDomenico and Gabriel Rebeiz in a paper they published in 1999.[11] The technology has several advantages. First, compared with standard maser beam technologies, it is very energy conserving since the transmitted microwave energy is restricted to a tight beam extending between the ground communication station and the spacecraft. Unlike a standard maser beam, very little radiation is lost into space. Second, the link is very secure since, unlike radio broadcasts, it is very difficult for anyone to eavesdrop unless he places his receiver in the path of the beam. Third, the phase conjugate beam locks on to the spacecraft and, hence, is able to automatically track the spacecraft, even though the spacecraft is moving. The DiDomenico and Rebeiz paper is interesting because it talks about phase conjugation of an incoming signal that has been Doppler shifted due to motion of the target. So, many of the same considerations involved in a vehicle propulsion system are already being discussed in the context of communication systems. The mathematics are somewhat involved, requiring the use of Heaviside operators or Laplace transforms.*

8.4 ■ TESLA'S MAGNIFYING TRANSMITTERS

Tesla's magnifying transmitters also functioned as phase conjugate resonators, and this was most likely known to him, although the specific concepts behind the phase conjugation phenomenon would not be developed for many decades. These devices were known for their ability to generate exceedingly high voltages and to occasionally produce violently destructive ball lightning sparks. The Wardenclyffe tower was the largest of his generators (see figure 8.6).[12] Tesla built it to show that it was possible to transmit megawatts of power over global distances to power entire cities and air vehicles as well.[13] Construction of the tower began in 1901 on Long Island near the town of Shoreham. The central part of its resonator consisted of a large, flat spiral coil mounted horizontally and shielded within a 68-foot-diameter mushroom-shaped dome electrode, this whole structure being perched 187 feet above the ground at the top of a wooden tower. The center of the coil was electrically connected to the dome electrode to form a resonator. Power inductively supplied to the coil would surge in resonance between the coil and the dome electrode, pulsing the dome with extremely high voltages. The dome electrode consisted of a honeycomb array of 1-foot-diameter parabolic shells whose small-curvature radii facilitated ionization of the surrounding atmosphere. Thus, when electrified, the entire dome would have become enveloped in an ion halo. The nitric oxide gases that would have formed in this halo have electric properties that are

---

*Those interested in preparing themselves for a career in field propulsion technology should consider going for an undergraduate degree in physics with a minor in electrical engineering. Then, go on to get either a master's or doctoral degree in electrical engineering, specializing in microwave or radar engineering. Make sure to take courses, among other things, in optical and microwave phase conjugation, Fourier analysis, and Laplace transforms. The best universities to pick are those at which professors are already working in the area of microwave phase conjugation applied to radar or communication systems. Examples that come to mind are the University of Michigan and the University of California Los Angeles. In particular, the California professors have in the past coauthored papers in this area with Rocketdyne scientists.

very nonlinear and would have served as an ideal phase-conjugating medium.[14]

We may deduce, then, that this ion halo would have phase-conjugated the tower's resonant oscillations, allowing it to function as a phase-conjugate resonator and to generate field powers far in excess of the power used to run it. Since the dome electrode would have been radiating longitudinal waves to the Earth's ionosphere, an immense soliton would have formed between the halo and the ionosphere, and since the halo would have phase-conjugated the waves returning from the ionosphere, the tower would have been able to draw on energy cohered from a vast region of space.

Unfortunately, this miraculous structure was never completed because Tesla's sponsor cut off the project's funding, but earlier, Tesla had built and operated smaller-scale versions. These magnifying

Figure 8.6. The Wardenclyffe tower. (Photo courtesy of C. Yost, from "Tesla's Tower [Wardenclyffe]" Electric Spacecraft Journal *[May/June 1991])*

transmitters emitted a repeating series of very high-voltage shock fronts from their dome-shaped negative electrodes. Unlike hertzian waves, these wave fronts had little transverse polarization. Their electric field profile was sawtooth-shaped with a very sharp rise in potential and a subsequent more-gradual decay, and the field gradients were oriented longitudinally in the direction of wave propagation. When operating, his transformer would gradually build up immense field potentials, approaching 100 million volts, as a result of the cumulative effect of repeating pulse cycles. The progressive amplification of these phase-conjugated waves is an example of what we earlier referred to as the field-induced soliton phenomenon. The high voltage that accumulated as a result of these repeating waves would have been apparent as a growing luminescence. In his book *Secrets of Cold War Technology*, Vassilatos wrote:

> He [Tesla] had already observed how the very air near these Transformers could be rendered strangely self-luminous. This was a light like no high frequency coil ever could produce, a corona of white brilliance which expanded to ever enlarging diameters. The light from Tesla Transformers continually expands. Tesla described the growing column of light which surrounds any elevated line which has been connected to his Transformers. Unlike common high frequency alternations, Tesla radiant energy effects grow with time. Tesla recognized the reason for this temporal growth process. There were no reversals in the source discharges. Therefore the radiant energy would never remove the work performed on any space or material so exposed. As with the unidirectional impulse discharges, the radiant electric effects were additive and accumulative. In this respect, Tesla observed energy magnifications which seemed totally anomalous to ordinary engineering convention.[15]

Further on, Vassilatos wrote:

> Tesla performed outdoor experimental tests of broadcast power in the northernmost reaches of Manhattan by night. Sending metallized balloons aloft, he raised conductive lines. These were connected to the

terminals of his Transformers, and activated. When properly adjusted, the white luminous columns began covering the vertical aerial line and expanded by the second. Enveloping Tesla, his assistants, and the surrounding trees, this strange white luminosity moved out into the countryside to an enormous volume of space. Tesla described this phenomenon in several of his power transmission patents, the obvious artifact of a non-electrical energy.[16]

Like Tesla's transformers, the Project Skyvault vehicle may similarly have made use of a nitric oxide halo to phase-conjugate its microwave beams. If part of the pump microwave beam signal was used to energize the outer surface of the craft, the resulting high-frequency, high-voltage field would have ionized the air immediately around the craft, enveloping it in a phase-conjugating layer of nitric oxide gas. Then, the entire lower surface of the craft would have served as a secondary mixer, and the three incoming ground-reflected probe beams would have become time-reversed (phase-conjugated) on the craft's hull rather than in its interior mixer diode. This could explain why the bodies of many UFOs are seen to be luminous and radiating microwave radiation in the 0.3-to-3-gigahertz frequency range. The observation that UFOs often visibly pulsate when hovering or taking off could be explained if they were pumping their surfaces with a low-Doppler beat frequency that modulated the brightness of their luminescence.

It is also possible that the hull of such a spacecraft could itself be fabricated out of a metamaterial such that any microwaves impinging on its lower surface would result in a propulsive force. The unusual, layered bismuth and magnesium metal known as Art's Parts, reportedly retrieved from the hull of a crashed UFO, could be an example of such a material (see chapter 9).

Such material could also have the dual purpose of functioning either as a radar-absorbing or as a radar-diverting material, one that would bend incident electromagnetic waves around an object in such a way as to give the impression that they had passed through the object completely unobstructed. For example, physicists David Smith, John Pendry, and David Schurig announced in May 2006 that within five years their team would be able to demonstrate a metamaterial cloaking shield that would render a craft within it completely invisible to electromagnetic waves at microwave frequencies.[17]

Ray's comment in chapter 4 that black-world antigravity propulsion technology was based partly on Tesla's work further supports the notion that phase-conjugate microwave resonators have been developed for aerospace propulsion. Ray was apparently quite familiar with phase-conjugate resonance technology because he said that one of the highly secret R&D projects he had been assigned to involved working on the application of this technology in the field of cryptography. Apparently, his group had developed a way to encrypt an electronic message by degrading it into unrecognizable noise and then later recovering it by using a phase-conjugate resonator to time-reverse that noise back to the original ordered message. He felt that this same technology provides some of the key concepts that can explain how these antigravity propulsion craft work the way they do. Discussing a version of this phase-conjugation technology, he stated, "I have seen demonstrations of this stuff—of the raw technology. One of them breaks the second law of thermodynamics, the law of entropy. It breaks that law!"[18] Statements that the first or second law of thermodynamics might be broken amount to blasphemy to the mind-set of the conventional academic physicist. In the world of black-project engineering, however, they are routine facts of life.

8.5 ■ BROWN'S PHASE-CONJUGATING MICROWAVE DISC

Brown's levitating disc would have operated much like the Project Skyvault vehicle. The disc-shaped antenna attached to the bottom of the conical dielectric (see figures 3.2, 3.3, and 3.8) would have radiated microwaves at a frequency of a few gigahertz and would have acted much like the Gunn diode in the Skyvault vehicle. The positive electrode, which would have had either a parabolic or a cone shape, would have served as a wave amplifier cavity since a portion of the microwave radiation reflected downward by the electrode would have been reflected back at the mouth of the cavity. As a result, its signal would have resonantly amplified and built up to a high intensity across the high-K piezoelectric dielectric cone located inside the hornlike cavity. This dielectric cone, which would have been polarized with high-voltage DC, would have had very nonlinear electrical characteristics and would have functioned in a manner similar to the mixer diode in the Skyvault vehicle. Since the dielectric was being pumped with the resonantly amplified microwave beam that was reflecting back and forth along its length, microwaves reflected up from the ground would have acted as a probe beam that would have interacted with the pump beam to produce a holographic grating pattern in the dielectric. At the same time, the pump beam would have produced a phase-conjugate beam that would have traveled downward to the ground, precisely retracing the pathway followed by the ground-reflected waves. The upward-reflected ordinary beam and the downward-propagating phase-conjugate beam would have been phase-locked to produce a soliton wave.

As in the Project Skyvault vehicle, the soliton wave extending between the ground and the mixer dielectric in Brown's saucer would have resonantly amplified to a very high intensity. This would be an example of the field-induced soliton phenomenon. Much of the microwave radiation radiated by the saucer's disc electrode, then, would have been bottled up in this beam.

It is possible that the ion discharge surrounding the positively charged umbrella electrode also served as a phase-conjugating medium, in addition to the barium titanate dielectric. As mentioned above, nitric oxide ions surrounding a high-voltage electrode would have very nonlinear electrical properties and could phase-conjugate waves, much like the glow discharge that surrounded the dome electrode of Tesla's magnifying transmitter.

With Brown's saucer, the DC polarization along the length of the dielectric would have progressively build up to a high voltage due to the dielectric's tendency to rectify some of the AC signal. As a result, a very steep potential gradient would have formed and would have exerted an upward thrust. This ramping up of the dielectric's potential gradient would have been helped by the tendency of the saucer to function as a phase-conjugate resonator with self-amplifying pump beams. In addition, if the oscillator could have been made to emit sawtooth-shaped waves, the saucer would have experienced more upward thrust due to the electrogravitic impulse effect.

Brown made no mention of phase conjugation in his electrokinetics patent, which he applied for in 1958. At that time, the phase-conjugation effect was unknown in the unclassified engineering world. The maser was invented in 1954 by Charles Townsend, and the rubidium optical laser was invented in 1960 by Ted Maiman at Hughes Aircraft. The field of laser holography began developing shortly after that, in the early 1960s, and it was not until 1972 that optical phase conjugation began to be discussed in the open literature. Thus, Brown was most likely unaware of the phase-conjugation principle behind the levitation effect he had experimentally discovered. Bahnson came close to the idea when he inferred that energy from the AC waves was being resonantly stored in the ether surrounding his saucer's electrodes. Indeed, the ether very likely also plays an important role, but the effect finds a ready explanation in terms of the phase-conjugating properties of the saucer's ceramic dielectric and its electrode's plasma sheath.

By 1957, when Brown had begun experimenting with the idea of pulsing dielectrics at microwave frequencies to get electrogravitic thrust, he was apparently rediscovering a microwave propulsion phenomenon that for some time had been known to Project Skyvault scientists and that by that time was in an advanced stage of secret development. His vertical-thrust apparatus would have functioned much like the homodyne version of the Project Skyvault vehicle. This is the version that incorporated the microwave transmitter and mixer in the same conduit. Since Brown was a latecomer with a history of conducting independent investigations that did not adhere to normal security protocols, he could have been regarded as a threat to efforts to maintain the secrecy of this area of R&D investigation. This may explain why his attempts to obtain military funding were continually rebuffed by the Pentagon and why Admiral Rickover had advised him to drop his work on electrogravitic propulsion.

8.6 ▪ THE RUNAWAY MODE

Phase conjugate resonance and its related field-induced soliton phenomenon appear to be key to understanding this futuristic aerospace technology. However, this technology is not without its hazards. One important problem that engineers have had to face is ensuring that their microwave-powered vehicle does not enter a runaway mode such that the energy of its soliton field increases exponentially and finally explodes.

Guy Obolensky, one of the early researchers in microwave phase conjugation, has observed firsthand this explosive resonant amplification phenomenon in the phase-conjugating systems he has worked with. He coined the term "faser phenomenon" to refer to this exponential energy increase, "faser" being an acronym that stands for "*force amplification by stimulated energy resonance*."[19] His term, in effect, describes the field-induced soliton phenomenon concept.

The phase-conjugate resonator Obolensky was testing in his laboratory in 1958 was so highly efficient that it entered this runaway energy-increase mode, which ended in a violent explosion.[20,21] The phase-conjugate resonator he had constructed, which he termed a "limit cycle faser," consisted of a long surface-waveguide resonator of a size that could be placed on top of a desk. The waveguide was made of an aluminum sheet approximately 0.25 millimeter thick laid over an aluminum slab and separated from it by an insulating Mylar film that was hermetically sealed on either side with layers of distilled water. The separation of the waveguide walls had to be accurate to within a few ten thousandths of an inch. At one end of the waveguide, a 17-kilovolt spark discharge was made to jump across a series of spark gaps, tuned so that their sparks were self-quenching. The resulting spark oscillations generated longitudinal microwaves that traveled down the waveguide, skimming between the top and bottom metal surfaces. Normally, waves reflecting back from the end of the waveguide would disturb the spark discharge, causing it to become noisy and have excessive energy losses. Yet by placing five evenly spaced strips of Permalloy tape at the far end of the waveguide, Obolensky was able to create a phase-conjugate mirror. The nonlinear electrical properties of these strips altered the waveguide's characteristics in such a way that they phase-conjugated the surface waves coming from the spark and reflected them back as time-reversed waves, thereby making the spark oscillations coherent, that is, totally ordered.* As a result, his resonator had a phenomenally high output efficiency—far over unity. A powerful soliton consisting of nine harmonics of the fundamental submillimeter wave was able to build up within it.

Apparently, Obolensky's waveguide was optimally tuned and its Permalloy magnetic grating was optimally configured, because the energy resonance process became self-reinforcing, causing the waveguide's stored energy to increase exponentially. The current gain was so enormous that ball lightning–like sparks began to erupt from the waveguide and actually perforate its aluminum wall. Finally, in a blinding flash, the whole resonator assembly explosively discharged its accumulated energy and fragmented itself. Surviving pieces showed that the waveguide's submillimeter-thick wall was perforated with clusters of tiny holes spaced apart by a certain precise distance-multiple of the planar waveguide's thickness to form a periodic pattern. The dendritic pattern connecting these holes traced out the branching path of the immense electrical discharge that had formerly traveled down the full length of the conduit. Judging from the amount of energy required to vaporize the quantity of aluminum that was missing from the perforated section, Obolensky concluded that it would have required several hundred thousand joules (~100 watt hours) of energy, about 100,000 times greater than the 7 joules (~2 calories) of coulombic energy in the DC charge that his power supply had fed into his waveguide.[22] In subsequent experiments, Obolensky found that this field amplification technique could be properly controlled by means of a feedback circuit that would temporarily detune the oscillator powering the resonator whenever an excessive energy began building up in the resonator. He found

---

*That is, the permaloy strips produced a series of impedance discontinuities that formed a current dependent electromagnetic grating pattern along the waveguide. The grating modulated the surface conduction of the waves in such a way that it would reflect them back as phase-conjugate waves. Tesla had also discovered the secret of making coherent spark discharges characterized by negative resistance; he once commented that an arc is not working properly when it is noisy: to be efficient it should "sing," i.e., be coherent.

that circuits employing normal hertzian signal conduction functioned much too slowly to effectively control the resonator, that only longitudinal shock front waves could travel fast enough.

In another experiment, Obolensky used such self-regulating faser circuits to achieve a 20 percent increase in light output from a 500-watt sodium vapor arc lamp, at the same time eliminating its flicker.[23,24] He attributed its increased efficiency to the nonlinear reactance element that he placed in series with the lamp that phase-conjugated the plasma oscillations of its arc.

Obolensky theorizes that his resonators derive their excess power by "cohering" incoherent energy present in their wave shape and possibly even entraining the zero-point energy in the surrounding ether.[25] He feels that this may have something to do with the resonator's ability to excite multiple harmonics of its fundamental frequency. Whereas a normal resonant electric circuit amplifies only the fundamental resonant frequency, these phase-conjugate resonators also exchange energy among and amplify up to nine harmonics.[26,27] Since these harmonics mutually interlink in the resonator's nonlinear elements, noise energy present in the environment that happens to excite certain of these harmonics would become entrained and cohered into this multimode resonance. That is, the phase conjugator's nonlinear elements would send time-reversed waves back to those "noise" fluctuations, creating a coherent soliton that would entrain incoherent energy into the self-amplifying energy resonance pattern, thereby reversing the entropy of that "noise." Since the soliton not only resides within the resonant circuit but also extends outward to surrounding space, its resonance would entrain the surrounding energy and cool its immediate environment.

Obolensky observed an environmental temperature drop when operating the 200-kilovolt pulse discharge magnifying transmitter described in chapter 6. He noticed that when he switched on his device, the temperature immediately dropped in the surrounding room. He attributed this to the ability of the ionized medium surrounding his dome electrode to phase-conjugate shock fronts reflected back from the environment, creating a soliton wave pattern that entrains environmental noise fluctuations. Like the dome of Tesla's magnifying transmitter, Obolensky's scaled-down replica creates a luminous aura that is ideal for phase conjugation.

Unlike Obolensky's limit cycle faser experiment, the energy entrainment rate of his magnifying transmitter was sufficiently low as to pose no danger of explosion. Oscillograms showed that the accumulated energy produced a 50-kilovolt negative bias on the dome electrode, which otherwise should have maintained a zero voltage. To control the energy buildup, he uses a series of high-ohm resistors immersed in an oil cooling bath to continually drain off power from his antenna dome to ground. He says that in so doing, he bleeds off excess power that his transmitter is cohering from the environment, a blatant example of entropy reversal. If Tesla's technology could be used on a large scale for power generation, the threat of global warming would indeed be a thing of the past.

It is possible to conceive that the Project Skyvault vehicle was similarly phase-conjugating and entraining environmental noise energy into its soliton pattern. If so, its Gunn diode may have initially been operated at full power to seed the microwave beam and get the soliton field established. Once deployed, the soliton beam would have drawn upon entrained energy as its supplementary power source.

Other researchers experimenting with nonlinear resonators also have reported observing environmental cooling effects. In the late 1980s, Russian physicists Vladimir Roshchin and Sergei Godin were testing a version of the Searl effect generator that they refer to as the magnetic energy converter (MEC; see chapter 10). They reported observing a seven-degree Centigrade drop in room air temperature when the MEC was in operation, with the temperature drop being confined to a series of concentric, shell-like cylinders surrounding the MEC's spinning rotor and spaced from one another at intervals equal to the rotor radius. This suggests that the MEC was setting up a radial stationary wave pattern, that is, a soliton. Like Tesla's dome electrode, their disc developed a luminous aura when operating, providing an ideal environment to phase-conjugate incoming waves reflected from the environment. The disc was likely entraining energy from the environment, because above a certain critical rotational velocity, the rotor was observed to self-accelerate and had to be forcefully restrained with a braking system.

The temperature drop in the environment was probably a consequence of this energy entrainment.

One day I received a telephone call from a physicist named Greg who wanted to discuss subquantum kinetics, but as the conversation turned to electrogravitics, I quickly learned that he had considerable inside knowledge about UFO propulsion technology. He told me his interest in this subject began as a young boy, since his father had served as a consultant on secret military projects that were attempting to reverse-engineer UFOs. Greg agreed that many of the antigravity vehicles being developed use microwaves to generate their propulsive force through interaction with certain kinds of nonlinear materials. However, he underscored the problem that these kinds of antigravity drive systems are inherently unstable. Referring to phase-conjugate technology, he said:

> I know why some of this stuff is dangerous and I agree with it being kept secret. Because, while achieving a desirable effect of free body levitation is relatively easy to do, . . . there is an energetic mode in addition to the force mode and the energetic mode has to be controlled with some finesse. It is far easier for the setup that they would create to blow up in their face and wipe them out, and perhaps their neighbors, before they would figure out that such a thing is a potential problem. Anything that has an exponential rise with a few microsecond time constant is not something to take trivially.[28]

Greg said someone would need a very sophisticated knowledge of mathematics to be able to design such a system so that it could operate safely. The reason is that linear mathematics, the kind most physicists use when they solve explicit function equations on the blackboard, does not adequately represent the behavior of the nonlinear interactions that characterize how individual parts of such a system mutually interact with one another and how they are affected by the system as a whole. He said:

> You have to be into nonlinear partial differential equations, and you have to be good with your numerical analysis. You can't go out and

use anybody's canned algorithm. You have to get all the auxiliary functions, analog solutions; anything that remotely smells of a linearization scheme to approximate what the nonlinear solution would be is likely to overlook the runaway solution, the one that's going to end up getting you. You can read about this in IEEE. They've come across this sort of thing in their microwave simulation studies before . . .

When you're doing high-frequency stuff . . . you will find that standard second-order linear differential equations are incapable of modeling the behavior. You will readily see that there are terms that you are neglecting of how the whole system is interacting with itself. It's wrong to think about it as "field" being separate from "material." You have to think of it as an implicit function system . . . Suppose you say that Z is a function of X and Y and Z, then you have to know what the Z function is before you can say what the answer to it is, that's an implicit system . . . For any of these nonlinear systems, especially the interesting ones, you end up with an implicit function system. So if you are making an approximation, guessing the behavior of the function in a nonrigorous way, and if you violate any of the convergence criteria, then what you'll end up with is a spurious solution. You have to go into the differential topology of the system. Chaos mathematics and stuff like that come in there.[29]

The phase-conjugation demonstration that Ray, the black-project scientist, had witnessed convinced him of the need to keep the details of this technology secret. In his 1992 phone conversation with me, he said, "When I saw the demonstration, it proved radically to me that this stuff has got to be kept under wraps. I agree with the secrecy."[30]

I said I had read about the weapons applications of phase-conjugate technology and had wondered, if this is true, are we really ready for this sociologically?

Ray responded, saying, "We're not. We are not. Let me tell you why. The engineering applications of this stuff are extremely simple, very fundamental, and there is no way to control it. What it amounts to is giving out the recipe to make an atomic bomb by going to the local drugstore. We don't want to broadcast this kind of stuff. At this point in time, its not good to do that."[31]

I felt that Ray may have been exaggerating a bit. The explosion that blew up Obolensky's limit cycle faser certainly could not have been more powerful than that produced by a cherry bomb or M-80 pyrotechnic. It appeared that this was more a concern for the personal safety of the experimenter than it was an issue of a destructive bomb that could potentially be used by a terrorist. Certainly it is nowhere near the hazardous potential of nuclear fission, which now is in common use worldwide. I then commented that, looking at the other side of the coin, there are many problems this technology could help to solve, such as providing an alternative to fossil fuels that could eliminate air pollution and ultimately do away with the global warming greenhouse effect.

Ray responded that there were economic considerations for introducing such a major shift in energy technology. He said, "But the problem exists also that we cannot switch from the way things are to the way things should be instantly, because one interferes with the other completely. You have to have a slow evolution with this."[32]

When I commented that this slow evolution did not seem to be happening since the technology was encapsulated within the black-R&D world, Ray said, "That's because there are political considerations at the moment. You are going to find a little bit more of this exposure beginning, of course, with some of the articles like the one in *Aviation Week & Space Technology* and other articles you're going to see. By the year 1995, you're going to hear a lot more about it, according to the grand plan, according to what I can tell. So it's coming out slowly but surely."[33]

However, 1995 has long passed and the existence of field propulsion technology is still being kept quiet.

9

UNCONVENTIONAL
FLYING OBJECTS

9.1 ■ SIGHTINGS

Information gathered from a variety of sightings suggests that many UFO disc craft support and propel themselves by means of phase-conjugate microwave beams similar to those used in Project Skyvault. In his book *Unconventional Flying Objects,* Paul Hill reviews a number of sightings of craft that propelled themselves by means of downward-directed force field beams. One example is a case that occurred in Norway in 1970 in which a 10-meter-diameter disc was hovering above a man standing next to his car.[1] The craft was steel blue and shimmered yellow all around its circumference. Suddenly, it began to leave, and as it did, an invisible force knocked the man to the ground and imploded and pulverized his car windshield. The man did not feel any pain from the impact of the force field, which suggests that it acted uniformly on every cell in his body.

In a similar fashion, the phase-conjugated microwave beam projected from a Project Skyvault craft would have exerted a repelling force on the ground and on ground-based objects or people as it supported the craft. Since the microwave beam would have been targeted over a large region of the ground and would have penetrated some distance into the objects it touched, its force would have been distributed diffusely, as was apparently the case in the Norwegian encounter.

Hill mentions several other sightings. The force field from an overhead UFO in one example gave a soft push to a moving vehicle; in another, rocked a vehicle from side to side; and in yet another, actually flipped over a stopped truck.[2] In another encounter, which happened in 1959 in the Greek villages of Digeliotica and Agio Apostolou, the field from a low-flying disc repelled several ceramic roof tiles off the roof of one house as the craft passed overhead. The village priest, Pappa Costas, who was inside the dwelling at the time, reported that the whole house seemed to shake, making him think there was an earthquake, but it could not have been an earthquake since other houses had not experienced a similar shaking. All of these force field effects would be expected if the UFO was projecting a microwave beam capable of exerting a repelling force on solid objects.

Downward forces have also been observed on underlying vegetation.[3] One bullet-shaped UFO, approximately 45 feet in diameter, was sighted in Maryland in 1958. As it moved at about thirty miles per hour at an altitude of 300 feet, it emitted a steady hum and its skin illuminated the surrounding terrain with a green glow. Tree branches lying along its flight path were bent down and in some cases broken. In another sighting, which occurred in 1974, four UFO discs were spotted hovering only a foot off the ground in a field of rape plants. Approaching to within 15 feet of one rotating craft, a man named Edwin Fuhr noticed that the grass below was being swirled down. The four craft departed vertically about fifteen minutes later, after which he noticed that the grass below where each had hovered was flattened in a clockwise swirl pattern, forming a ring with the grass in the center being left standing upright.

Generally, UFOs are observed to sit level when they hover and to tilt when they perform all other maneuvers. For example, they tilt forward to move forward, tilt backward to stop, bank to the left to turn left, and so on. All of these tilting maneuvers are the kind that would be performed by a craft driven by a matter-repelling microwave soliton beam.

Another common characteristic of UFOs is their penetrating humming, buzzing, or whining sound. In his book, Hill describes one case in which a man reported a UFO casting a greenish light into his cabin as a throbbing hum shook its walls.[4] In another case, the observers "felt" a high-pitched intense sound as a 5-meter-diameter UFO took off. In yet another encounter, a UFO hovered 1.5 meters above the surface of a mountain lake and was seen to excite the water below to dance in thousands of sharp-pointed waves. Hill concludes that the propulsive fields that UFOs project downward are oscillatory and that the energy they transport to the ground and objects below excites oscillations at the same frequency and induces sound to radiate from the objects themselves. A craft levitated by a phase-conjugated microwave soliton beam having a beat frequency φ in the audio range would produce precisely these effects.

Also, UFOs have been observed to extend luminescent beams to the ground. Hill reviews one sighting that was made in Bahia, Brazil, in 1958 in which a 70-foot-diameter UFO disc was observed to emit a silver-blue glow.[5] As it hovered 90 feet above the ground, its luminosity was seen to extend like a curtain all the way to the ground, creating an illuminated area on the ground that was about twice the diameter of the UFO. After climbing to an altitude of about 600 feet, it made a tight circle in the sky, and as it banked for this turn, its luminous focus on the ground traced out a much larger circle. Hill concludes that the luminosity surrounding UFOs and coming from their beams must be caused by their field energy ionizing the air and producing a cool, luminous plasma. He reasons that the plasma must be cool because in one case a UFO that looked like a ball of fire had passed very close to foliage without burning it.

Although Hill suggests that X-rays might be producing the ionization, the same effect could also be produced by an intense microwave beam. In particular, a phase-conjugated soliton beam would store an enormous amount of energy and build up very high electric potentials capable of ionizing the air and exciting these ions to become luminescent, much like the gas molecules inside a fluorescent lamp. Recall that Tom's boss had said that the Project Skyvault vehicle supported itself on a microwave beam that gave off a greenish blue glow.

9.2 ▪ THE CASH-LANDRUM ENCOUNTER

Evidence that the Air Force was test-flying an antigravity craft surfaced on the night of December 29, 1980. Betty Cash, age fifty-one, her friend Vickie Landrum, age fifty-seven, and Vickie's seven-year-old grandson Colby had been driving through the Pinewoods area near the Houston, Texas, suburb of Huffman, located about twenty miles north of Johnson Space Center.[6,7] About 9 p.m., they spotted a fiery object high in the sky that quickly descended to treetop level. Eventually, it came to hover above the road. They drove to within 130 feet of it and got out of their car for several minutes to watch it. The craft was hovering about 70 feet off the ground. It was diamond-shaped, tapering to rounded points at the top and bottom, and was about the size of a city water tower (about 20 feet in diameter; figure 9.1). Every so often, a reddish orange cone of flames would roar out of its bottom, as if from a giant blowtorch or rocket. At such times, the craft would loft into the air about 25 feet, only to gradually descend once again. The flames brightly lit the surrounding pine woods and bathed them in an intense heat, turning the nearby pine branches brown and badly damaging the road's blacktop surface.

Frightened by what they saw, Landrum and her grandson got back in the car, and were joined some time later by Cash. The car door became so hot from the radiation that Cash could not touch it with her bare hands, but instead used her coat to grab the handle. After about ten minutes, the object rose up and once again hovered over the trees. At

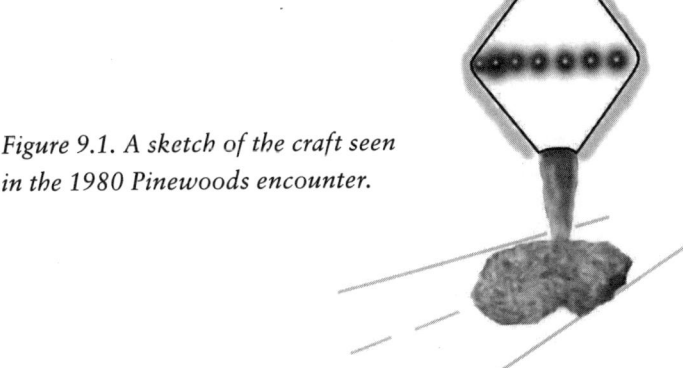

Figure 9.1. A sketch of the craft seen in the 1980 Pinewoods encounter.

that point, the three witnesses noticed that the vehicle was approached by almost two dozen twin-rotor military helicopters, later identified as CH-47 Chinooks and some of the Bell-Huey type. They appeared to be escorting the craft. The three concluded that they had witnessed a test flight of some kind of advanced antigravity military aircraft. One year later, Cash met a Chinook helicopter pilot who admitted to her in front of a witness that on the night of the encounter he had been called to fly to the area to check on a UFO that was in trouble near Huffman.

The description Cash and Landrum gave suggests they had observed a test flight of a prototype microwave vehicle similar in some respects to the Project Skyvault craft. The highly incandescent reddish orange "flames" were likely the exhaust from a flame-jet high-voltage generator adjusted for incomplete combustion. As mentioned in chapter 2, Brown had proposed a 10-foot-diameter saucer with a downward-pointing flame-jet generator as one version of the vehicle he had planned to research as part of Project Winterhaven. The Winterhaven design may have looked something like the sketch shown in figure 2.11, chapter 2.

The high voltage from this flame-jet generator may have been used to energize high-power Gunn diode oscillators to generate a downward-directed microwave beam of a kind similar to that used in the Project Skyvault saucer. After their encounter, Cash, Landrum, and Colby experienced radiation burn symptoms such as hair loss and inflamed eyes, the sort produced by exposure to an intense beam of microwave radiation. All of them became extremely sick within the next few hours. Of the three, Cash had spent the longest time out of the car (about ten minutes) and, not surprisingly, she had the worst symptoms. Her head and neck were blistered, and soon her eyes swelled shut, fluid seeped from welts on her head and scalp, and she suffered from severe headaches, nausea, vomiting, diarrhea, and body pains. After a couple of days being cared for at Landrum's home, Cash checked into a hospital, where she was treated as a burn victim, remaining for fifteen days. She began losing large patches of skin from her face, her hair began to fall out, and her eyes swelled so badly that she could not see for about a week. After a month in the hospital, she still showed no improvement. Then she developed breast cancer and had to have a mastectomy. She later died at the young age of sixty-nine. Landrum was also losing her hair and her scalp was numb and painful. Colby had problems with his eyes. All three of the victims were treated for radiation poisoning, and doctors listed their condition as life threatening.

Cash and Landrum sued the U.S. government for $20 million in damages, but after dragging on for many years, their case was finally dismissed on the grounds that no such object was owned, operated, or in the inventory of the Air Force, the Army, the Navy, or NASA. The ABC television show *Nightline* in 1987 broadcast a recorded statement made by Richard Doty, a special agent with the U.S. Air Force Office of Special Investigations who then named himself "Falcon." Doty claimed that the object Cash and Landrum saw was a captured alien UFO that was being test-flown and had temporarily experienced some flight problems. Quite likely, Doty was dispensing misinformation. A more plausible explanation is that Cash and Landrum encountered a prototype unmanned electrogravitic craft built for the military by an aerospace corporation. Possibly the craft was remotely controlled, and the helicopters were there to observe it and provide military security should the need arise. Had information about Project Skyvault been made public, along with the existence of black projects in microwave phase-conjugate propulsion, perhaps Cash and Landrum would have won their suit.

The severe effects that Cash, Lundrum, and Colby sustained in their encounter suggest they were exposed to a very intense microwave beam. This could have occurred if the vehicle's microwave propulsion beam was confined to a narrow angle and had mistakenly "locked on" to the observers and their car. They would have then been exposed to its full intensity. An incident similar to the Cash-Landrum encounter occurred in the late 1980s in the vicinity of Fort Hood, which lies about sixty miles north of Austin, Texas. A woman and her daughters, who had been observing a glowing, hovering object, became badly burned and suffered serious health effects. The victims subsequently sued the military for damages.

If the propulsion beam from these craft was being properly controlled to fan out to a wide enough area on the ground so that its radiation level per unit area was at a safe level, then a brief exposure would not be hazardous. Even so, a pilot should not fly a beam propulsion vehicle in a populated area so that accidents of this nature are avoided. If anything, these casualties of microwave exposure should be a warning to hobbyists that they are taking a serious health risk when they experiment with high-intensity microwave beams in the kilowatt range.

9.3 ■ TRIANGULAR CRAFT

During the late 1980s, there were numerous sightings of hovering vehicles that resembled the B-2 bomber. For example, in 1987, a year prior to the B-2's unveiling, hundreds of people living in Wythe County, Virginia, claimed that on several occasions they had seen a triangular-shaped black craft hovering in the night sky, mostly between the hours of eight and ten. Many who had seen the craft concluded that there was military involvement. One resident said that low-flying jets flew until 7 p.m. and then the "saucers" took over. Another witness had observed these objects flying a certain pattern at night and said that at the first crack of dawn, helicopters would fly the same pattern. Observers said the strange flat craft made no noise and in some cases hovered motionless in the air.

Danny Gordon, news director of radio station WYVE in Wytheville, Virginia, was one witness to the phenomenon. Regarding the similarity to the B-2 bomber, he stated, "Unequivocally, undoubtedly it was the same aircraft. I saw it, a lady in Fort Chiswell saw it . . . the same aircraft, flat-wing V-shape. This is not all the UFOs we've seen here, but this is one type, and I believe it's connected."[8]

Gordon said that several nights earlier he had paced a similar aircraft while driving his car at a speed of twenty-five miles per hour. He said he did not know how anything that big could travel at such a slow speed and not fall out of the sky. He concluded that the stealth bomber was being tested in their area by the Air Force and that other aircraft also observed might be part of the experiment. While Air Force authorities have acknowledged that the B-2 is a relatively slow-moving aircraft, such sightings lead us to believe that it also has the ability to hover totally motionless. If so, then this may have been accomplished along the lines suggested in chapter 5, that is, by applying high-voltage

AC across ceramic dielectrics oriented vertically at spaced intervals within its wing. The propulsion method would have been similar to that used in Brown's vertical-lift electrokinetic apparatus, and as concluded earlier, this type of AC field propulsion would have generated soliton beams beneath the craft.

These sightings might not all be B-2 craft. U.S. Air Force officials acknowledge that diamond- and triangular-shaped vehicles are "the trend now."[9] According to *Aviation Week*, one of these high-altitude military craft has earned the name Pulser because it is seen as a single bright light that sometimes pulses. The craft emits no engine noise or sonic boom, yet it has been seen crossing the night sky at extremely high velocities, exceeding the speed of conventional aircraft. Speaking about aircraft under development in U.S. black projects, the magazine reported in 1990:

In addition, there is substantial evidence that another family of craft exists that relies on exotic propulsion and aerodynamic schemes not fully understood at this time . . . Over the past 13 months, large, triangular wing-shaped aircraft characterized by a relatively quiet propulsion system have been the object of at least 11 sightings near Edwards Air Force Base, Calif., and one near Fresno, Calif. These are supported by additional reports of similar vehicles seen and heard around remote central Nevada communities near government ranges operated by the Energy Dept. and the Air Force.

Possibly prototypes or concept demonstrators of the Air Force B-2 or Navy A-12, the fairly flat, triangular-shaped vehicles have a rounded nose, rounded wingtips and probably no vertical tail surfaces. The flying wings' trailing edges may be slightly curved, but definitely are not sawtooth-shaped like those of the Air Force's B-2 bomber, according to reports received so far. One observer in Nevada described the shape as "like a manta ray."[10]

Very large aircraft whose shape fits this general description have been seen in the Hudson Valley region, thirty to sixty miles northeast of New York City.[11] Beginning in the spring of 1983 and continuing for a period of several years, tens of thousands of people on various occasions saw an immense craft described as looking like a boomerang-shaped flying wing with a rounded prow and measuring about 300 feet from wingtip to wingtip. It was usually seen flying at night with lights along its leading wing edge and at various locations beneath its body. These would periodically turn off or sometimes change in color. The craft was often seen hovering noiselessly or moving very slowly, about twenty to forty miles per hour, but occasionally it would accelerate to enormous speeds, disappearing to a point on the horizon in the blink of an eye. Either this was an exotic craft that the United States was secretly developing or one must presume it was an alien vessel. Clearly, to be able to hover noiselessly and undergo such enormous accelerations, the craft does not use a conventional means of propulsion.

Triangular-shaped craft have been sited hovering over various parts of Belgium on numerous occasions since 1989, with witnesses also numbering into the tens of thousands. Eyewitness accounts and photographs suggest the shape shown in figure 9.2. On top, the craft have a dome fitted with several windows. Viewed from beneath, they have bright white circular regions at each corner and a single red light near their center. Could these corner "lights" be luminous emissions from microwave horns that are part of a microwave phase-conjugation resonator system projecting down beams that support and propel the craft?

The craft were observed to hover, sometimes to move slowly horizontally, and at other times to accelerate vertically or horizontally to great velocities. On one occasion, after one of the craft was detected by radar, the Belgian government scrambled two F-16 fighters, but they were unable to apprehend it. The craft exhibited erratic changes

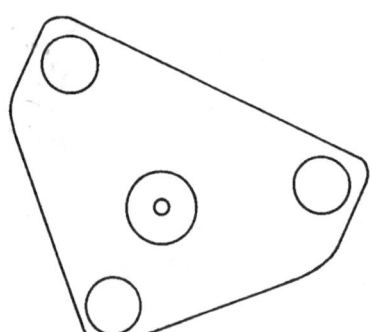

Figure 9.2. Drawing of a spacecraft seen over Belgium as viewed from below.

in direction and very rapid linear accelerations. In one case, a craft changed its altitude by 9,500 feet in just five seconds, an acceleration of more than 24 g. This would have been fatal to any pilot if it had been a vehicle operating on the conventional jet-thrust principle. Although the spacecraft attained a velocity of nearly twice the speed of sound, no sonic boom was heard. Quite possibly, this was a military test of a beam propulsion vehicle of the kind developed in Project Skyvault.

9.4 ■ CRASH RECOVERY OPERATIONS

The early research efforts leading to Project Skyvault began not long after July 1947, when an extraterrestrial spacecraft is reputed to have crashed near Roswell, New Mexico. Shortly after the crash, the site was secured by a top-secret military task force, and the scattered wreckage, including the vehicle's occupants, was subsequently removed to U.S. military laboratories for analysis. Similar operations were conducted in subsequent years to recover other downed alien vehicles. Although the military has made a concerted effort to keep knowledge of these incidents from the public, much information has since come to light as a result of research by dedicated investigators. This has been summarized in books such as *Behind the Flying Saucers, The Roswell Incident, UFO Crash at Aztec, Above Top Secret, Alien Contact,* and *The Truth About the UFO Crash at Roswell.*[12-17] These recovery operations are said to have resulted in an accelerated program to develop antigravity propulsion technology under projects code-named Y and Redlight. Since Project Skyvault and other advanced propulsion technology development programs were initiated around this time, it should be helpful to review something about these UFO crashes.

The first incident is believed to have taken place on the night of July 4, 1947, when a wedge-shaped spacecraft measuring about 15 by 25 feet crashed about thirty-five miles north of Roswell, New Mexico. Some say that two crashes actually took place at the same time but separated by some hundreds of miles. A few months later, in October, a 36-foot-diameter, dome-shaped craft is said to have crashed in Paradise Valley, Arizona. Then in March 1948, a 100-foot-diameter disc reportedly crashed in Aztec, New Mexico, and in July of that same year, a

90-foot-diameter disc is said to have come down near the Sabinas River in Mexico, thirty miles southwest of Laredo, Texas. In *UFO Crash at Aztec,* UFO researchers William S. Steinman and Wendelle C. Stevens estimate that as many as sixteen UFO craft may have crashed at various locations around the world between 1947 and 1986.[18]

Although the U.S. government has officially maintained that UFOs do not exist, a secret memo dated September 23, 1947, indicates that the military was taking this matter very seriously. It was written by General Nathan Twining, commander of the Army Air Force's Air Materiel Command at Wright Field, to the Air Technical Intelligence Command in Dayton, Ohio. Excerpts from this memo, printed in *Aviation Week & Space Technology,* read as follows:

1. As requested by AC/AS-2 there is presented below the considered opinion of this Command concerning the so-called "Flying Discs" . . .
2. It is the opinion that:
    a. The phenomenon reported is something real and not visionary or fictitious.
    b. There are objects probably approximately the shape of a disc, of such appreciable size as to appear to be as large as man-made aircraft.
    c. There is a possibility that some of the incidents may be caused by natural phenomenon, such as meteors.
    d. The reported operating characteristics such as extreme rates of climb, maneuverability (particularly in roll), and action which must be considered evasive when sighted or contacted by friendly aircraft and radar, lend belief to the possibility that some of the objects are controlled either manually, automatically or remotely.[19]

Also, a twenty-six-page classified report issued in 1948 by the Air Technical Intelligence Command stated:

It must be accepted that some type of flying objects have been observed, although their identification and origin are not discernible. In the interests of national defense it would be unwise to overlook the possibility that some of these objects may be of foreign origin . . . if it is firmly indicated that there is no domestic explanation, the objects are a threat and warrant more active efforts of identification and interception.[20]

Additional confirmation about the saucer crashes and the government's secret R&D involvement in the matter has come from a conversation that Wilbert B. Smith had in September 1950 with electrical engineer Dr. Robert Sarbacher, who was then serving as a consultant to the Research and Development Board, headed by Dr. Vannevar Bush. Smith, who was a senior radio engineer with the Canadian Department of Transportation, had read the account in Scully's book about the Aztec, New Mexico, saucer crash and secret government retrieval operation and wanted to verify if there was any truth to it, so he contacted Sarbacher through the Canadian embassy in Washington. According to Smith's handwritten notes, their conversation went as follows:

> Smith: . . . I have read Scully's book on the saucers and would like to know how much of it is true.
> Sarbacher: The facts reported in the book are substantially correct.
> Smith: Then the saucers do exist?
> Sarbacher: Yes: they exist.
> Smith: Do they operate as Scully suggests on magnetic principles?
> Sarbacher: We have not been able to duplicate their performance.
> Smith: Do they come from some other planet?
> Sarbacher: All we know is, we didn't make them, and it's pretty certain they didn't originate on earth.
> Smith: I understand the whole subject is classified.
> Sarbacher: Yes, it is classified two points higher even than the H-bomb. In fact, it is the most highly classified subject in the US government at the present time.
> Smith: May I ask the reason for the classification?
> Sarbacher: You may ask, but I can't tell you.[21]

Note that Sarbacher's comment about the classification level was made two years before the H-bomb had been detonated. On November 21, 1950, Smith sent an intra-departmental memo to the Controller of Telecommunications of the Canadian Department of Transportation that summarized some of what he had learned from Sarbacher. The memo, which is reproduced in appendix F, was marked TOP SECRET but was downgraded to "confidential" in September 15, 1969. It stated:

a. The matter is the most highly classified subject in the United States government, rating two points higher than the H-bomb.
b. Flying saucers exist.
c. Their modus operandi is unknown but a concentrated effort is being made by a small group headed by Doctor Vannevar Bush.
d. The entire matter is considered by the United States authorities to be of tremendous significance.[22]

In a response to inquiries made in 1983 by UFO researcher William Steinman, Sarbacher wrote a letter confirming that the U.S. government had recovered crashed flying saucers, along with the bodies of their occupants.[23]

The top-secret administrative group that Sarbacher said was headed by Bush was later discovered to have the code name MJ-12, or Majestic 12. This group of twelve individuals is said to have been formed September 24, 1947, under the authority of President Harry S Truman for the purpose of investigating UFOs, reporting the findings to the president, forming policies on the basis of those findings, and implementing policies that had received presidential approval. A photocopy of the memorandum Truman wrote to authorize its formation is displayed in appendix F.

Programs to analyze the crashed saucers and to attempt to duplicate their technology came under the direction of the Research and Development Board, which, in turn, reported directly to MJ-12. The Research and Development Board, which Sarbacher was consulting for in 1950, was organized by Bush in 1947 at the time that MJ-12 was formed. Under Bush's direction, this board headed up the R&D organizations of three branches of the military: Army Research and Development, Air Force Research and Development, and the Office of Naval Research. Brown's electrogravitics technology, which was evaluated by the Office of Naval Research in 1952, was probably closely scrutinized by this overseeing organization. Bush had previously headed the Office of Scientific Research and Development, which administered the Manhattan Project and other top-secret wartime efforts, such as the development of radar and the proximity fuse.

On November 18, 1952, two weeks after his election, President-elect Dwight D. Eisenhower is said to have been briefed on MJ-12 and the crashed saucer retrieval operations. In 1984, television producer Jaime Shandera and UFOlogist William Moore obtained a document through intelligence contacts that they believed to be the top-secret "eyes only" document used in that briefing. Although some doubt whether the document is in fact genuine,[24-26] a *Washington Post* article does confirm that President-elect Eisenhower had received a military briefing on November 18, 1952, the same date stated on the MJ-12 briefing.[27] Moreover, the existence of a subsequent briefing with MJ-12 is corroborated by a memo that Moore and UFO researcher and physicist Dr. Stanton Friedman obtained from the National Archives through a Freedom of Information Act request. The memo, dated July 14, 1954, was written by Robert Cutler, special assistant to President Eisenhower, and sent to General Nathan Twining, one of the individuals claimed to belong to MJ-12. The memo, which concerned the NSC/MJ-12 Special Studies Project, states, "The President has decided that the MJ-12/SSP briefing should take place during the already scheduled White House meeting of July 16 rather than following it as previously intended."[28] The NSC designation refers to the presidential office's National Security Council, which was also created in 1947. MJ-12 is said to operate under the NSC as an unacknowledged subcommittee called the Special Studies Group, with a current membership of thirty-three.

9.5 ▪ ART'S PARTS REVERSE ENGINEERED

The *Coast to Coast* radio talk show, and in particular Art Bell, who served as its host for many years, is well known to many. The show's favorite topics have been UFOs and alien encounters. In April 1996, one of the show's listeners, a man who asked to remain anonymous, mailed to Bell a number of metallic artifacts that he said had been retrieved from the exterior of an alien spacecraft that crashed in 1947 between White Sands and Socorro, New Mexico.[29] He said that his grandfather had gathered the materials while he was a member of the military security team connected with the retrieval cleanup operation and had given them to him before he died in 1974.

The parts, which have come to be known as Art's Parts, were extensively discussed on *Coast to Coast* and pictures of them for some time had been posted on Bell's webpage. One of the two shipments of the alleged alien artifacts that were sent consisted of two irregularly shaped pieces of metal measuring approximately 6 by 3 centimeters and 5 by 2 centimeters, respectively, both having a thickness of 3 to 4 millimeters. These were alleged to have been taken from the exterior underside of the craft and were believed to have formed a shell-like shield of sorts.

In the following months, the fragments were analyzed using a scanning electron microscope outfitted for energy-dispersive spectroscopy.[30] The results for the two hull fragments were quite interesting. Analysis showed that they consisted of twenty-five well-defined layers alternating between a thick layer of magnesium-zinc (97 to 97.5 percent magnesium and 2 to 3 percent zinc) and a thin layer of pure bismuth. The metals were of exceptional purity. The magnesium-zinc layers ranged in thickness from 100 to 200 microns (0.1 to 0.2 mm), and the bismuth layers ranged in thickness from 1 to 4 microns. When examined in cross-section, it was apparent that the layer interfaces were not even but rather contained microscopic undulations.

Some researchers found it unusual that the material would jump around when exposed to the high-voltage field of a Van de Graaff generator. However, American research technologist Nicholas Reiter, who conducted a similar test, says there is nothing unusual about this since any metal fragment would similarly dance around in a 200-kilovolt AC field.[31] To check for any electrogravitic force effects, he exposed the artifacts to DC voltage potentials in the range of 15 to 50 kilovolts, but observed no weight change as measured with a laboratory digital milligram balance. Thus, contrary to widely publicized claims, there is no evidence to date that the fragments might lose weight when subjected to high voltage potentials.

Linda Moulton Howe, an American investigative journalist and documentary producer who was investigating the nature of the fragments, interviewed a large number of metallurgic experts from various companies and scientific institutes, including aerospace and defense companies. None had heard of such a material, and they didn't understand its purpose. Howe also wrote letters to various agencies like the National Science Foundation to get information about the material. A foundation scientist working in the Division of Materials Research said that he was unaware of any research into such a material. A computer search of the foundation's database on materials consisting of bismuth, magnesium, and zinc turned up nothing.

However, insights into the nature of this bismuth-layered material may be gained by considering recent investigations into negative index of refraction materials. In 2005, Professors Victor Podolskiy and Evgenii Narimanov and graduate student Leonid Alekseyev, working at Oregon State University and Princeton University, respectively, announced their discovery that a thin layer of monocrystalline bismuth exhibited a negative index of refraction at microwave frequencies, making it the only known, naturally occurring substance to exhibit such a property.[32-34] They sandwiched a 4.5-micron-thick layer of monocrystalline bismuth between two metal plates, as shown in figure 9.3. In this arrangement, the semimetallic bismuth acts as a dielectric and the flanking metal layers act as waveguide walls. When a 5,000-gigahertz microwave beam (60-micron wavelength) was directed into the bismuth layer, the beam was found to refract negatively. That is, they found that over a narrow band of wavelengths, ranging from about 53 to 63 microns, the bismuth exhibited a negative index of refraction.

Bismuth achieves negative refraction in a manner very different from that of the metamaterials described in chapter 7. Recall that such materials exhibited negative refraction because they had magnetic and

Figure 9.3. Waveguide made of monocrystalline bismuth sandwiched between two metal layers and used to demonstrate negative refraction of a 5,000-gigahertz microwave beam.

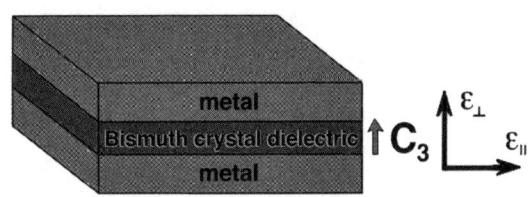

electric resonances near the same frequency, creating a frequency range over which their permittivity, ε, and permeability, μ, would simultaneously attain negative values. With ε and μ simultaneously negative, the refractive index would also be simultaneously negative. However, bismuth is nonmagnetic and, hence, has no magnetic resonances. However, theory shows that a dielectric can exhibit a negative refraction index if the material has a permittivity anisotropy, that is, different values of permittivity for different wave propagation directions relative to the dielectric crystal axis, and if permittivity in one of these directions becomes negative over a specific frequency range while the permittivity in the other directions remains positive. Bismuth has such a property (see text box).

Why Bismuth Exhibits a Negative Index of Refraction

Although its population of free electrons is much smaller than that of most metals, bismuth has what is termed an *electron mass anisotropy*, in which the effective mass of its free electrons is lower parallel to its trigonal axis as opposed to perpendicular to its axis. Since these free electrons behave as a plasma, which has a specific resonant frequency, this anisotropy causes the plasma frequency to be lower in the direction parallel to the bismuth layer, as compared with the plasma frequency for oscillations perpendicular to the plane of this layer (parallel to the trigonal axis; i.e., $f_\parallel < f_{perp}$). Since the dielectric constant for bismuth is determined both by the value of its electron plasma frequency and the frequency of the exciting beam, this differing plasma frequency causes the permittivity parallel to the bismuth layer ($\varepsilon_\parallel$) to be more negative than the permittivity perpendicular to the bismuth layer (ε_{perp}). Consequently, when the bismuth is excited at frequencies between these two plasma frequencies, the permittivity in the direction parallel to the layer will be negative when the permittivity perpendicular to the layer is still positive (i.e., $\varepsilon_\parallel < 0$, $\varepsilon_{perp} > 0$), which provides the necessary condition for the index of refraction to be negative (see figure 9.4).

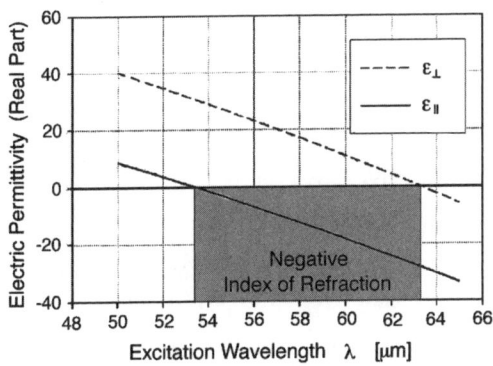

Figure 9.4. Electric permittivity (real part) plotted as a function of excitation wavelength. The solid line charts the component parallel to the bismuth layer and the dashed line charts the component perpendicular to this layer. Negative index of refraction is exhibited in the range in which $\varepsilon_{\parallel} < 0$ and $\varepsilon_{perp} > 0$.

Interestingly, the "hull fragments" among Art's Parts consisted of bismuth layers having a thickness range (1–4 microns) only slightly thinner than that tested by the Oregon State–Princeton team. One might then guess that the magnesium-zinc layers in the Art's Parts fragments had the function of acting as metal waveguide walls around the bismuth layers. This would ensure that microwaves propagating within the bismuth layers would be confined to those layers. Magnesium is a relatively good electric conductor, so it would serve as a good metal to use for a waveguide wall. It also has the advantage that it is lighter and stronger than aluminum. Negative refraction should characterize bismuth layers that are even as thin as 1 micron. Although the layer thickness is a factor in determining the exact value of the permittivity, it does not affect the values of the electron plasma frequencies along the two bismuth crystal axes. Thus, the bismuth-layered samples in the Art's Parts collection would also be expected to exhibit negative refraction of a 5,000-gigahertz beam.

As mentioned in chapter 7, metamaterials having a negative index of refraction are also capable of developing a strong repulsive force when exposed to a microwave beam. The same may be true of bismuth films, so it is possible that the layered material would develop a thrust when excited with 5,000-gigahertz microwave radiation. This could be easily checked in a laboratory. An Art's Parts hull fragment could be mounted on a pendulum or on a torsion balance and exposed to a high-power terahertz beam. A 100-watt free-electron laser that has been tuned to a 5,000-gigahertz frequency might serve as an adequate beam source. If a thrust effect was found, it would be the first time that such an effect has been discovered, since to date no research group has considered looking for such a beam thrust effect in layered bismuth. A positive result could help validate the claim that the Art's Parts layered-metal fragments were part of a shield that once covered the underside of a spacecraft and may have been a lofting material that was part of the craft's exotic electro-gravitic field propulsion system.*

One would expect a thrust response to occur in a direction perpendicular to the bismuth C_3 trigonal axis. In the case of the sample tested by the Oregon State–Princeton team, such a thrust would occur parallel to the layer plane. If the intention was instead to produce a force perpendicular to the layer plane, that is, perpendicular to the spacecraft hull, the C_3 trigonal axis would need to be oriented in the plane of the bismuth layer instead of perpendicular to its plane. It would be interesting to discover how the trigonal crystal axis is oriented in the Art's Parts hull artifacts.

9.6 ▪ PROJECT REDLIGHT

In the years immediately following the first saucer recoveries, MJ-12 ran a super-secret investigation program that concentrated primarily on analysis of the saucers, with the hope of learning something about their power source, mode of propulsion, instrumentation, and weaponry. In parallel with this effort, autopsies were performed on the saucer's occupants to learn something about their physiology. Information obtained from reliable eyewitnesses indicates that the recovered discs and occupants have been stored and analyzed at a number of secret facilities that

---

*In August 2006, I contacted members of the Oregon State–Princeton team to see if they knew of anyone who had observed a microwave beam repulsion phenomenon in layered bismuth. They said they did not know of anyone who had investigated this, but indicated interest in my suggestion that such an effect might be present. They requested I send anything I may have published on this. However, after I outlined to them how an experiment might be easily carried out and suggested mutual collaboration on the project, they mysteriously broke off contact. My subsequent e-mails to them went unanswered.

include Wright-Patterson Air Force Base, in Dayton, Ohio; Kirtland Air Force Base and the Sandia Laboratory complex in Albuquerque, New Mexico; and a highly restricted area on the Atomic Energy Commission (AEC) Reservation, in southern Nevada. Here again, we encounter the Wright-Patterson connection, although artifacts that were in storage there are believed to have been transferred to the Kirtland-Sandia complex when the Department of Special Studies at Wright Air Development Center was moved there in 1956.[35] Substantial information about a program to analyze and even test-fly some captured saucers is found in the "above top secret" Report No. 13 of Project Blue Book, Blue Book being a U.S. Air Force project established to document and analyze UFO sightings. Steinman and Stevens, who summarize this report in *UFO Crash at Aztec,* received this information from a witness who inadvertently reviewed the document in 1977 while working as an information analyst at a highly secure Royal Air Force/U.S. Air Force radio espionage facility in Chicsands, England.[36] The front cover of this 624-page bound document was dated 1953 with a 1963 date in parentheses, indicating that it was later updated with penciled annotations. A length of red tape indicating code-red security measures was stretched diagonally across its front from corner to corner, and the cover was stamped in red ink, TOP SECRET—NEED TO KNOW ONLY—CRYPTO CLEARANCE 14 REQUIRED. The top-secret compartmentalized clearance the report demanded was higher than that of the Blue Book management office staff, who were cleared only up to the secret level. This would explain why Project Blue Book itself has no record of Report No. 13, even though its inventory includes status reports numbered 12 and 14.

According to the Air Force analyst, Report No. 13 reviewed the U.S. government's official procedures concerning downed UFOs and UFO close encounters. It also summarized what the Air Force knew about crashed discs, their power systems, and their weaponry, and it included photographs of alien craft, crash debris, and the bodies of some of the craft's occupants. Also of interest, the report described a project called Redlight, whose purpose was to test the propulsion and weapons systems of retrieved saucers and to examine various pieces of hardware recovered from the crafts. This operation was carried out in the highly restricted one-hundred-square-mile UFO research facility located in

Nevada in the north-central part of the three-thousand-square-mile AEC Reservation. The facility was said to harbor at least three alien saucers. One was dismantled, and the other two were in good enough shape that they could be made flyable, although one of the two was said to have later exploded in flight with two U.S. pilots aboard.

Based on information given in Report No. 13 and from eyewitness accounts, the following is known about the highly secret Nevada facility.[37] Variously known as Area 51 or Dreamland, it is situated in the Groom Lake area northwest of Las Vegas. It is the most heavily secured area in the United States. It lies inside the guarded perimeter of the existing AEC nuclear-testing reservation and Air Force weapons practice range. This dry lake site is screened on all sides by a mountain range, and this is ringed with electronic detectors, including infrared, motion, and ammonia detectors, which are so-called people sniffers. The area itself is surrounded by three additional defense perimeters. Security teams in helicopters and planes are on twenty-four-hour alert to respond to any intrusion.

The site was originally a Navy air field installation that was being used as a nuclear weapons storage base. In 1951, the base was put on alert, and all personnel were evacuated except for the medical personnel, who were restricted to the hospital facility. The Navy then brought in a Seabee construction battalion and, over a six- to eight-month period, dismantled the base, built underground work facilities, and surmounted them with large aboveground hangars. At the end of 1951, after the work was completed, the Seabees moved out and the Project Redlight personnel moved in. Their ranks grew to eight hundred to one thousand personnel permanently on duty and all living on the site. A large but undetermined number of top scientists having very high security clearances were reported to come and go from this maximum security area. Some had been formerly associated with the Manhattan Project.

Nevada residents living in the vicinity of Area 51 have seen disc-shaped craft being tested there from the 1950s to the present. In their book, Steinman and Stevens described several cases in which hovering, disc-shaped aircraft were seen to be test-flown in the Area 51 vicinity.[38] One story concerns a Navaho Indian who was backpacking in a canyon that ran down into the AEC Reservation area (date unknown).

After having camped the previous night in the canyon, he had awoken and had just finished preparing his pack for the hike ahead when a helicopter approached. It was broadcasting warnings from a loudspeaker cautioning anybody in the canyon area to make his presence known so that he could be moved to safety and explaining that a military test was scheduled to be conducted that would be very dangerous. The helicopter returned fifteen minutes later broadcasting the same message. Feeling safe among the rocks, the Indian remained hidden and waited to see what would happen. About a half hour later, two helicopters came into view flying up the canyon about 500 feet apart. They escorted between them a dark gray, metallic, disc-shaped craft that had a raised dome at its center. Ten minutes after the three had passed, the two helicopters flew back the way they had come, but without the saucer. The saucer appeared some time later as it flew very fast and silently down the middle of the canyon, retracing its original path of entry.

Another story concerns an Air Force fighter pilot who was part of the Tactical Air Command Combat Squadron and had been taking part in a "red flag" war game exercise that was being conducted in an area adjacent to the AEC Reservation. The pilot accidentally flew his jet across a corner of the reservation and happened to pass just north of the Area 51 region. At that time, he saw below him to the south a 60-foot-diameter, circular, disc-shaped craft in flight. At that moment, he was hailed on the open emergency channel of his radio, told to abandon his mission, and ordered to fly directly to Nellis Air Force Base, where he was told to land. Once on the ground, he was taken into custody and escorted to a security office for interrogation. He was released two days later, only after pretending to be convinced that the disc-shaped object he had seen was merely a water tower.

Yet another story concerns a man who during the early 1960s performed top-secret radio work for the Air Force at Area 51. He reported seeing one unconventional aircraft that was being flight-tested there under Project Redlight. The craft was 20 to 30 feet in diameter and pewter colored. He didn't see the craft in operation since at those times he was brought indoors for security reasons. However, he did note that, unlike conventional craft, it made no engine sound when it took off or landed.

Aircraft of nonconventional design such as the stealth bomber, stealth fighter, and Navy A-12 fighter have all been air-tested at Area 51. So, quite possibly, there is a close relation between Project Redlight and the development of such advanced aircraft. This area also may have served as the test site for the discs developed in Project Skyvault.

Another sighting of a very large, 200-foot-diameter craft was reported by Frank Batts of Santa Barbara, California. Writer George Balanus summarized Batts's story as follows.[39] On the evening of April 30, 1997, Batts and his friend Joe had set out on an expedition to Area 51 in the hope of viewing some of the unusual hypersonic vehicles often seen in that vicinity by UFO watchers. About 10:20 p.m., they had set off to find the landmark known as "the black mailbox," which is located along Highway 375 about twenty miles southwest of Rachel, Nevada. The area where most viewing enthusiasts camped out was about eight miles from there, away from the main highway, but Batts and Joe lost their way and did not find the black mailbox. Instead, they ended up at a spot on the north side of the Area 51 range, opposite from where UFO buffs normally congregated. This northern spot was known as the "back door" of the installation. It lay much closer, about seven miles from the edge of the Area 51 facility. Flight testing had been conducted in that area in the past, but locals reported that tests had supposedly ceased there for some time.

Batts and Joe had parked on the shoulder of the road and were facing out into the desert. After about an hour, they saw a blue ball appear above the mountainside, hover, and then dance about for about two minutes before disappearing below the ridge. This sounds very much like a plasma ball test that was sighted in 1993 in the Superstition Mountains twenty-five miles east of Phoenix.[40]

After the blue ball had vanished, they noted red, yellow, white, and blue lights still glowing out in the desert 175 to 200 yards from their location. Sometimes the red lights flashed and sometimes they stayed constant like the other lights. The two men thought they were observing a building at the base, but after about an hour and a half, what they thought was a building suddenly lifted slowly off the ground and hovered. At that point, they realized that what they were seeing was not a building. White light reflecting off the desert floor illuminated the underside of the craft, revealing it to be a very large silver saucer. They estimated that the disc was about 200 feet across. It had curved upper and lower hulls with a bank of red and orange lights around its middle. For about an hour and fifteen minutes, they watched it maneuver from side to side and up and down. Eventually, it receded and finally disappeared over a distant mountain range.

During the sighting, Batts tried to operate his camera, but it wouldn't work. They also tried the car radio but got only a high-pitched whining sound, whereas before and after the sighting they were able to tune in a radio station.

9.7 ■ THE SPORT MODEL

Robert Lazar, a former employee of Los Alamos National Laboratory, claims that in December 1988, the Office of Naval Intelligence gave him a compartmentalized clearance thirty-four levels above a top-secret "Q" clearance and employed him at the highly secret "S-4" test facility located about fifteen miles south of Area 51. He says that he was hired to study the power source of a captured alien flying saucer and try to figure out how it functioned.[41,42] Four months later, having become disenchanted with his work and concerned that such important scientific discoveries were being kept secret from the American public, he broke his vows of secrecy and began describing his experiences to friends. He led them on night outings to remote spots near Area 51 to view some of these captured UFOs being test-flown. Later, he appeared on a local Las Vegas television news broadcast to relate his experiences and present some insights into the propulsion hardware on the craft he had been assigned to. Subsequently, he lectured at a number of UFO conferences and also put up a Web site on the subject. His description of the propulsion unit is of particular interest because it sounds in many ways similar to the microwave propulsion system developed in Project Skyvault.

Gene Huff, who has socialized with Lazar since the late 1980s and knew him during the period when he was hired to work at S-4, has written an interesting biography that corroborates many aspects of Lazar's story.[43] However, others have come to mistrust Lazar's claims, considering the large number of contradictions in his story as well as statements he has made that appear to indicate a substantial lack of knowledge of basic physics. Several of these critiques appear on the Internet website www.dreamlandresort.com/area51/lazar/index.html.[44] Nevertheless, the gravity wave propulsion beam technology that Lazar refers to comes sufficiently close to the field propulsion ideas reportedly developed in Project Skyvault, so it is worth summarizing his story, although, as will be pointed out, many of his claims appear to be disinformation that may have been planted to protect the technology's secrecy.

Lazar says that while working at the S-4 test site near the dry bed of Papoose Lake, he was shown a 52-foot-diameter spacecraft that he nicknamed the Sport Model (see figure 9.5). He says he was told that the craft was powered by an "antimatter reactor" located at its center. He claims that the reactor was designed to emit bursts of positrons 7.46 times per second, which, in turn, would generate bursts of type-A "gravitational" microwaves that he terms Gravity A waves. He says that these gravity waves would travel up the vertical conduit attached to the top of the reactor, where they would become amplified in intensity. This conduit, which is said to be about 8 centimeters in diameter, could act as a microwave waveguide and could serve as a microwave amplifier, just as Lazar claims, provided that its length was properly matched to the microwave wavelength. However, from Lazar's description, it is not entirely clear whether he believes these to be pure gravity waves or electromagnetic waves that have gravitational effects. Indeed, a waveguide would be unable to contain a pure gravity wave of the sort commonly known to physics. Such waves should freely pass through waveguide walls without reflecting from them, much like Podkletnov's gravity impulse beam did. If the microwave emissions from the Sport Model's reactor are able to be contained by a waveguide, then they cannot be considered exclusively gravitational.

In fact, at one UFO seminar in 1993, Lazar disclosed his belief that gravity is electromagnetic in nature but that it is an electromagnetic wave of a particular microwave frequency, which he did not wish to disclose at the time. Yet, in my opinion, it is a major error to assume that gravity is electromagnetic in nature or to suggest that the electric or magnetic field itself produces gravitational effects. It would instead be more reasonable to postulate that electric and gravity potential

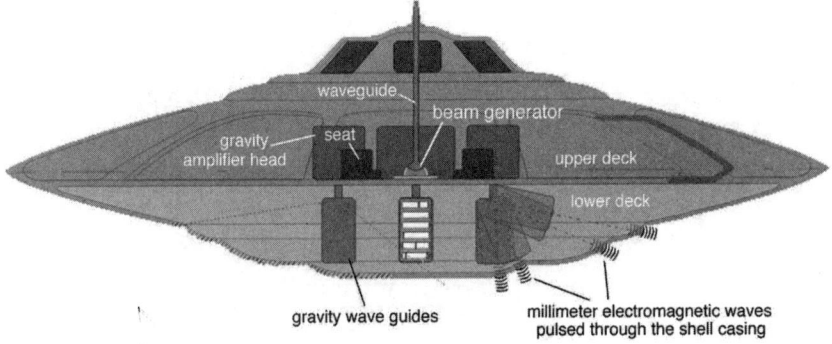

Figure 9.5. A cross-sectional view of the Sport Model, a flying disc of alien origin. Based on Robert Lazar's description. (After P. Potter)

fields are coupled and that electromagnetic waves and electric shock discharges are accompanied by a distinct gravity wave component. To refer to the craft's microwave emissions as gravity waves per se and to claim that such gravitational effects manifest only at a specific frequency, in my view is rather outlandish. Recall that in chapter 7 we learned that Project Skyvault scientists found that when microwave beams were tuned to specific frequencies, they were able to strongly interact with matter and produce strong electrogravitic repelling forces. Also, we learned that metamaterials that have resonant frequencies in the microwave range exhibit strong repulsion forces when beams are tuned to frequencies slightly above these resonances. Thus, frequency is critical to obtaining a maximal repelling force from a microwave beam, but not the way Lazar seems to imply.*

So if we discount Lazar's contention that the so-called reactor is generating gravity waves (his Gravity A waves)—and we will discover that there is good reason to ignore such an assertion—we are then left with the idea that this is essentially a microwave generator, hence, the equivalent of the Gunn diode oscillator cavity on the Project Skyvault vehicle. Lazar did not describe the inside of the vertical waveguide, but

---

*Lazar suggested that one could infer the microwave frequency from the dimensions of the waveguide tube. Theorizing that its inside diameter is somewhat less than its 8-centimeter outside diameter, say 5 centimeters, and that a full wavelength fits across this dimension, this would imply a frequency of about 6 gigahertz (or 3 gigahertz for a half-wavelength fit).

if it were to have a series of properly spaced ringlike cavities along its length, it could function as a klystron, that is, a linear microwave generator. As charged particles (positrons or electrons) would shoot along the length of the waveguide and move past the cavities, they would set up microwave oscillations in the cavities that would become progressively more intense as the end of the tube was approached.

However, one has great difficulty believing Lazar's story that this gravity wave generator is powered by bursts of high-energy positrons emitted from the radioactive decay of a slab of exotic metal located at the bottom of the reactor. For instance, in one radio interview, he described having been present when this reactor was being bench-tested with its waveguide tube removed and said that he had been allowed to put his hand over the top of it. If this had been a real high-energy positron beam, no reasonable physicist or engineer would have let him put his hand over it to feel its matter-repelling force field. If Lazar had done so, his hand would have received a severe radiation burn and a dangerously high dose of radiation. Clearly, this would have upset the health and safety people at the laboratory, if the story was true. Also, when positrons annihilate, they produce 1-million-electron-volt gamma rays. Lazar made no mention of such energetic by-products or precautions taken to shield them. More likely, what was coming out of the reactor was a microwave beam emitted by a crystal oscillator, without any accompanying beam of high-energy particles. But even if it had been a microwave beam, the beam power would have had to be sufficiently low so as not to harm him.

The other part of this reactor story that raises doubts is Lazar's claim that this slab of exotic stable metal was composed of element 115 and that these matter-repelling waves were emitted when this element was bombarded with protons, inducing it to transmute into element 116, which then immediately decayed by emitting a positron. He says element 115 is a stable element that does not exist on Earth but is generated in the cores of massive stars many light-years from Earth and that the only available supply of this material on Earth is held under tight security in a supersecret facility in Area 51. He says that only discharges of this normally inaccessible material are able to generate his so-called Gravity A waves used for propelling the disc.

However, in 2004, about fifteen years after Lazar began making claims about this rare element, Russian scientists successfully synthesized small amounts of elements 113 and 115, and four years earlier had created element 116. The problem is that both isotopes of element 115 that they were able to synthesize were unstable. They lasted only for several tens of milliseconds, and they decayed not into element 116 as Lazar claimed, but into element 113, with the emission of an alpha particle. Also, element 116, which Lazar claimed could be momentarily created by bombarding element 115 with a proton, was found to decay, but not by emitting a positron. It too decayed by emitting an alpha particle.[45] It seems, then, that Lazar's antimatter reactor idea is on the rocks, as is his idea for the existence of Gravity B waves, which he claims can be produced only through this 115-to-116 decay process.

This antimatter reactor part of the story was most likely a fabrication included to make the Sport Model field propulsion technology sound very exotic and, hence, not easily duplicable in most laboratories. If we are willing to accept that Lazar actually had worked at this secret Nevada test site, engineers there most probably misinformed him on purpose as to how this microwave oscillator functioned. In this way, the technical details of the disc's operation would be protected in case he decided to rebel and speak out, as he has done. Alternatively, some have suggested that agents may have employed brainwashing techniques to manipulate his mind.[46] Such black-project research facilities are likely in possession of technologies that would allow them to erase specific memories and replace them with false ones, similar to what is portrayed in the movies *Total Recall* and *The Bourne Identity*. Area 51 security agents may have felt that they were not breaking the law to use such methods since Lazar says that prior to his employment on the project he had to sign a secrecy agreement waiving his constitutional rights.

Conversations I have had with several people lead me to believe that such abhorrent brainwashing techniques are unfortunately being used. One case concerns an individual who claims to have worked on a highly classified time travel project supposedly conducted on Long Island, near Montauk, as an extension of research done on the highly secret Philadelphia Experiment. I had a chance to talk to him after he had just finished giving a lecture about his unusual experience. He believed that he had formerly been director of this time travel project and that, after having had some disagreements with people there, he was subjected to a mind-altering therapy that attempted to erase from his memory experiences he had on the project. He says that he later began to recall these memories after he had a chance street encounter with someone who he claims had once worked under him on that project. He said this began to trigger the suppressed memories.

In his mind, this individual thoroughly believed that what he was saying was the truth, that he had directed this project and that it had been capable of temporarily transporting people into other time periods. However, it was obvious to me that what he was saying was a complete fabrication. I believe he was correct in stating that he had been the unwilling subject of mind control, but I believe that during his brainwashing session, false memories were implanted in his mind about the existence of this fictitious time travel project and how he had once served as its director. While under hypnotic inducement, he was most likely instructed that he would forget about the mind-control session, but that meeting a certain person in the street would be a trigger that would allow him to begin to remember the implanted memories as well as some of the mind-control session. However, the implanted memories were to be remembered as being events that had actually taken place. Also, he was instructed to believe that the purpose of the mind-control session had been to erase his recollection of those supposed real events (the implanted memories). Perhaps he also was given a subliminal suggestion to write about his implanted experience and to lecture about it. A person who actually believed in his heart that what he was saying was the truth would be the ultimate in misinformation dissemination. Could Robert Lazar have actually had some experiences working on the Sport Model and later undergone similar techniques to implant confusing ideas about the technology?

Lazar's diagram of the reactor depicts a conical structure at the bottom of the chamber that he identifies as the element 115 fuel source, which he says is made up of a stack of thin wafers. In view of the element 115 fiasco, much of what he has described about the reactor should be discounted. If there was such a microwave-emitting structure in the reactor chamber, it would be better construed as a solid-state oscillator similar to an IMPATT or Gunn diode. Its enclosing hemispherical chamber and capping waveguide tube would then form the microwave amplifier duct.

We might guess this crystal oscillator was made of a material dense enough that it could give an unsuspecting person the impression that it was in fact made of an element having an atomic weight of 115. For example, recall from chapter 4 the disclosure by black-project engineers of the development of very dense, radar-absorbing materials containing uranium. Lazar says he was not involved in measuring the atomic mass of the material he worked with but only correlated the data taken by others who reportedly worked on the project before him. These data, then, could have been "cooked" to give Lazar the impression that he was discovering something with an atomic weight of 115. The trick apparently worked, because he wholeheartedly believed them.

Lazar commented that when the reactor was bench-tested with its waveguide tube removed and he was allowed to place his hand over its mouth, he could feel the pressure of the field, which he described as being similar to the repulsion one feels when two like poles of a magnet are brought together. He said that they also played around with the repulsion field by bouncing golf balls off it. The force that he refers to sounds very much like what Tesla says he felt from the radiant energy shocks discharged from his magnifying transmitter. Podkletnov also says that he was able to feel the repulsive field generated by the momentary discharges of his gravity impulse beam. However, both Tesla and Podkletnov were feeling the repulsive force of sawtooth-shaped waves produced by electron shock discharges. Subquantum kinetics predicts that if these were positron shock discharges, they should instead have produced an attractive force. If anything, Lazar's hand should have been sucked into the reactor if it was actually emitting positron discharges. His positron pulse claim is, in my opinion, misinformation that he is perhaps unwittingly disseminating.

To continue the story, Lazar said the gravity wave (i.e., microwave emission) generated in the reactor was piped into three "gravity amplifiers" (i.e., microwave amplifiers) located in the lower compartment of this vehicle, each amplifier measuring 2 feet in diameter and 4 feet in length (figure 9.6). These were equally spaced from one another in a triad arrangement and could be swiveled to aim in any direction. He said the gravity wave (electromagnetic microwave) from the reactor was of too low an amplitude to be effective for propulsion and that it became amplified in the gravity amplifiers into waves powerful enough to propel the craft. Each of these gravity amplifiers would emit a microwave beam downward, which was used to buoy the craft upward. The craft would sit on these beams and tend to bounce around on them in a flight mode he referred to as the "omicron configuration." He said the disc would move forward by focusing one or more of the beams behind it, which would cause the craft to fall forward.

More specifically, Lazar said the gravity amplifiers achieve their lifting force by sending out a microwave beam, the Gravity A wave, toward the Earth's surface and by phase-shifting this wave relative to the microwave propagating up from the Earth, which he termed the "Gravity B wave." His description sounds a lot like that for a microwave phase-conjugate resonator, although described in very vague terms. In other words, his outgoing Gravity A wave would correspond to the outgoing phase-conjugate microwave beam and his incoming Gravity B wave would correspond to the incoming ground-reflected probe beam, which would actually consist of microwaves that had previously originated from the disc's microwave generator (its reactor). The energy in outgoing A wave would be locked in phase with the incoming ground-reflected B wave and would eventually retrace the B wave's scattering path back to the beam's ground target point and then back to the vehicle's central source oscillator. As in the Project Skyvault craft, these incoming and

Figure 9.6. One of the gravity amplifiers. Based on Robert Lazar's description. (After P. Potter)

outgoing beams would be focused by some sort of microwave lens.

Clearly, considering the controversy surrounding the veracity of Lazar's statements, it is difficult to pick out fact from fiction. All we can say is that many features that he describes bear a strong resemblance to the Project Skyvault technology. Moreover, we are compelled to accept that the Sport Model or something like it exists since many people claim to have seen from a great distance some sort of unusual levitating craft being test-flown in the vicinity of Area 51. Consequently, to understand how this craft might work, it makes more sense if we disregard the gravity wave mumbo jumbo and reframe Lazar's dialogue in terms of what is already known about microwave phase conjugation.

The gravity amplifier would be the equivalent of the mixer diode cavity in the Skyvault vehicle. Like the Skyvault mixer cavity, each such amplifier is reportedly energized by microwaves from the craft's central microwave source, that is, its reactor and waveguide resonator. Provided that each gravity amplifier contains a polarized dielectric medium, these piped-in microwaves would serve as pump beams, which would interact with the probe beam entering the amplifier (Lazar's gravity B wave) to generate a holographic grating pattern in the dielectric. The pump beams would then interact with the dielectric's grating pattern to produce an outgoing microwave beam that would be the phase conjugate of the incoming probe beam. In describing the amplifiers, Lazar makes no mention of any internal dielectric but does say they contain a series of plates. Perhaps these are dielectrics.

Lazar's gravity amplifiers, then, most likely function as phase-conjugate resonators that allow microwaves from the craft's central microwave source to self-amplify and create powerful soliton faser beams between the craft's mixer diodes and the ground. Provided that the microwaves consist of sawtoothlike shock discharge waveforms, as one may infer from Lazar's description, the soliton beams should create a repulsive force both on their ground surface target and on the craft, which would tend to buoy the craft upward.

Lazar makes a number of statements concerning the nature of gravity that seem to be nonsense and that some have had issues with.[47] For example, he contends that the two gravity waves the craft phases relative to each other to obtain its propulsion, the Gravity A and Gravity

B waves, are actually the results of two very different types of gravity. He identifies the Gravity A wave with the strong nuclear force, the very short-range force that binds protons and neutrons together in the atomic nucleus, and claims that only through the element 115-to-116 antimatter reaction is it possible to release this in the form of a traveling wave having a specific wavelength in the microwave frequency range. He identifies the Gravity B wave with the gravitational force of standard physics, that is, the field that causes celestial bodies to attract one another.

However, to suggest there are two types of gravity and to identify one of them with the strong force sounds absurd, even to the most open-minded of physicists. The idea has no basis in standard physics, nor is such a concept compatible with subquantum kinetics. In subquantum kinetics, nuclear binding arises because nucleons have wavelike electrostatic potential fields in their cores that interlock with one another when the particles are in close proximity, that is, close enough to form an atomic nucleus. This subquantum kinetics model of the strong force has been confirmed by particle-scattering experiments.[48] Lazar, however, offers no experimental evidence to support his odd theory of gravity except vague references to the operation of a flying saucer kept in the supersecret S-4 test facility. Considering that one key aspect of his theory has now been disproved, his claim that the Gravity A wave is produced by the spontaneous positron decay of element 116, we may conclude that his gravity wave theory should not be taken seriously or, at least, we should regard it as disinformation that was purposely disseminated by black-project security staff. Along these same lines, we may safely disregard Lazar's statements that the Gravity A wave entering the craft's gravity amplifiers bends space around the disc as it becomes amplified.

Lazar claims that gravity, as observed in nature, which he refers to as the Gravity B field, is in essence electromagnetic and that it specifically involves oscillations in the microwave range. This again is nonsense, apparently interjected to create confusion. The Earth's gravitational field does not oscillate at microwave frequencies. If it did, this would have been widely known to the physics community, since various gravity wave antenna experiments have been conducted over the past thirty-five years, and if we include Brown's gravitoelectric detectors, this work would date back more than seventy years. Lazar's assertion of the Gravity B wave being electromagnetic in character and oscillating at microwave frequencies would make better sense if it was not part of the Earth's natural gravity field, but instead consisted of gravity microwaves that originated from the craft itself; that is, if they radiated from its central reactor, or wave amplifier unit, and then reflected from the ground and happened to return to the craft. Lazar reports that before beginning work on the Sport Model, he was allowed to watch a demonstration of it taking off, ascending about 30 feet, moving to the left and right, and then returning to the ground. He says that just prior to and during its liftoff, it gave off a hissing sound similar to that coming from the coronal discharge from a high-voltage line. Its bottom gave off a blue glow, which he says was due to air atoms being excited by the craft's electromagnetic emissions. Again, these characteristics seem to suggest a field propulsion technique similar to the one used in Project Skyvault.

Lazar relates that the gravity wave generators operate in two modes—the omicron and delta configurations. In the omicron configuration, mentioned earlier, one or more generators (microwave mixers) are directed downward to form a supporting beam for the craft (see figure 9.7a). In the delta configuration (see figure 9.7b), all three beams are intersected at a distant location to achieve propulsion relative to that intersection region, which measures on the order of a meter in diameter. The beams then are said to develop an attractive force, rather than a repulsive force, which causes the craft to suddenly jump to that location. Lazar maintains that the beams gravitationally warp space-time at the intersection zone and that the resulting warping is what pulls the spacecraft toward that point. Here it seems he relies on standard general relativity concepts while pursuing an unusual theory of gravitation that has nothing to do with general relativity. His critics, however, rightly contest that if space-time was indeed warped at that distant location to the extent Lazar claims, every other object in the vicinity of that intersection zone should also be suddenly sucked in, as though toward a miniature black hole, which would result in a major collisional catastrophe. So again we encounter a baloney factor that appears to cast a shadow over the whole matter. Also, note that Lazar does not invoke

Figure 9.7. The beam configurations for flying the Sport Model, as described by Robert Lazar: (a) the omicron configuration; (b) the delta configuration.

general relativity concepts in explaining the beam repulsion effect produced in the omicron configuration, probably because standard general relativity theory does not allow gravitational repulsion.

To pursue an independent line of thought in regard to the delta configuration, one might imagine that these microwave soliton beams are intersected at a particular location to generate a plasma that acts as an anchoring point for wave scattering to take place. If the polarity of the beam's sawtooth-shaped waves are reversed to produce a sharp decline of negative electric potential at the beginning of each wave cycle or microwave burst rather than a sharp rise in potential, then it is possible that the beam could function as a tractor beam instead of a repulsion beam, as was theorized in chapter 6 in the discussion of the Podkletnov experiment. In that case, the soliton field the craft would be creating would generate an attractive force on the craft that would be vectored toward the target region and result in propulsion of the craft in that direction.

Although so much of Lazar's story sounds like nonsense, it is difficult to dismiss entirely when so many features of the Sport Model resemble Project Skyvault technology. This raises a lot of questions. If this was really one of several captured alien discs, did Project Skyvault begin as an attempt to reverse-engineer alien technology? One can make a strong case for microwave phase conjugation having originated as an outgrowth of the World War II development of radar. Also, there is the work of Townsend Brown, who apparently had gotten quite close to these ideas on his own without having any captured alien discs at hand to reverse-engineer. If we are to believe that the discs were given to us by Zeta Reticulans as part of some technology-exchange program, as Lazar claims, then perhaps these beings did so because they knew that our own research had already advanced to the point that they would have little to lose to let us look at what they had.

Then there is the issue of the misinformation that was apparently given to Lazar during his employment at S-4, with the outside possibility that he also went through some sort of mind-manipulation session, as some have suggested. If Lazar worked at S-4, was his recruitment planned from the start as a disinformation campaign to create confusion about how field propulsion vehicles might operate?

10

THE SEARL EFFECT

10.1 ■ THE SEARL EFFECT GENERATOR

It was during an indoor test that British inventor John Searl watched his permanent magnet generator levitate off its bench and ultimately bump into the ceiling. As he had envisioned in his dreams, his generator was not only able to propel itself with an overunity power efficiency, but it was also able to defy gravity. In its more advanced design, his generator consisted of three concentric ring magnets flanked by three sets of roller magnets that revolved in a clockwise direction about their circumference (see figure 10.1).[1] The innermost ring consisted of a stationary stator magnet, which he called a plate, having its magnetic north pole pointing down perpendicular to the plane of the ring. Twelve or more roller magnets, called runners, were spaced around the plate's periphery and allowed to roll around its circumference (see figure 10.2). The runners were placed with their magnetic fields oriented north-pole up, so that they would be magnetically attracted to the plate's rim as they rolled around it. However, they were spaced so that there would always be a small air gap between their surfaces and the plate. The diameter of the runners was such that they would revolve on their own axis a whole number of times for each revolution around the plate. This allowed the revolving magnets to establish a condition of resonance, with each revolution reinforcing the previous to build up a stationary wave magnetic field oscillation.

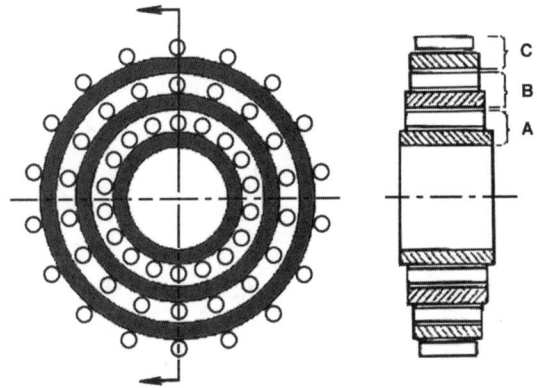

Figure 10.1. The Searl effect generator, showing the arrangement of the concentric ring and runner magnets. Ideally, the middle ring would have twenty-two or more runners and the outer ring thirty-two or more runners. (After S. Sandberg)

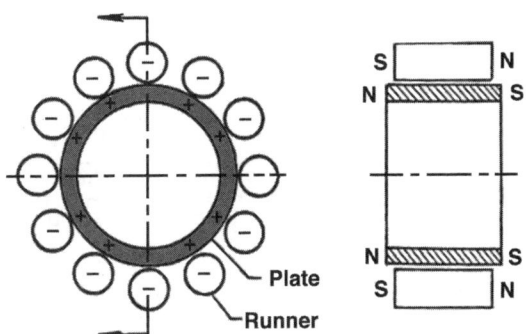

Figure 10.2. Schematic of the Searl effect generator's inner ring and its runner magnets, illustrating the magnetic polarity and electric polarity developed during operation. (After S. Sandberg)

This inner roller ring was, in turn, surrounded by a second ring-shaped magnetic stator plate with an adjacent set of at least twenty-two runner magnets. This was surrounded by a third ring-shaped magnetic stator and an adjacent set of at least thirty-two runner magnets. Previous tests of smaller-sized units indicated that a negative voltage was produced at the rim of the runner magnets, allowing a current to be collected by stationary brushes contacting the runner's rim.

In 1952, Searl and a friend conducted the first outdoor test of one of his 3-foot-diameter multi-ring generators.[2] They set the rotor in motion using a small engine that applied torque through a clutch mechanism. Even though it was rotating at a relatively slow speed, the device produced an unexpectedly high potential at the rotor's periphery, on the order of 100 kilovolts or more. The large potential was indicated by the field's characteristic crackling sound and ozone smell and the effect it had on surrounding objects.*

After the rotor had passed a particular threshold speed, its rate of rotation began to accelerate. Operating on its own power, it began to lift off and break the union to its starter engine. It is said to have risen to a height of some 50 feet, where it hovered for a while, still speeding up. It surrounded itself in a bluish pink halo similar to the glow discharge phenomenon seen when air is ionized in a moderate vacuum. Its pulsing field caused nearby radios to turn on. Finally, the whole generator is said to have accelerated upward at a fantastic rate and disappeared, presumably flying into space.

Searl's unusual device has since come to be known as the Searl effect generator (SEG) or the Searl levity disc. In the years that followed, he and his team members built and flew some dozen or more generators, some of which similarly became lost in flight until Searl found a way of controlling the rotor speed. He also built some generators that were 12 feet in diameter and two that were 30 feet in diameter; see figure 10.3.

There appeared to be a positive-feedback loop involved in the operation of Searl's generators since beyond a critical speed they would begin to accelerate on their own, without any mechanical assistance from the driving motor. In the beginning, Searl had not figured out how to control the effect. Later, he found that it was possible to reduce the speed by electrically loading the generators so that power was siphoned off from them.

Searl found that when the generator was running, there was a dramatic drop in air temperature in the immediate vicinity of the generator and in its interior.[3] Also, there was an accompanying drop in air pressure in its interior, with air being found to move outward from the generator's rim. This air expulsion effect was attributed to the high-

---

*Some claim even higher potentials on the order of 10 million megavolts!

Figure 10.3. John Searl (left) and coworkers constructing a Searl effect generator. (Photo courtesy of John Thomas)

voltage ionization present along the generator's rim, the effect being most pronounced when the rim voltage exceeded 30 kilovolts. The high-voltage emission was believed to cause the generator to be enveloped in a vacuum. Electrons expelled from the rim as negative ions would have arced to the positive plate, exciting a glow discharge in the low-pressure environment.

The electric fields produced by the device were apparently quite strong, since Searl noticed that after he had been working near the generators when they were operating, he had a "cobweb" static electricity sensation on his skin and found that his clothes clung to him. The fields were apparently strong enough to leave a residual electric polarity in his body tissue. When the generator was made to hover low above the ground for an extended period, the soil was seen to become burned because of the heat from the electric currents induced in the ground.

Objects placed inside the generator ring were found to lose weight. Also, a lit candle placed at its center became extinguished. Whether this was due to lack of oxygen from the lowered air pressure or to lack of convection from the reduced gravity field is not certain. Outside the generator, the field was such that objects approaching it were diverted before actually colliding.

In his 1968 paper on the Searl effect, P. L. Barrett presented a sketch showing the approximate direction of the gravity field in the vicinity of an operating Searl generator (figure 10.4).[4] The hatched regions indicate gravitational neutral zones, one being centered below the generator and a ringlike region positioned above the generator. Barrett wrote that any objects that entered those regions would tend to be held there. Also, it was found that upon taking off, a Searl craft would lift up a chunk of earth along with it. As a result, the craft would often leave behind a large hole in the ground. A similar lifting phenomenon has been observed to occur beneath UFOs. Numerous reliable witnesses who observed UFOs hovering over a body of water have noted that the water peaked beneath the craft, as shown in figure 10.5.[5]

The takeoff effect of a Searl-type disc might explain how a massive chunk of soil was torn up and displaced to a spot 73 feet away on a farm in north-central Washington.[6] A farmer's two sons discovered the chunk on October 18, 1984, while rounding up cattle. Passing through a wheat field that had been harvested about a month earlier, they discovered an irregularly shaped hole about 10 feet long and 7 feet wide,

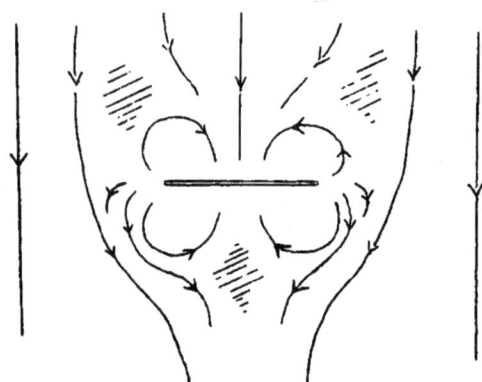

Figure 10.4. Sketch showing how the gravity field flux vectors (e.g., G-on flux vectors) are oriented in the vicinity of a hovering Searl disc. The vectors, which are represented as arrows, map the Earth's gravity field combined with that induced by the disc. Shaded regions indicate neutral gravity zones. (From Barrett, "Searl Effect," fig. IV)

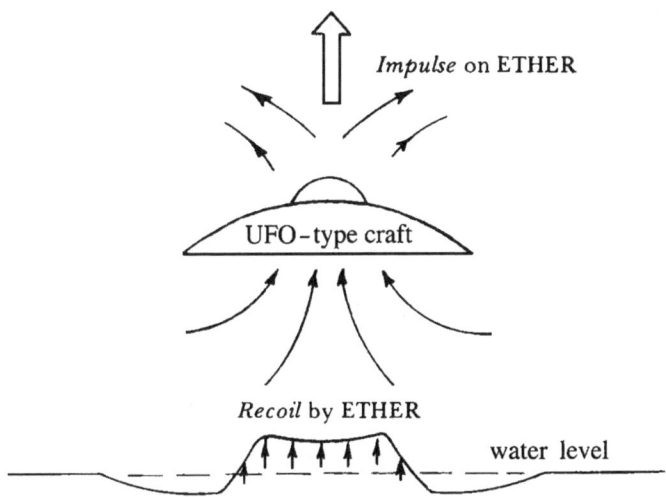

Figure 10.5. How water peaks beneath a low-hovering UFO, as based on numerous eyewitness reports. (After Sigma, Ether Technology 80)

and from 1½ to 2 feet in depth. The hole was not a meteorite crater, as it had vertical walls and a fairly flat bottom, looking as though it had been cut by a giant cookie cutter. The removed piece, still intact, was found right-side up and rotated about 20 degrees counterclockwise from its original position in the hole. An arc of dribblings traced a path from the hole to where the chunk was found. By one estimate, this chunk of soil must have weighed about 5 metric tons. The dense root mass in the soil apparently held the clump together, but some of its roots looked as if they had been torn out at the time the piece was removed, since they were seen still dangling from the vertical walls of the hole. This feat of transportation could not have been executed by any kind of excavation machine. Besides, there was no evidence that the removal had been man-made, nor were any machinery marks found.

One geologist speculated that the shock waves from a local 3.0-Richter earthquake spontaneously focused their energy to this spot and caused the massive chunk of soil to pop out, but seismologists found this unlikely. Besides, for this to happen, nature would have to violate the law of entropy. For this multi-ton piece of soil to move intact without breaking apart, it must have been subject to a uniform force pulling all of its mass upward, countering a downward gravitational pull of at least 90 grams per square centimeter. This would be about two hundred times greater than the lift force developed beneath Brown's 6-inch-diameter (15-centimeter) AC electrokinetic test apparatus.

Since this event took place on a remote farm field at a time when no observers were present, no UFO sighting was reported. Nevertheless, the physical evidence at the site was consistent with the UFO soil-displacement scenario that Schaffranke had documented seven years earlier. The craft must have hovered near the ground, with its electro-gravitic field penetrating the soil below. Then as the craft rose, this extended field gradient would have moved up as well, inadvertently drawing up this chunk of soil with it. As the craft slowly swung in an arc-like trajectory across the field, pieces of soil dribbling from the bottom of this hovering chunk would have floated to the ground. Eventually, the craft must have accelerated, leaving the soil chunk to break away from its towing field and thump to the ground at its new location. What would otherwise be a mystery becomes easily explainable when one has a foreknowledge of gravity field propulsion technologies. Nevertheless, it leaves us to marvel at the enormous lift that such an antigravity field can generate, one capable of overpowering the natural force of gravity to allow tons of soil and a massive spacecraft to float freely. Whether it was made by us or came from somewhere else maybe we will never know.

10.2 ■ THE MAGNETIC ENERGY CONVERTER

The Russian physicists Vladimir Roshchin and Serge Godin in the mid-1990s built and tested a version of the SEG, which they named the magnetic energy converter (MEC). It resembled one of Searl's earlier generators, consisting of a single magnetized stator ring measuring 1 meter in diameter flanked by twenty-three roller magnets, each having a diameter of 7.4 centimeters; see figure 10.6.[7,8]

Roshchin and Godin's generator differed from Searl's in several respects, the principal difference being in the design of the runner magnets. They refer to the runners as rollers, so I will use their term instead. Searl used a special magnetization technique to create magnetic spoke domains in the cylindrical sides of his runner magnets

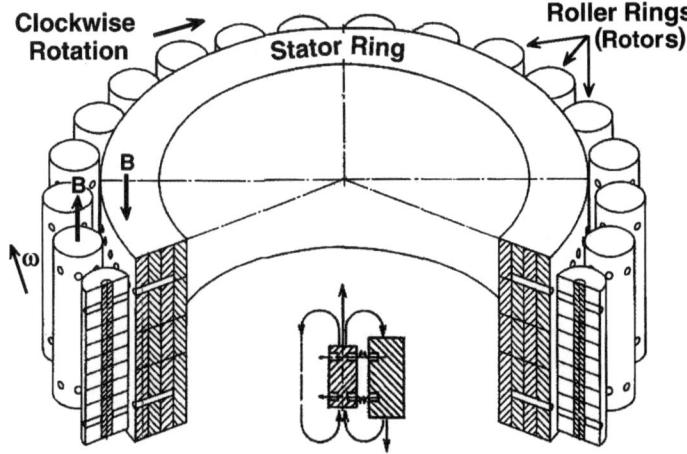

Figure 10.6. The magnetic energy converter. (Based on a drawing by V. Roshchin and S. Godin, 2000)

oriented perpendicular to the runner's axis of rotation. The MEC accomplished the same thing with dipole magnets implanted perpendicular to the surface of each roller magnet so that the north pole of each magnet was oriented perpendicular to the stator plate. The stator had a complementary set of dipole spokes with their north pole pointing out toward the rollers. As in Searl's SEG, these mutually repulsive magnetic domains helped to orient the rollers as they revolved around the circumference of the stator plate and helped to keep the rollers from contacting the plate.

The axles of the roller magnets were secured via bearings to a common rotor frame that kept each roller at its proper relative spacing as they rotated around the stator ring. The rotor had a shaft that was connected to an electric motor via a clutch mechanism. As in Searl's experiments, the motor was used to get the generator up to speed. Roshchin and Godin found that the MEC had a specific rotational speed at which it began to partially power itself, with its drive motor accordingly consuming less power. This occurred when the rotor speed surpassed about 200 revolutions per minute (3.3 hertz). After about one and a half minutes, the rotor had accelerated to 550 revolutions per minute (rpm), at which point the starter motor current consumption had reached zero and was beginning to go negative. The clutch assembly was then made to disconnect the motor from the rotor shaft and in its place connect a generator. As the rotor continued to accelerate, an increasing load was applied to the generator, reaching 7 kilowatts by the time the rotor had a reached a speed of 595 rpm. Greater loads caused the rotor speed to subsequently decline. At greater than 590 rpm, the MEC made "an unpleasant high frequency whistling sound" that damped out as soon as an increased electric load was placed on the generator.

Roshchin and Godin also found that the weight of the generator assembly decreased as the rotor accelerated. Weight loss first became noticeable when the rotor speed reached its critical value of 200 rpm, and by the time its speed had reached 450 rpm, weight loss began to increase exponentially (figure 10.7). At 550 rpm, there was an inflection pause in the weight-loss curve and after 590 rpm, an extremely steep rise in weight loss with increasing rotor speed. By the time the rotor speed had reached 595 rpm, the apparatus as a whole had become 35 percent lighter, with the weight of the generator alone dropping 50 percent. There was substantial hysteresis in this weight loss–rotation rate curve. As rotor speed decreased from 595 to 400 rpm, weight loss remained constant at the 35 percent reduction level. Measurements were not made at greater than 600 rpm for fear that the MEC would enter an uncontrollable supercritical regime in which positive feedback would cause an exponential rise in rotor speed. A similar positive feed-

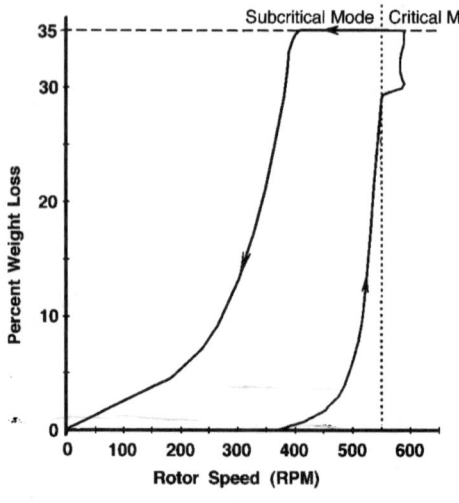

Figure 10.7. Weight loss as a function of rotor speed when a 20-kilovolt potential was applied between the plate and the rollers of an MEC. (After Roshchin and Godin, 2000)

back mode had resulted in Searl losing several of his devices through uncontrolled levitation.

The MEC was found to produce this weight-loss effect only in clockwise rotation (for rollers with their north pole oriented up). Counterclockwise rotation produced a weight-gain effect, with the gain in weight being proportional to the rotor speed. Roshchin and Godin found that application of a 20,000-volt bias potential applied between the plate and the roller magnets improved the performance of the MEC, allowing its weight reduction to begin to take effect at a lower rotor speed. This potential would have ionized the air around the MEC, allowing the electron current to flow more easily across the 1-centimeter gap between the stator and the rollers.

Like the SEG, the MEC generated a luminescence around itself when seen operating in the dark. The bluish pink ionization cloud was observed to cover both the stator ring and the roller magnet ring. Also, when looking at the edge of the rotating roller magnet ring, Roshchin and Godin saw, superimposed on this emission, a series of horizontal, yellowish white luminescent bands (four or five) spaced along the height of the roller magnet's cylindrical surface. This luminescence suggests a possible high-voltage electron discharge from the surface of the roller magnets, although it was not accompanied by sounds characteristic of arc discharge. This silent emission could occur because the emission was coming from a large surface area, rather than from a point source. They compare it to high-voltage, microwave-induced luminescence observed prior to the point of electric breakdown.

Roshchin and Godin also found that when in operation, the MEC surrounded itself with a stationary magnetic-wave pattern consisting of a nested series of cylindrical "magnetic walls" (see figures 10.8 and 10.9).[9] Magnetometer measurements indicated the presence of a magnetic-field flux inside the walls of 0.05 tesla, with the field orientation being the same as that of the roller magnets and at about 6 percent of their 0.85 tesla magnetic flux. No magnetic flux was detectable outside of the walls. There is no evidence to suggest that the magnetic-wall pattern is hazardous in any way. Roshchin and Godin did not notice any harmful effects. On the contrary, Searl has reported that the field emissions of his SEG actually had healing effects.

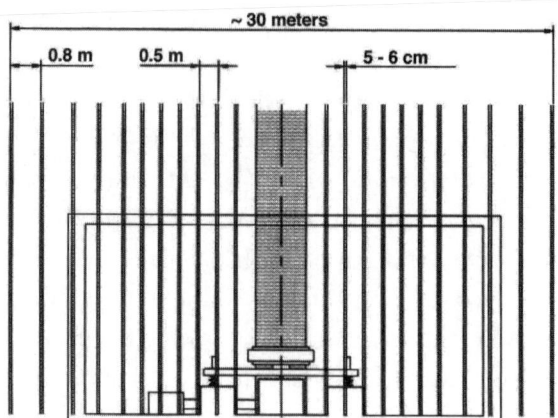

Figure 10.8. A side view of an MEC and its research lab, showing the positions of the nested magnetic walls. (After Roshchin and Godin, 2000)

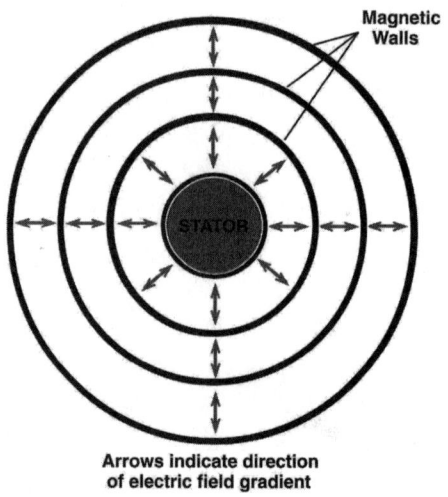

Figure 10.9. A top-down view of the layout of the cylindrical magnetic walls that together formed a stationary wave pattern around an MEC stator ring. (P. LaViolette, © 2006)

The magnetic-wall pattern extended out from the MEC's stator ring for a distance of up to 15 meters, beyond which the magnetic flux within successive walls rapidly decreased in intensity. The field pattern could be sensed on the second floor of the laboratory, indicating that the walls extended upward at least to a height of 6 meters and possibly downward into the ground for a similar distance. The innermost walls were spaced from one another by approximately 0.5 meter (one stator radius), and this spacing increased 0.8 meter as distance from the MEC increased. Also, the walls had a thickness of about 5 to 6 centimeters, approximating the 7.4-centimeter-diameter of the roller magnets.

The proximity of these dimensions to that of the ring of roller magnets led Roshchin and Godin to conclude that there was a direct connection between this stationary field pattern and the circumferential movement of the magnetic rollers. This conclusion was also supported by their observation that there was a decrease in air temperature within the magnetic walls, which could be immediately felt by placing one's hand within a wall. Moreover, they found that the temperature decreased in proportion to the rate of roller-ring rotation, the decline becoming noticeable above a rotor speed of 200 rpm and reaching –7.5°C by the time the rotor speed had reached 550 rpm. Consequently, they concluded that energy from the environment was somehow being transferred to the ring of roller magnets to assist their rotation.

Below, I will attempt to explain the principle of how the MEC generator operates, with the understanding that the same explanation should apply equally well to the SEG. These ideas were first presented in June 2001 at the Conference on New Hydrogen Technologies and Space Drives.[10] Briefly, clockwise rotation of the rollers causes a current to flow radially outward from the MEC's plate to its rollers due to the Faraday disc dynamo effect. As a result of the "ball-bearing motor effect," this current then creates a torque that induces the rollers to continue their clockwise rotation. The rotating rollers create resonant extremely low-frequency (ELF) oscillations that phase-conjugate to form an extended soliton wave pattern. Energy entrained from this soliton helps to propel the rollers. Let us begin by examining the Faraday effect.

10.3 ▪ THE FARADAY DISC DYNAMO EFFECT

The Faraday disc generator, also known as the homopolar generator, was first built by physicist Michael Faraday in the late nineteenth century. Faraday placed a copper disc between the poles of two cylindrical magnets so that the magnets' field ran perpendicular to the plane of the disc. He found that when he spun the disc, a current was induced to flow, with electrons moving outward from the disc's center toward its periphery. This was attributable to the $v \times B$ rule, in which v represents a conductor's velocity vector and B represents the magnetic force field vector, the two being vector-multiplied with one another.

Faraday also discovered that if the copper disc was cemented to the magnets and the magnets and disc were rotated together as a unit, electrons would still flow to the disc's periphery (figure 10.10).[11] This led him to conclude that the magnetic field produced by the magnets did not rotate with the magnets, but rather was anchored in space (i.e., in the local ether rest frame). Researchers such as Bruce DePalma, Adam Trombley, and Paramahansa Tewari built various versions of this cemented-disc homopolar generator with the hope of developing a motor generator having an overunity efficiency.

As in the classical Faraday disc generator, in this cemented-disc version the magnitude and direction of the electron-current flow is determined by the $v \times B$ rule. Remember to use the left-hand rule instead of the right-hand rule when dealing with electron current flows. That is, if you point the index finger of your left hand in the magnetic-flux direction (south to north) and your thumb in the direction of rotary movement, then your middle finger will indicate the direction of electron flow.

The runners in the SEG and the rollers in the MEC are essentially little Faraday disc generators. They do not have a cemented-copper disc, but the magnets themselves are conductive—not as conductive as copper, but nevertheless they conduct electricity. Thus, like the Faraday disc generator, they should generate a radial current. If we consider a single roller magnet rotating in a clockwise direction with its north pole

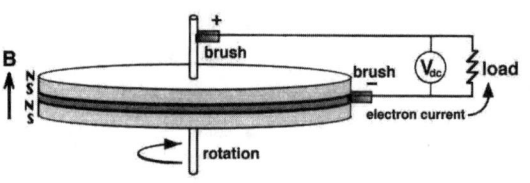

Figure 10.10.
A Faraday disc generator with a copper disc cemented to the magnetic pole pieces. (After Archer Energy Systems)

pointing up, as shown in figure 10.6, the left-hand rule indicates that the roller should produce an electron current flow from the roller into the stator ring plate. However, when operating, the Searl disc and the MEC produce an electron current that flows outward from the plate to the rollers, not inward.

This discrepancy arises because we have not accounted for the effect of the entire system in motion. In addition to the Faraday effect due to the rotations of the individual roller magnets, we must also account for the Faraday effect produced by the collective translational movement of those magnets around the circumference of the stator plate, that is, about the stator's central axis. In considering this effect, we may treat the roller magnets collectively as composing a single ring magnet whose radial thickness is equal to the diameter of the rollers and whose magnetic field is in the same direction as that in the individual roller magnets.* Once again applying the left-hand rule, we see that clockwise rotation of this ring produces an electron-current flow outward from the stator plate to the roller magnets, opposing the current flows arising individually from the Faraday effect of each roller. As it turns out, the Faraday-effect voltage induced by this collective translation is much greater than the opposing voltage polarity that arises from the individual magnet rotations. The net result is that the electron current should flow outward from the plate to the roller magnets, just as is observed. For illustration, the relative magnitude of these two opposing Faraday disc effects is calculated in the accompanying text box.

Faraday Effect Potential Induced by Roller Ring Rotation

Let us first consider the voltage generated by the rotation of each individual roller. The induced voltage may be calculated using the equation applicable to a Faraday disc dynamo:[12]

$$V = \tfrac{1}{2}\omega B(b^2 - a^2) \tag{9}$$

in which ω is the magnet's angular velocity and b and a are its outer and inner radii. In this case of the MEC prototype, $b = 0.037$ meter, the

---

*Here we assume that most of the return magnetic flux from the individual roller magnets (south-pole-seeking flux) returns either outside or inside the ring of roller magnets and to a lesser extent in the space between the roller magnets.

roller diameter, and a = 0.005 meter, the diameter of the roller's central shaft hole. When the ring of rollers is moving about the circumference of the stator plate at 550 rpm, the individual rollers will be rotating twenty-three times faster, hence, ω = 23 x 550/60 x 2π = 1325 radians per second. Taking B = 0.85 tesla, equation 9 predicts a voltage of V = 0.76 volt, in which the center of the roller is positive and its periphery is negative.

Now let us calculate the voltage generated by the displacement of the entire ring of rollers. The outer and inner radii of the ring are taken as b = 0.574 meter and a = 0.5 meter, which are the distances from the MEC's central axis to the outer and inner circumferences bounding the roller ring. If this roller ring revolves about the MEC's central axis at 550 rpm, or at an angular velocity of ω = 57.6 radians per second, then equation 9 predicts that it would generate a voltage of V = 1.95 volts, with the potential being negative at the edge farthest from the stator plate. Consequently, even though the roller ring rotates far slower than the individual rollers, it produces a far larger voltage.

On the side of the roller nearest the stator plate, the voltage generated by the clockwise displacement of the roller ring will be opposed by the voltage being generated by each roller. Yet on the opposite side of the roller, the side farthest from the stator plate, the roller's Faraday effect will produce a voltage polarity in the same direction as that produced by the displacement of the roller ring; hence, the two will add to one another. The net result is that the Faraday effect arising from the rotation of each roller cancels out, leaving just that arising from the roller ring's clockwise displacement. In the example presented in the text box, this would leave a net voltage of 1.95 volts, inducing an electron current to flow toward the MEC's periphery.

Roshchin and Godin used neodymium iron boron magnets in their MEC (typically 58 percent iron, 37 percent neodymium, 4 percent ferrous sulfate, and 1 percent boron). This alloy has an electric resistance of about 144 micro-ohms per centimeter, so a 7.4-centimeter-diameter roller would offer a resistance of 1066 μΩ. Roshchin and Godin did not state how much current was flowing into each roller at a given rotor speed, but as a guess, we might suppose that at an operating speed of 550 rpm, this was on the order of 800 amps for each roller.

As this electron current passes radially outward into the rollers, it generates a motive force on the rollers that assists their rotation. This production of circumferential torque from a radial current flow is known as the *ball-bearing motor effect*.

10.4 ▪ THE BALL-BEARING MOTOR EFFECT

I first saw the ball-bearing motor effect in action in the spring of 1999. An alternative-energy political demonstration was being held in Washington, D.C., on the western Mall. Many there were preparing themselves for the predicted breakdown of society that was expected to occur the following year, when the year-2000 computer-calendar glitch was to occur. Among the attendees sprawled on the grassy lawn was Mark Gubrud, a University of Maryland physics student. He was demonstrating an interesting motor that consisted of a shaft mounted in a ball-bearing race (figure 10.11). He applied DC voltage from a

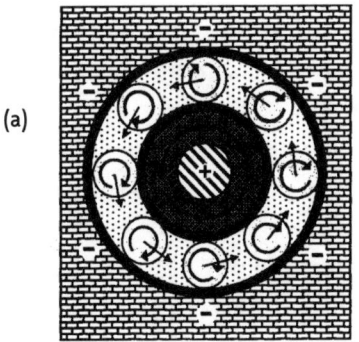
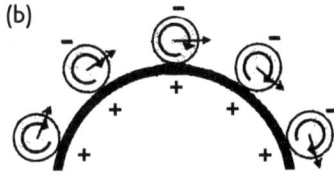

(a)

(b)

Figure 10.11. Comparison of the ball-bearing motor to the MEC's stator and roller ring. (a) The ball-bearing motor. (After M. Gubrud, in T. Valone, Homopolar Handbook, *54–55) The central shaft rotates relative to a stationary cylinder. (b) The lower portion of the above diagram, unfolded to show equivalence to the geometry of the roller magnets traveling around the MEC stator. In each case, the electron flow moves from the stator toward the ball bearing (or roller magnet). Charge polarity in (b) is reversed since the MEC functions instead as a generator, with the electron flow inducing the buildup of charge whereas in the case of (a) an applied charge instead induces electron flow.*

small battery pack between the shaft and the outer casing of the ball-bearing race. Then, when he gave the shaft a starting torque, it continued to rotate in the direction of the applied torque and continued to turn as long as voltage was applied. If the shaft was given a starting torque in the opposite direction, it again continued to rotate, but in that new direction. Gubrud had with him an explanatory write-up of the motor's principle of operation, which he gave me. I gave a copy to Tom Valone, and he subsequently published it in his *Homopolar Handbook*.[13]

After learning about this phenomenon, I realized that the same effect powered the rotary motion of the runners in the SEG and the rollers in the MEC. Using Gubrud's explanation, let us review how an applied radial-current flow through a ball bearing induces torque forces on the bearing, causing it to rotate around its bearing race (see figure 10.12). A current passing through the ball bearing at time t_1, flowing from the bearing casing to the central drive shaft, will magnetize the bearing. The bearing, though, will retain a residual field in this same magnetization direction at time t_2, although the direction of this residual field has changed because of the bearing's rotation. At time t_2, this residual magnetic field will be directed at some angle α to the direction of current flow, which always occurs through the points where the ball bearing contacts the axle and bearing race. The current component i that lies perpendicular to this residual field will then induce a force ($F = i \times B$) that produces torques on either side of the bearing, which induces it to keep revolving in the direction of its initial rotation.*

The same principle applies to the rollers of an MEC or the runners of an SEG. Consider the ball-bearing motor in figure 10.11a. The outer ball-bearing race remains stationary as the axle turns. If we take this outer-race circumference and fold it back so that its inside faces out, we get the geometry shown in figure 10.11b. Imagine that the ball bearings are roller magnets rolling around the stator plate. The two mechanisms are then seen to be equivalent. The electric polarities are reversed in each

---

*Note that the current flow illustrated in figure 10.12 follows the engineering convention that current flows in the opposite direction from electron flow. Electron current would be in the opposite direction, and one would then use the left-hand rule instead of right-hand rule to determine the directions of the torques. The result, however, comes out the same.

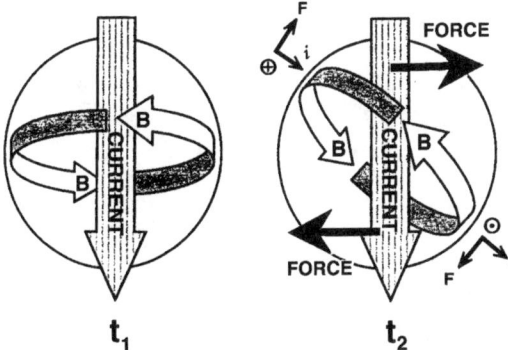

Figure 10.12. A ball bearing shown magnetized at time t_1 that retains a residual field in that same direction at time t_2, even though it has rotated in a clockwise direction. Vectors show the clockwise torques developed by the applied current. (Based on M. Gubrud's diagram, in T. Valone, Homopolar Handbook, *54–55)*

case because the motor in figure 10.11a is powered by dissipated power (electrons flowing from minus to plus) and the motor in figure 10.11b is powered by generated power (electrons flowing from plus to minus). As in the ball-bearing motor, if an electron current flows from the plate outward through each roller, this current will produce a torque on each roller, assisting it to move in the direction of its established rotation.

These two processes together, the Faraday disc dynamo effect and the ball-bearing motor effect, form a positive feedback loop in which a clockwise displacement of the roller-ring rotor produces a radial electric current that induces roller rotation and greater clockwise displacement of the rotor (see figure 10.13, upper left).

The high voltage the MEC induces at its periphery could be due to the sudden change in the electric resistance the electrons encounter in the course of their radial outward movement. Upon leaving the low-resistance environment of the magnets and continuing their push through the high-resistance environment of the surrounding air, the electrons' voltage potential associated with their current flow would have shot up proportionately, since $E = iR$. That is, for the same current value, voltage will increase in direct proportion to resistance.

Roshchin and Godin found it necessary to use an external motor to apply mechanical torque to the MEC to start it and keep it going in

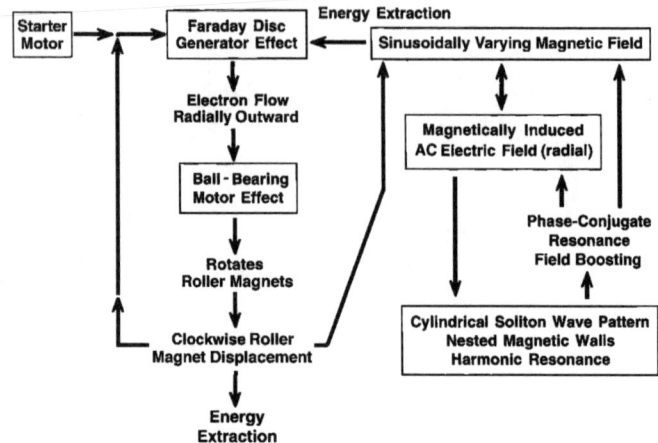

*Figure 10.13. An energy flow analysis chart of the MEC. (P. LaViolette, ©
2006)*

the low-revolutions-per-minute regime. Nevertheless, as in Searl's earlier generators, they found that the roller ring began to spontaneously
accelerate once it had been spun up to a critical threshold speed. For the
MEC, this critical rate of rotation was around 200 rpm, although Searl
succeeded in designing generators that would spontaneously accelerate,
even from rest. This acceleration phenomenon suggests that the MEC,
like the SEG, must have been receiving an additional input of energy
from some unknown source. Let us next consider where this energy may
have been coming from.

10.5 ■ THE CYLINDRICAL SOLITON

This additional, unaccounted-for energy input that was assisting the
roller ring's movement around the stator plate most likely was being
entrained from the MEC's immediate environment. This hypothesis
finds support in the observation that during its operation, the MEC
caused a drop in air temperature within a series of nested cylindrical
zones established in its immediate vicinity (see figures 10.8 and 10.9).
Here we will attempt to understand how this magnetic wall pattern is
generated, and later we will examine how energy from this field pattern
might be entrained to power the roller-magnet rotor.

As the roller magnets travel clockwise around the stator plate, they generate an oscillating magnetic field in the stator's frame of reference. That is, as a roller magnet coincides with a given reference point on the plate's circumference, in the stator reference frame the magnetic field strength at that point would increase to a maximum since the flux lines between the oppositely oriented roller and stator magnets attract and complement one another. However, as the roller makes a half turn in its clockwise motion, its downward return field, which is south-pole seeking, would now occupy the space that formerly held the roller's north-oriented field. This same space would also be filled with the stator's upward-directed south-seeking field. Since these two fields are in opposition, the net magnetic field strength there would add up to a minimum value. As a result of the roller magnet's clockwise motion, the magnetic field in the stator reference frame would sinusoidally vary in magnitude, and would do so without reversing its magnetic polarity (figure 10.14).

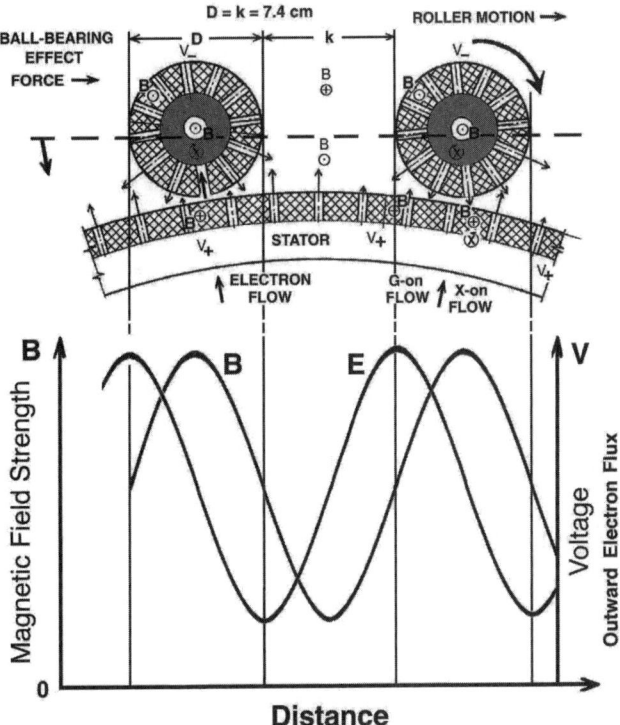

Figure 10.14. *The resonant variation of magnetic field strength (B) and electric field intensity (E) in the stator reference frame excited by the circumferential displacement of the MEC's roller ring. (P. LaViolette, © 2006)*

Because the diameter of the roller magnets (D) and the space between the roller magnets (k) in figure 10.14 are approximately the same, the rotary displacement of this succession of equally spaced roller magnets would set up a resonant oscillation in the stator reference frame. As an example, when a ring of twenty-three roller magnets revolves at 550 rpm, the generated frequency would equal 211 hertz. This time-varying field (B), in turn, would induce a sinusoidal electric-potential wave oscillation at the same frequency that is directed radially inward and outward from the MEC's center in the stator's nonrotating reference frame. However, compared with the magnetic field maxima, the electric potential maxima would be displaced 90 degrees in phase, that is, by one roller radius in the clockwise direction (see lower trace in figure 10.14).

Also, as discussed earlier, the Faraday disc dynamo effect would be generating a radial electric field gradient across the diameter of the roller magnets with the potential being more negative at the outer periphery of each magnet, thus inducing an electric current to flow radially from the stator through the roller magnets. Consequently, the sinusoidal AC electric field that the roller magnets generate would modulate this DC-negative potential in the vicinity of the roller ring, causing its amplitude to vary sinusoidally with time. This is analogous to what was happening in Brown's electrokinetic apparatus when the high-voltage DC field across the capacitor dielectric was modulated with an AC field from the apparatus's negative antenna electrode, but in the case of the MEC, the field geometry would be radial rather than axial.

As a result, the electron flow from the MEC's stator toward its rotor would pulsate at the frequency induced by the translating roller magnets. When a roller magnet is tangential to a given location on the stator, that is, most proximal to that location, the electric potential that induces current to flow radially outward to the roller would at that moment only be at its median potential. The potential at that particular location would reach its maximum only after the roller magnet has passed and become displaced by a distance of one roller-magnet radius, so there would be a tendency for arcing to occur on the trailing side of each roller. This would be especially apparent when a high-voltage gradient was present in the vicinity of the MEC's rollers, thereby allowing negative ion discharges from the stator to cross large air gaps.

This recurrent outward pulsing of negative charges in some ways resembles the asymmetrical, longitudinal waves radiated from Tesla's transformers. The oscillating, nonreversing magnetic field and the oscillating, nonreversing electric field produced around the circumference of the roller magnet ring would induce a longitudinal ELF electromagnetic wave to propagate radially outward from the stator ring in the stator ring plane. This wave would have a cylindrical wave-propagation geometry, with its magnetic field aligned to the cylinder wall that parallels the roller magnet field and its electric field component oriented perpendicular to the cylinder wall, oscillating radially outward and inward with respect to the center of the MEC. Roshchin and Godin, however, did not report having observed such radial electric field oscillations in these walls.

Also, this radiated ELF wave would have a circumferential wave component. Since the stationary wave produced by the rotating roller ring contains a total of twenty-three wavelengths that fit a whole number of times around the roller-ring circumference, a similar whole number of wavelengths would be required to fit around the circumference of field oscillations appearing outside the roller ring. These would be able to manifest only at multiples of the 0.5-meter stator radius. At a radius of 1 meter, equal to two stator radii, forty-six wavelengths would fit around the wave's cylindrical circumference. At 1.5 meters, equal to three stator radii, sixty-nine wavelengths would fit around the wave pattern circumference, and so on. No oscillations would manifest at intermediary radial distances because a whole number of wavelengths would be unable to fit into the cylinder circumference at those other radial distances. This is because Roshchin and Godin used twenty-three magnets in their roller ring, 23 being a prime number. Hence, oscillating potentials are able to build up to detectable levels only at radii that are whole-number multiples of the stator radius. At any intermediary radial distances, the oscillations would destructively interfere, hence preventing resonant oscillations from building up there.

The radial ELF oscillations would manifest as a soliton pattern consisting of a series of concentric cylindrical wall nodes similar to those shown in figures 10.8 and 10.9. Each wall would have a depth of about one roller radius and would be separated from the next wall by a distance of about one stator radius. The symmetry axis of this cylindrical soliton pattern would be centered on the MEC's axis of rotation. Hence, when the MEC's field patterns are analyzed in this way, it becomes understandable why Roshchin and Godin observed their MEC to be surrounded by a cylindrical, shell-like field pattern. Although stationary wave patterns that vary in smooth-sine-wave fashion are more commonly observed, patterns with sharp transition boundaries are also known to occur. For example, such sharp-edged boundaries have been observed in stationary waves produced in the laboratory with Tesla-wave-type oscillations.[14]

The waves within each of the soliton's walls would revolve around the circumference of their walls in synch with the rotation of the roller-magnet ring, remaining stationary with respect to the rotating frame and with respect to circumferential oscillations occurring in the other wall-like nodes. Hence, the waves in rings at successively greater distances would circuit around the circumference of their walls at successively higher velocities. The waves circuiting in the 1-meter-radius node, whose circumference would be twice the roller-ring radius, would oscillate at twice the roller-ring fundamental frequency (i.e., at $2f_o$). More-distant walls spaced at one-stator-radius intervals would support oscillations circuiting at progressively higher rates, at frequencies of $3f_o$, $4f_o$, and so on. Thus, as one proceeds outward through this soliton wave pattern, moving radially outward from the center of the MEC, one encounters progressively higher harmonic modes, whole-number multiples of the roller magnet's fundamental frequency f_o. We might also speculate that a whole series of harmonic modes might coexist within each of these walls but at intensities below that of the main harmonic mode circulating in that wall.

Since 23 is a prime number, lower-frequency subharmonic modes, such as $f_o/2$ and $f_o/3$, situated at radii smaller than the roller-ring radius, would not form. The roller ring would itself be the lowest-frequency resonance in this stationary cylindrical-wave pattern. Given that Roshchin and Godin observed the walls to extend about 15 meters out from the generator, this indicates that the outermost wall had a frequency of about $30 f_o$. This would have produced a frequency of about 6,300 hertz for a ring-rotation rate of 550 rpm. It is remarkable that the soliton pattern was stable enough to produce such a high harmonic. This 15-meter soliton pattern, though, may limit how close MECs can be placed to one another. For example, if an aircraft was to employ more than one MEC or SEG for lofting, the units would need to be spaced apart by at least 50 meters.

If Searl had used twenty-two magnets for his inner magnetic ring, then he also would have observed a cylindrical wave pattern forming around his discs, but with twice as many nodes because 22 is not a prime number. But when divided by two, it yields the prime number 11, so such a disc would have generated nodes at half multiples of the fundamental frequency, with a lowest-frequency node at $f_o/2$ positioned at half a stator radius and with nodes repeating at radial distance intervals of half a stator radius.

10.6 ■ ENERGY ENTRAINMENT

The MEC's AC electric and magnetic-field oscillation is induced in a nonlinear medium. That is, the ferromagnetic material making up the stator and roller magnets has nonlinear electric and magnetic properties. In addition, the high-voltage discharge that surrounds the stator plate and roller ring when the MEC is operating at high rpm would generate nitric oxide, which also has nonlinear dielectric properties. As explained in chapter 7, the nonlinear dielectric properties of such plasmas make them good phase-conjugating media. For example, the phase-conjugating characteristics of the plasma surrounding the dome of Tesla's magnifying transmitter could explain how his tower was able to "beam" radio-frequency energy to distant locations without appreciable attenuation. Also, as mentioned earlier, Obolensky has succeeded in phase-conjugating arc fluctuations in an arc lamp by placing a nonlinear reactance element in series with the lamp.[15] This demonstrates that phase conjugation can take place even in the ELF frequency range. The notion that the MEC was phase-conjugating the ELF waves it was generating, then, seems plausible.

We might presume that some of the MEC's outgoing ELF radiation was reflected back to the MEC from surfaces in its environment and that these reflected waves were then phase-conjugated by the MEC's nonlinear media. The phase-conjugate wave would have retraced the path of the ordinary wave to the remote reflection site and then back to the MEC's oscillating field. The ordinary and phase-conjugate ELF waves would have interlocked and constructively reinforced one another to produce phase-locked field potentials that would have manifested at each node of the cylindrical soliton pattern.

One interesting characteristic of a phase-conjugate resonator is its ability to decrease entropy. For example, in the case of optical and microwave phase conjugation considered in chapter 8, when an outgoing laser (or maser) beam scatters from the environment, its entropy increases—it becomes more disorganized. The probe beam scattering back toward the phase-conjugate mirror therefore is more disordered than the original outgoing beam. This state of disorder, however, may be reversed through the emission of the phase conjugate of the probe beam. That is, the emitted phase-conjugate waves precisely retrace the path of the scattered ordinary waves, causing the entropy of the wave system to decrease as the wave regains its original ordered state. As a result, energy that normally would be lost through scattering to the environment becomes bottled up in the soliton wave pattern.

As in the case in which microwave phase conjugation is used to generate intense beams for spacecraft propulsion, so too would phase-conjugate resonance occurring in the vicinity of the MEC generator bottle up back-scattered ELF waves, storing their energy in the soliton wave pattern. The repeating AC oscillations that the roller magnets would generate in the reference frame of the stator plate would then add to one another, causing the soliton's ELF field oscillations to progressively increase in magnitude. Experiments have shown that an optical phase-conjugate resonator can self-excite to intensities sixty times that of the input signal beam without any additional energy input, and it has been suggested that even higher amplification coefficients should be achievable.[16,17] We might speculate that even-greater signal amplification occurs in the MEC.

The faser effect produced by phase conjugation of the ELF waves should amplify not only the MEC's electric field pulsations in the stator reference frame, but also the associated magnetic field soliton pattern rotating with the roller-magnet ring. Consequently, since the direction of the soliton's B field matches that of the roller magnets, the soliton magnetic field should reinforce the roller-magnet fields, causing the ambient magnetic field in the vicinity of the roller magnets to progressively increase over time to far exceed the 0.85-tesla strength produced by the roller magnets alone. The maximal amount of this induced increase would depend on the rpm of the roller ring, higher ambient field strengths being achieved at higher revolution speeds.

As explained earlier, the radial electric field and current flow induced by the circumferential displacement of the roller magnets depends not only on the roller ring's rpm but also on the field strength of the magnets. Since phase-conjugate resonance would boost this field strength, one would expect that the potential developed at a given rotation rate would far exceed the voltage that we calculated earlier on the basis of the Faraday disc effect. In addition, the radial electric field component of the soliton's ELF waves would amplify through phase-conjugate resonance, and in the MEC's nonlinear environment, a portion of this amplified AC potential would likely transfer over to boost the MEC's DC electric field component. Consequently, it is possible to imagine that these effects in combination could produce a DC voltage drop in the vicinity of the MEC of tens of thousands of volts or more, rather than just of a few volts. This would explain the origin of the high-voltage glow discharge that Roshchin and Godin observed even when their 20-kilovolt external bias voltage was switched off. It would also account for the pink, high-voltage glow discharge that Searl observed around his discs. We are here reminded of Tesla's towers, which similarly developed potentials so high that nearby objects were excited to luminesce.

Earlier, we concluded that the circumferential displacement of the B fields of the roller magnets induces a radial electron flow from the stator plate to the rollers that, in turn, generates a ball-bearing-motor torque that aids the clockwise rotation of the rollers. Consequently, the boosting of the radial electric potential gradient due to these various faser effects would proportionately increase the induced-current flow and thereby aid the rotation of the roller ring. In this way, energy from the soliton-field pattern would be continuously converted into mechanical energy, inducing the roller ring to accelerate in spite of resistive losses.

This additional energy input into the MEC is illustrated by the feedback loop on the right side of figure 10.13.

Since this array of magnetic-wall solitons is part of a single resonance phenomenon, these resonant modes should exchange energy with one another. Hence, the buildup of potential energy in the vicinity of the roller ring (the inner soliton cylindrical wall that was assisting roller propulsion) could easily be conveyed from the other cylindrical wall nodes whose frequencies resonated harmonically with this base frequency.

The temperature drop observed in the magnetic walls may be a direct result of thermal energy being extracted from the air and being entrained into the soliton throughout its harmonic range. If so, the soliton field must then somehow be physically interacting with the air molecules in the magnetic walls and possibly be drawing energy from their Brownian motion. For example, the magnetic field oscillating in the cylindrical walls might slightly magnetize the air or solid objects that a wall happened to intercept, and magnetized molecules whose Brownian-motion oscillation happened to match any of the wall's ELF harmonics would then have their energy entrained into the soliton. Consider, for example, an inner wall located 1 meter from the MEC's axis. It would support an ELF frequency of $2f_o$, twice the fundamental frequency. However, this particular harmonic could be excited by an entire spectrum of harmonics existing in the ambient molecular movement: f_o, $2f_o$, $3f_o$, $4f_o$, $5f_o$, and so on, up to the ninth harmonic; recall the discussion in chapter 8. Thus, there are abundant opportunities for frequency matches to develop.

Using the heat capacity of air, Roshchin and Godin have estimated that heat was being lost at the rate of 1,700 calories per second from the eight innermost magnetic walls residing within 4 meters of the MEC's center.* This loss rate equals 7 kilowatts, which slightly exceeds the 6 kilowatts of electric power that the MEC was mysteriously generating without any mechanical input. If the caloric loss of the entire soliton pattern is taken into account out to a radius of 15 meters, then there is more than enough energy loss to account for the MEC's source of

---

*Roshchin and Godin estimated that the walls measured 12 meters in height by 5 centimeters in breadth and calculated that the air passing through that volume underwent a 6°C temperature drop when the MEC was operating at 550 rpm.

power. Thus, the MEC may have a clearly identifiable energy source in its immediate environment. It would be partially propelled by heat flowing into the magnetic walls from the ambient air and laboratory structures, with this energy being subsequently entrained into the soliton pattern. A generator such as this that cools its environment while it generates power appears to be the ultimate solution for the global warming problem.

It is also possible that the MEC is entraining background electromagnetic energy. The universe is permeated by energy spanning a wide range of frequencies, including frequencies in the ELF range. Also, there is the well-known ionospheric Schumann resonance, which is excited at approximately 8 hertz by solar particle bombardment of the ionosphere. Higher harmonics of this resonance might match some of the soliton harmonics.

The MEC may also be cohering energy from the omnipresent zero-point energy background that extends throughout space. As was noted in chapter 4, the reactive ether of subquantum kinetics is conceived to have X, Y, and G reactant concentrations that continually fluctuate in magnitude in seemingly random fashion. The X and Y ether fluctuations correspond to spontaneous pulsations of the ambient electric potential field, and the G ether fluctuations correspond to spontaneous pulsations of the ambient gravity potential field. Together they make up the zero-point energy continuum. It seems plausible that the zero-point energy background would locally transfer a portion of its energy to material bodies it surrounds, so that if there was a decrease in the energy density of the zero-point energy background, then one might observe a corresponding local decrease in air temperature. Thus, if the MEC was cohering energy from the zero-point energy background, the observed drop in air temperature might be a collateral effect and not the actual source of the MEC's energy. If it was able to extract energy directly from the zero-point energy background, then the MEC would be able to continue powering itself in empty space.

The subquantum kinetics zero-point energy spectrum differs from that of conventional physics in that it spans all frequencies, including ELF frequencies. The zero-point energy concept of quantum mechanics and quantum field theory, for example, predicts that such fluctuations should instead occur primarily at high energies since such fluctuations are theorized to arise in the simultaneous appearance of a virtual sub-atomic particle and its virtual antiparticle. Most of the energy in the conventional zero-point energy spectrum then would be at frequencies greater than 10^{20} hertz (the MeV range) and would extend on up to the Planck limit of 10^{43} hertz (10^{28} electron volts). Conventional physics, then, would be off by at least eighteen orders of magnitude in provid-ing an appropriate frequency match for the MEC's energy extraction. Subquantum kinetics, on the other hand, allows energy to be extracted from the fluctuating ether at frequencies in this ELF range and even lower. In fact, in subquantum kinetics, the probability of a fluctuation occurring increases as frequency decreases. High-energy fluctuations in the MeV range and greater that would be potentially large enough to nucleate the materialization of a subatomic particle would be exceed-ingly rare events.

If we define our system boundary so that it surrounds both the MEC and the magnetic-wall soliton pattern that it creates in its immedi-ate vicinity, we find that the Searl effect does not violate the first law of thermodynamics, but it does violate the second law of thermodynamics. However, the violation of the second law is the expected norm whenever phase-conjugate resonance is occurring.

10.7 ▪ EXPLAINING THE WEIGHT-LOSS EFFECT

The MEC's weight-loss effect is not easily explained in terms of stan-dard physical theory, but it is understandable within the framework of subquantum kinetics.[18,19] The MEC would develop a negative electric potential at the periphery of its roller ring and positive potential near its central axis. The resonant oscillations would cause this field to fluctu-ate in magnitude at its negatively biased periphery. From an electrody-namic standpoint, the Searl disc's oscillating field is analogous to the nonreversing AC field that Brown was exciting across the dielectric of his AC electrokinetic apparatus. Hence, the analysis illustrated in figure 4.5 for Brown's apparatus should apply equally well to the Searl disc. That analysis assumed that gravitational thrust was being produced as a result of the creation of a virtual-charge gradient across the capacitor dielectric. Furthermore, on the basis of the electrogravitic coupling prediction (equation 7 from chapter 4), we would expect that this virtual-charge gradient would induce a corresponding gravity potential field. As figure 4.5c shows, the resulting oscillating gravity field gradient would induce a gravitational thrust in the negative-to-positive electric pole direction. A similar thrust would be predicted for the MEC, directed from the MEC's periphery toward its center.

In the subquantum kinetics ether concept, this radial gravity potential gradient is envisioned as a G-on concentration gradient that angles downward toward the MEC's center and whose slope varies cyclically with time. This concentration gradient would induce G-ons to diffuse radially inward at a rate that just compensates the rate at which G-ons are being added to the MEC's periphery as a result of the electrons and negative virtual-charge densities that are being pumped there and that act as G-on sources. Thus the revolving ring of roller magnets would act as an ether pump, pumping G-on sources (electrons and negative virtual-charge densities) toward the MEC's periphery, thereby lowering the G-on concentration at the MEC's center.* This outward G-on flux would likely have a rotary component aligned in the clockwise direction of magnetic ring rotation, in which case a clockwise ether vortex would be produced.

The above analysis suggests that while in operation, the MEC or Searl disc would generate a gravity field in its generator's interior where *up* would be oriented toward the generator's periphery and *down* would be oriented toward its center. Thus, the induced internal gravity gradient would act as a centripetal force that would counteract the centrifugal force of rotation. This disagrees with Barrett's inference that the gravity field in the Searl generator would be oriented with the center being up and the rim being down.[20] In the same paragraph, Barrett commented that "side effect electromagnetic forces help to keep the Searl generator

---

*In addition to these G-on fluxes, X-on and Y-on fluxes would be induced by the radial X-on and Y-on concentration gradients, which in conventional physics are the counterparts of the electric field. Thus, at the MEC's negatively charged rim, X-on concentration would be elevated and Y-on concentration depressed, while at the MEC's center, X-on concentration would be depressed and Y-on concentration would be elevated. Consequently, accompanying the centripetal flow of G-ons there would be a centripetal flow of X-ons and a centrifugal flow of Y-ons.

together," that is, counteracting the centrifugal force of rotation. Here he is partially correct; the Searl generator induces a "side-effect" force that helps to keep it from flying apart, but this force is electrogravitic, not electromagnetic as Barrett infers. On the other hand, if the Searl generator was spinning counterclockwise, the gravity and electric field gradients would be just the reverse, and in that case the gravitational field would act to pull outward in the same direction as the mechanical centrifugal force.

Let us consider what effect this radial G-on flux would have on the Earth's gravitational field in the vicinity of the MEC. According to sub-quantum kinetics, the Earth is a net consumer of G-ons. Hence, it forms a radial concentration gradient in the G ether that extends out into space, with G-on concentration progressively rising with increasing distance from the surface of Earth. More specifically, this ether gradient, which corresponds to Earth's gravity potential gradient, diminishes according to the inverse of increasing radial distance, with the G-on concentration gradient progressively decreasing with increasing distance. This gradient induces G-ons in space to continually diffuse downward into Earth, where they are reactively consumed at a higher rate. This downward flux is illustrated by the large gray arrows in figure 10.15. This environmental gravity gradient extends vertically through the MEC and tends to exert a downward force on it; see my book *Subquantum Kinetics* for an explanation of how etheron gradients induce movement.

Figure 10.15, which displays a side view of the MEC stator and roller ring, also shows the directions in which the MEC induces G-on movement. Thus, when the MEC is operating, G-ons that normally would diffuse downward toward Earth, forming Earth's gravity field gradient, would instead be induced to move in a perpendicular direction, parallel to the MEC's rotational plane. G-ons would be drawn from above the MEC as well as from below, so G-ons residing below the MEC that normally would flow away from the MEC downward toward Earth now would diffuse upward toward the MEC's center, which establishes a low G-on concentration, or G well, when the MEC is operating. Note that the ether flux pattern mapped out here is similar to that mapped by Barrett (see figure 10.4), which was inferred from experiments with the Searl disc.

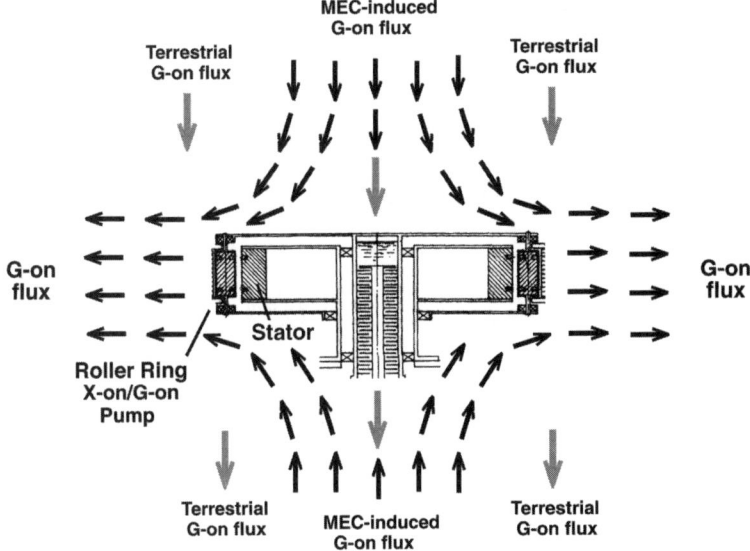

Figure 10.15. Vertical cross-section of the MEC showing how it alters the G-on flux in its vicinity when operating. (P. LaViolette, © 2006)

The alteration of the G-on trajectories correspondingly alters the gravity field gradient passing vertically through the MEC. Figure 10.16 shows how Earth's gravity potential gradient would be altered across vertical sections taken near the MEC's peripheral roller ring (right profile) and near its center (left profile). The dashed line indicates the gravity potential gradient that Earth normally produces. This figure shows that gravity potential is boosted in the vicinity of the roller ring because the outward flow of electrons (negative gravity mass) increases the G-on concentration near the periphery of the MEC and decreases it near its center. Objects approaching near this peripheral region would be gravitationally repelled. The opposite situation would occur near the MEC's center. Gravity potential at the center of the MEC would be decreased relative to environmental levels. Hence, objects approaching the MEC near its center would be drawn inward.

Along the MEC's midplane, at both its center and its periphery, the gravity gradient would approach a zero-gradient condition, giving the MEC a condition of weightlessness at a sufficiently high rate of rotation. Even so, the MEC would experience a net force repelling it

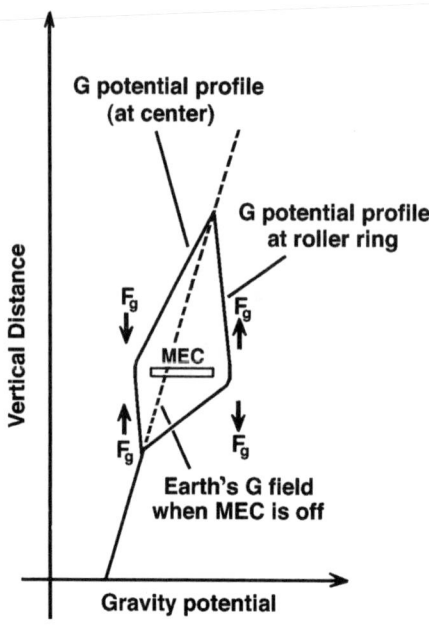

Figure 10.16. How Earth's gravity potential field is altered when the MEC is operating. Arrows indicate the direction of gravitational force outside the MEC. Along the plane of the disc, where the G profile becomes vertical, there would be no gravity gradient, hence a state of weightlessness.

from Earth, just like Searl's SEGs, which would suddenly rush upward after a brief period of hovering. The reason for this is that the MEC would surround itself with a gravity potential hill that would tend to screen the gravity potential well being generated at its hub. This lenticular gravity potential hill would behave just like a body of negative gravitational mass polarity. That is, under the influence of Earth's gravity potential gradient, it would spontaneously move up this gradient from a lower toward a higher potential. Consequently, the MEC's gravity potential hill field would migrate away from Earth's surface, the MEC moving along with it.* This scenario, which follows from subquantum kinetics, is consistent with Barrett's analogy that when the SEG is in the drive condition, its surrounding field is such that "the craft is shot out of the earth's field like a wet orange pip from between the fingers."[21]

If the ring of roller magnets was instead made to revolve in a counter-

---

*If a spacecraft was built with the MEC or Searl generator as its central engine, it would be beneficial to convey the electrons emitted from the generator's rim to the hull of the craft, which would be electrically insulated from the generator's positive plate. This would cause the entire craft to be surrounded by a high-voltage negative potential and a matter-repelling gravity potential hill.

clockwise direction, G-ons (and electrons) would be propelled radially inward, toward the center of the stator. In this case, this auxiliary vortical effect would surround itself with a gravity potential well that would tend to screen the gravity potential hill accumulated at the MEC's core. Thus, the MEC's outer field pattern would behave as though it was produced by a mass of positive gravitational polarity. Its field, then, would be attracted toward Earth and would pull the MEC downward, thereby increasing its overall weight. This is consistent with the observations of Roshchin and Godin.

The MEC's weight-reduction hysteresis may also have a ready explanation. This concerns the observation that when the MEC's rate of rotation passes a critical threshold, its weight decreasing with increasing rotor speed, if rotor speed is subsequently decreased, the MEC maintains its lowest attained weight even when rotor speed has been dramatically reduced. This may be an indication that much of the MEC's G-on pumping action comes from the field oscillations that produce its soliton pattern. Even though the rotor speed drops, this pattern would still be supplying energy to these oscillating fields and would continue to pump etherons.

Another prediction that emerges from subquantum kinetics is that the inertial mass of the MEC or SEG should decrease when it is operated in the levity mode (clockwise rotation). That is, in subquantum kinetics, an increase in G-on concentration (higher gravity potential) would affect the Model G ether reactions in such a manner as to cause a lengthening of all photon and particle wavelengths. That is, a rise of gravity potential would increase the Compton wavelength λ_0 of the electric potential wave pattern that characterizes the particle's field pattern.[22] This wavelength is related to the particle's inertial mass by the formula $\lambda_0 = h/m_0 c$, in which h is Planck's constant and m_0 is the particle's inertial mass. So an increase in Compton wavelength is equivalent to a decrease of inertial mass. This supports Searl's claim that his SEG becomes inertia-free during operation.[23,24]

These interrelated effects of gravity-induced wavelength change and gravity-induced inertial mass change, which emerge as a necessary outcome of subquantum kinetics, allow subquantum kinetics to account for well-known astronomical phenomena such as the gravitational lensing of starlight by a massive body and orbital precession. Gravitational lensing would arise from the wavelength-alteration effect, and orbital precession would arise from the inertial-mass-alteration effect. In the past, general relativity has attempted to explain both of these as outcomes of the supposed warping of space-time by massive bodies. In the twenty-first century, the outmoded theory of general relativity will be forced to relinquish its ownership of these astronomical phenomena as subquantum kinetics enters to fill its vacuum.

This inertial-mass-alteration effect has been demonstrated in the laboratory. Kazakhstan physicist Valery Mikhailov conducted two experiments in which he observed the oscillation frequency of an electron located within a charged sphere. In one, the sphere's potential was varied between −3,000 volts and +3,000 volts, and in the other, the sphere's potential was varied between −125 kilovolts and +125 kilovolts.[25,26] He found that the electron's effective inertial mass changed in direct proportion to the applied voltage potential—decreasing with increasing negative potential and increasing with increasing positive potential. This is in agreement with the predictions of subquantum kinetics. That is, by negatively charging the sphere, the gravity potential in the interior of the sphere should be raised; hence, the inertial mass of particles in the sphere's interior should be decreased, as was observed.

Apparently unaware of subquantum kinetics, Mikhailov cited his results as confirmation of a different theoretical prediction, made by Brazilian physicist André Assis on the basis of force field interactions predicted by the electrodynamic approach of the ninteenth-century German physicist Wilhelm Weber.[27] Assis's interpretation of Weber's electrodynamics theory, however, makes the different prediction that the sign of this inertial effect should depend on the particle's electric charge; hence, a proton's inertial mass should increase when the sphere is negatively charged. Subquantum kinetics, on the other hand, predicts that the proton's inertial mass, like that of the electron, should decrease with increasing negative charge. Apparently, another experiment should be performed on the oscillation frequency of a positively charged ion to determine which of these two theories is correct. Meanwhile, we currently have only anecdotal reports on the behavior of the SEG to suggest that the subquantum kinetics alternative may be the correct outcome.

Yet another prediction coming out of subquantum kinetics concerns the phenomenon of dematerialization. It predicts that if a spacecraft's gravity potential was to rise very high, the craft could become invisible or, in some cases, could dematerialize all together. In such a case, an increase in gravity potential above a certain critical threshold, g_c, would dictate subcritical conditions in the ether reaction system, which in the extreme would cause energy waves and matter to ultimately dissipate, leaving behind uniform concentrations (i.e., the vacuum state). Further experimental evidence is needed to determine whether this could account for observations of spacecraft invisibility.

The analysis presented above for the MEC would apply equally well to Searl's multiring SEGs. Since a second added ring would rotate in the same direction as the first but would be rotating twice as fast and clockwise relative to the innermost runner ring, the gravitic effects of the two rings would be additive, effectively doubling the weight loss of the apparatus at a given rotation rate. This can be compared to operating two water pumps in series. The inner ring of roller magnets would pump G-ons outward and the outer roller magnet ring would further assist this pumping action. Adding a third ring would boost this effect even more.

Compared with the beam propulsion technology discussed in chapters 7 and 8, the SEG and the MEC appear to offer a simpler approach to gravitational levitation. However, it may not be as desirable from a weaponry standpoint in that a craft using such levitation would not be able to abruptly change its direction of flight. This may explain why the military has preferred the beam propulsion technology, since high-speed maneuverability would give a craft a distinct advantage in combat. For more-peaceful applications, such as for high-speed personal transport across the globe or beyond to other planets, the SEG and MEC technology appears to be the better choice. It also has the benefit of offering a virtually limitless energy source that could ultimately eliminate global warming. Power companies, however, will need to reeducate themselves rapidly on basic physics and engineering so as not to follow the police-state tactics of their predecessors. For example, in May 1982, government agents broke into Searl's home, confiscated an SEG unit that was

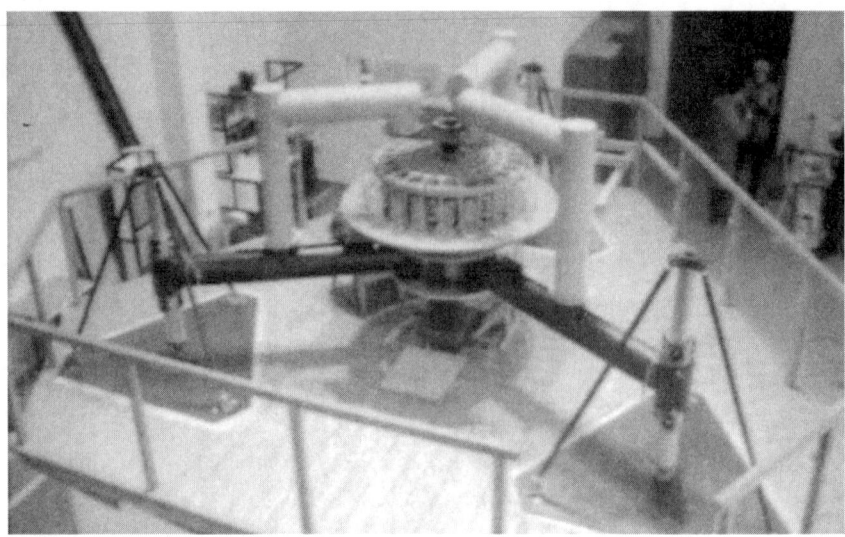

Figure 10.17. The second magnetic energy converter prototype in the process of being assembled by Roschin and Godin in Moscow.

under test supplying electricity to his house, and tore out all the electrical wiring from his house. Citing as evidence a sequence of unusually low metered electrical bills, the Southern Electricity Board then prosecuted him on trumped-up charges of "stealing electricity by means of a unique device," and sued him for a large sum of money. As a result, Searl's family broke up and he became very depressed. The Court had him confined to jail for about a year and while so detained an arsonist set his house on fire, destroying most of his records and equipment.[28]

Roshchin and Godin do not presently have access to their earlier generator. Reminiscent of Searl's plight, they report that their first prototype was stolen from their laboratory. However, they are currently working on building a second prototype at the Glushko "NPO Energomash" Company in Moscow with the intention of duplicating their earlier results (figure 10.17). Groups led by Searl in the United Kingdom and by American researcher John Thomas in the United States are also undertaking projects that intend to duplicate Searl's earlier work.

11

ELECTROGRAVITIC

WAVE EXPERIMENTS

11.1 ▪ THE DIMITRIOU GRAVITY SHIELD

Stavros Dimitriou, a professor of electrical engineering, has performed an experiment that may demonstrate some degree of gravity shielding. Like the SEG, the weight-loss effect produced by Dimitriou's apparatus appears to arise because electrons are induced to move radially in and out from a central point in a plane oriented perpendicular to the Earth's gravity field. Dimitriou arranged a set of eight wire loops in a radial pattern, the loops' ends being joined at upper- and lower-hub junctures (see figure 11.1).[1] The entire "antenna," which has a diameter of about 90 centimeters, was suspended from the ceiling by means of a thread.

A 15-volt, square-wave pulse signal having a frequency of about 75 megahertz was applied across the antenna's upper and lower hubs to excite currents to oscillate back and forth through the wire loops. These currents would flow radially with respect to the hubs and for the most part parallel to the Earth's surface. Dimitriou was supplying a radio-frequency power of only 2.5 watts to his antenna. In order to maximize the current flowing in the loops, the excitation frequency was chosen to match the antenna's resonant frequency, at which a quarter of a wavelength would fit across the loop's approximately 1-meter

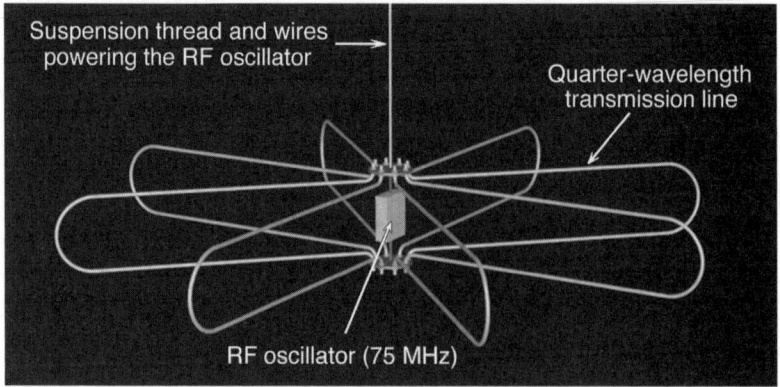

Suspension thread and wires
powering the RF oscillator

Quarter-wavelength
transmission line

RF oscillator (75 MHz)

Figure 11.1. The gravity screening radio-frequency antenna array tested by Stavros Dimitriou. (Adapted from a drawing on Jean-Louis Naudin's website, http://jnaudin.free.fr/)

radius. At resonance, the currents circulating in the loops would be far higher than the AC exciting current, with maximum values being reached at the antenna's periphery. For example, if his loop system had a Q-resonance quality factor of 90, the 100-milliamp AC signal would have excited currents reaching up to 9 amperes.

Dimitriou measured the antenna's ability to alter gravity by swinging it back and forth on its suspending thread and measuring its period of oscillation, with the AC turned both on and off. Comparing the periods, he found that the antenna's swing period was slightly greater when the AC was on. Since the length of the antenna's suspending cord did not change, he concluded that the antenna was able to locally reduce the gravitational accelerating force by 1.3 percent, with the period of any swing of the pendulum being determined by its length and the force of gravity acting on its mass.* He believed that this reduction arose because currents induced in the loops were somehow creating a local gravity-shielding effect. He also swung a small pendulum bob that he held near the radio-frequency antenna and, by timing its swing, found that its period also increased when the antenna was excited.[2] Consequently, he concluded that objects in the immediate vicinity of the antenna were similarly affected by the gravity reduction.

---

*Period $= 2\pi\sqrt{\frac{L}{g}}$, in which L is the length of the pendulum cord and g is the gravitational acceleration.

The question that arises, however, is whether the reported percentage of reduction in the gravitational acceleration, g, is a real effect or a statistical artifact. There is considerable error involved in measuring the period of a pendulum. Even with no variation of the gravitational force or the pendulum length, the clocked period of the bob's swing can vary from one clocked period to the next due to friction in the pendulum string, room air currents, and timing inaccuracies. Consequently, such a pendulum period measurement has meaning only if many sets of measurements are made with the AC signal both turned on and turned off, for which period averages and period standard deviations are calculated. If the difference between the average length of the "on" periods and the average length of the "off" periods is shown to be significantly greater than the data standard deviation, then one might surmise that a real effect on gravity is present. But past reports of the existence of a g reduction effect may instead be misinterpreting statistical artifacts of the measurement process as evidence of a real effect.

French researcher Jean-Louis Naudin attempted to duplicate Dimitriou's pendulum antenna experiment.[3] He built a similar wire loop array and excited it at a resonant frequency of 83 megahertz. Whereas Dimitriou had used a handheld stopwatch to clock the antenna swing period, Naudin allowed the antenna pendulum to cut the path of a laser beam as it swung and recorded the swing intervals electronically. Measuring the period increase of the pendulum swing, he calculated a much greater, 7 percent reduction in gravitational acceleration when the oscillator was switched on. In another version of the experiment, Naudin tested the period swing of an antenna having wire loops that each included an extra turn at its outer extremities, with the extra loops being oriented in the horizontal plane. With this design, Naudin measured an 11 percent reduction in gravitic force, for a power consumption of about 3.5 watts. Both test rigs were found to have a Q value of about 10.*

---

*Naudin's Q value was likely lower than that of Dimitriou's antenna because Naudin made his loops from uncoated wire that was subject to surface oxidation and, hence, would have had a greater surface resistance. At megahertz frequencies, such as were used in this experiment, current flows mainly on the wire surface. Hence, surface oxides can diminish the antenna's Q value.

However, like Dimitriou, Naudin reported only an estimated percentage of gravitational reduction, without providing the standard deviation values to allow one to judge the accuracy of the method of measurement. Thus, until a test of the radio frequency (RF) pendulum antenna is properly performed giving the purported RF-induced period change along with the variance of the period measurement, it is not possible to determine for certain if the effect is real. All that can be said at present is that it is an interesting subject for investigation.

Dimitriou also proposed another version of his RF pendulum antenna experiment in which the wire loops of his antenna are replaced by a disc-shaped printed circuit board having copper-clad upper and lower surfaces electrically joined at its periphery (figure 11.2).[4] In this way, the two surfaces would form a pancake-shaped RF cavity that would have a beneficial high Q value. He proposed that the lift effect would be more easily seen by testing a 30-degree pie section of the disc and observing its tilt when energized with radio-frequency power, given that a greater weight loss would be obtained at the sector's periphery than at its center.

Naudin has attempted to duplicate Dimitriou's RF pie-wedge experiment. The wedge shown in figure 11.3 was balanced around its center of gravity and checked for evidence of any tilt when energized with a radio-frequency-exciting circuit connected at its apex. Naudin reported that he was able to get some weak upward movements of the wedge's rim, but that these movements were not easily reproducible.

In 2007, I witnessed several tests of the disc version of the Dimitriou RF pendulum antenna being resonantly excited with 20-volt pulses, but in my opinion there was no change of the pendulum's period above the

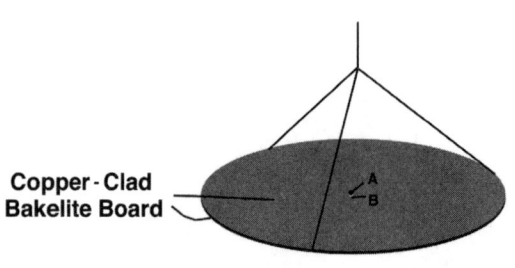

Copper - Clad Bakelite Board

Figure 11.2. The disc-shaped RF antenna proposed by Dimitriou in 2001. The upper- and lower-disc surfaces would be energized with radio-frequency power near points A and B.

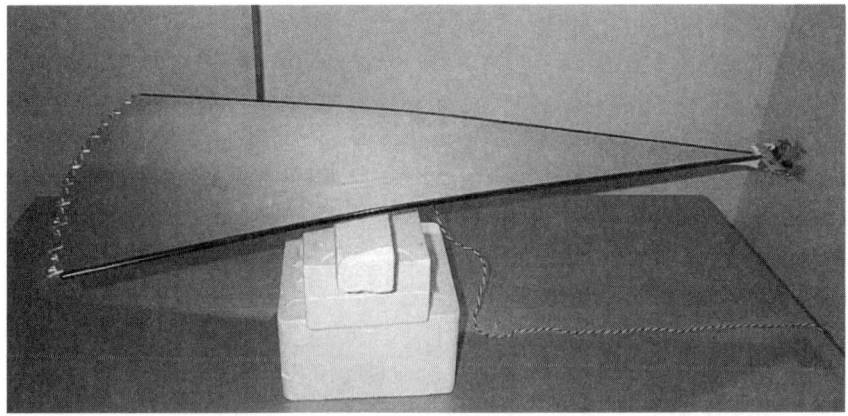

Figure 11.3. A wedge-shaped printed circuit board constructed by Naudin as a test of Dimitriou's experiment. (Courtesy of J.-L. Naudin, from his website, http://jnaudin.free.fr/)

margin of measurement error; hence, no clear indication that gravity was being reduced when the disc was energized. If there was an effect, it was too slight to be detected. By accurately recording large numbers of pendulum swing periods, it should be possible to substantially reduce the variance in the period measurement data, in which case the sensitivity of the test to detecting small alterations in gravity might be improved. If future tests demonstrate that a gravity-screening effect is produced, this would be one of the most promising among electrogravitic technologies, not only because of its simplicity, but also because of its high energy efficiency.

Although it may be too early to theorize about how Dimitriou's wire and disc antennae might produce a gravity-screening effect, we might venture an explanation similar to that given for the operation of the Searl disc. That is, one might expect a gravity-screening effect to arise because G-ons are being induced to move in a horizontal direction (i.e., perpendicular to the Earth's gravity field gradient). The AC resonance in the antenna would set up oscillating electric potential field gradients directed radially inward and outward along the length of the upper or lower conductor wires. Alternatively, in the case of the RF disc, these would be directed radially to and from the centers of upper or lower disc surfaces. These oscillating field gradients would be steepest near the periphery of the wire loops (or disc surfaces), where the induced currents would also be maximal. According to subquantum kinetics, these oscillating electric fields would be accompanied by oscillating gravity potential fields whose field gradients would also be maximal at the loop or disc periphery. Consequently, the radial back-and-forth movement of electrons would be accompanied by a horizontal radial back-and-forth movement of G-ons. The oscillating gravity field propelling these G-ons in the horizontal plane would also entrain G-ons associated with the vertical G-on flux that naturally flows toward the Earth in response to the Earth's gravity field gradient. Consequently, in the vicinity of this oscillating electrode, the ambient G-on flux and its associated G-on concentration gradient would be deflected from their normal vertical orientation toward the horizontal. As a result, objects in the vicinity of this disc would no longer feel the full downward pull of the Earth's gravity, being effectively screened from this field by the disrupting effect of this AC resonator.

In the Searl disc, the pulsing fields were nonreversing and always induced G-ons (and electrons) to move radially outward toward the periphery of the disc. In that case, then, an occupant of such a vehicle would feel a centripetal inward-pulling gravitational force. In a gravity shield created by radio-frequency excitation of a Dimitriou antenna, on the other hand, the direction of this radial G-on flux would reverse 100 million times per second or so. Consequently, the horizontal gravity field component would have a net-zero value.

Brown may have inadvertently been producing such a gravity-screening effect in his vertical-lift electrokinetic apparatus. His electrokinetics patent proposed applying AC to a negatively charged "half-wave radiator" disc electrode positioned at the tip of a conical dielectric (see figure 3.8 in chapter 3). Like Dimitriou's disc, this would have propelled G-ons radially inward and outward in the plane of the disc and set up a gravity-screening field in the electrode's immediate vicinity. The induced G-on flux would have locally redirected the G-on flux that normally moves downward toward the Earth to flow in a horizontal direction, thus decreasing the gravity field gradient across the saucer and locally reducing the g-value affecting the saucer's mass. The AC oscillations induced in the positively charged canopy electrode would also have contributed to the gravity-screening effect. Thus, part of the lift that Brown was getting might be attributed to an induced gravity-screening effect.

11.2 ■ LOW-VOLTAGE SAWTOOTH-WAVE EXPERIMENTS

In the mid-1990s, Dimitriou conducted an experiment in which he radiated sawtooth waves from the end of a specially configured dipole antenna.[5] The antenna, which is shown in figure 11.4, measured 18.5 by 12.5 centimeters. A detailed explanation of why it was constructed in this fashion may be found in his master's thesis. He excited the antenna with a 1.1-megahertz, 15.5-volt peak-to-peak RC-Norton signal of the sort graphed in figure 7.6b. The oscillating current would have reached its maximum value along its central wire axis and attained lower values in the two outlying wires, each of which was capacitively loaded with a total of 53 picofarads.

Dimitriou discovered that the antenna created a gravitational force in line with its central axis when it was being excited with this sawtooth wave. He suspended a 4.1-gram, 1.5-centimeter-diameter glass sphere at the end of a 2-meter cotton string from the ceiling, positioning the sphere close to one end of the central antenna wire. A grounded copper

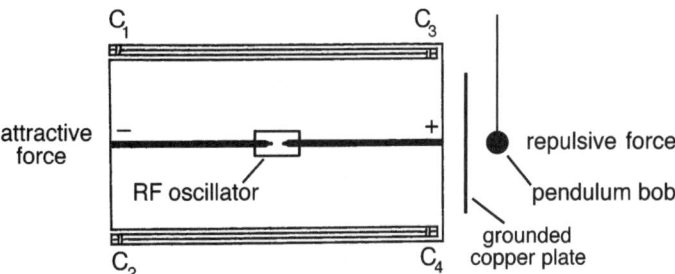

Figure 11.4. An antenna constructed by Stavros Dimitriou that radiates longitudinal gravity waves from either end of its central conductor. C_1, C_2, C_3, and C_4 are loading capacitances. The pendulum bob was hung 3 centimeters from the antenna in line with the central conductor.

plate was placed between the antenna and the pendulum to screen any electromagnetic effects. He observed that when the antenna was energized, a longitudinal force was exerted on the nearby test pendulum. When placed near the negative end of the central antenna wire, the bob was attracted with a force of 4 dynes, indicating that it was subject to a gravitational attraction of 0.1 percent. When placed near the positive end of the central wire, it was repelled with a force of 3 dynes, that is, repelled with a gravitic force of 0.08 percent g. Dimitriou theorized that this gravitic force was exerted on the test mass by a beam of gravity waves emitted from the end of the central wire.

In yet another experiment, Dimitriou demonstrated that sawtooth waveforms produced frequency shifts in light being emitted from the junction of a light-emitting diode (LED). He reasoned that the LED's junction functioned as a miniature capacitor and that the sawtooth wave created a gravitational force that induced it to move, with the motion producing a Doppler shift of the LED's frequency. He measured the resulting velocity change of the junction by observing the amount and sign of the frequency Doppler shifting that this motion induced in the LED's light. A blueshift (frequency increase) indicated a forward thrust of the LED's light-emitting junction, and a redshift (frequency decrease) indicated a reverse thrust of the junction.

Dimitriou excited the LED with a 1.85-megahertz sawtooth wave having an amplitude of about 2 volts peak-to-peak. The voltage was adjusted so that the LED began to emit its light just when the voltage reached its peak value. This was done because as the LED junction reaches full luminance, it loses its capacitive characteristics and, hence, no longer functions electrogravitically.

He studied the effect of two types of waveforms. One was an RC-RC waveform of the type pictured in figure 7.5b, with an exponential voltage rise that lasted one-third as long as its exponential voltage decline. The other sawtooth wave was a ramp-type wave having a linear rise and linear fall, also in a one-to-three duration ratio. The leading edge of the exponential waveform produced a frequency blueshift equal to 8.16 millimeters per second, and its trailing edge produced a frequency redshift equal to 2.85 millimeters per second, which was 2.86 times less. The ramp sawtooth waveform surprisingly produced frequency shifts as well, with the leading-edge velocity shift being 2.57 times the trailing-edge velocity shift. Also, the ramp wave produced velocity shifts that were about 2.7 times less than those produced by the exponential waveform.

The velocity change cannot be attributed solely to the electrogravitic effects of virtual charges since, if such were the case, the ramp waveform, whose potential varied linearly with distance, should have induced no velocity change in the LED junction. Alternatively, it is possible that the frequency shift of the LED junction arose because the sawtooth waves induced changes in the electrical characteristics of the junction through some unknown effect.

Dimitriou also performed a series of experiments in which he repeatedly charged and discharged a parallel plate capacitor with an RC-Norton sawtooth wave similar to that shown in figure 11.5 and looked for evidence of whether the capacitor was experiencing a gravitational thrust.[6-8] He experimented with sawtooth-wave frequencies ranging from several hundred kilohertz up to slightly more than 1 megahertz, having a comparably low peak voltage of up to 12.4 volts. He produced this waveform using the circuit shown in figure 11.6, which consists of a 7555 integrated circuit chip, two capacitors, and a charging resistor R_1 in series with the test capacitor. The value of this resistor was chosen to be 2,367 ohms, $2\pi Z_o$, in which Z_o is the free space impedance of 376.7 ohms.

Dimitriou applied this waveform to two capacitors attached to either end of a 38-centimeter-long rotor arm, repeatedly charging and discharging them at a 238-kilohertz frequency (see figure 11.7). Each capacitor measured 8 centimeters and consisted of a 1-centimeter-thick

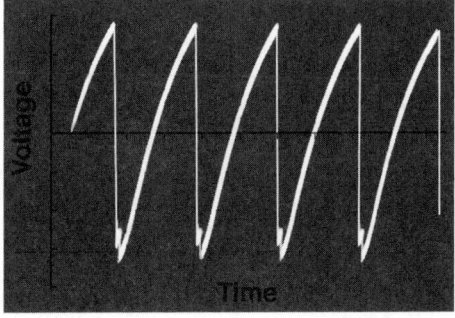

Figure 11.5. An RC-Norton sawtooth wave having a gradual exponential voltage rise and rapid linear voltage decline. The ramp voltage drop was made to last about 3 percent of the duration of the voltage rise phase.

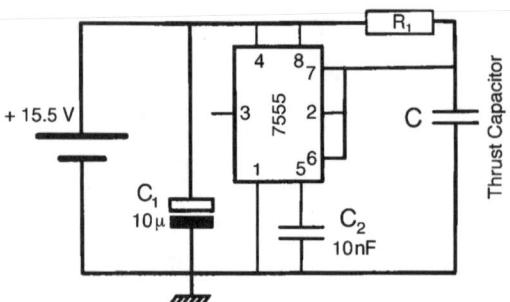

Figure 11.6. Circuit diagram for producing the RC-Norton-type sawtooth wave.

slab of copper flanked by two thin sheets of dielectric that, in turn, were flanked by 0.5-millimeter-thick bronze end plates. The outer bronze plates served as the capacitor's positive and negative electrodes. Dimitriou reported that when he energized them with this sawtooth wave, the capacitors developed a thrust in the negative-to-positive direction, causing the rotor arm direction to twist by about 5 millimeters, which was equivalent to a one-degree rotation. He also ran a similar test made with an air gap instead of a dielectric between the capacitor's plates and reported that it also produced a thrust. Thus he concluded that the effect did not depend on the presence of a dielectric.

In March 2007, I conducted my own tests of Dimitriou's thrust capacitor effect. I constructed printed circuit board capacitors measuring about 3.5 by 5 centimeters and having a capacitance of about 410 picofarads. A piece of aluminum foil formed one plate and the copper-clad printed circuit board formed the other plate, both separated by a layer of double-stick tape. I also built an RC-Norton oscillator based on the circuit diagram shown in figure 11.6 and used it to energize the capacitor with 15-volt sawtooth waves having a frequency of about 1 megahertz. I hung an 80-centimeter-long pendulum bob near one face of the capacitor, but observed no deflection when the capacitor was energized (i.e., $\Delta x < 1$ mm). This indicated that any lateral gravitational acceleration produced in the immediate vicinity of the capacitor would have had to be smaller than 0.1 percent g. I also used a waveform generator built by Dimitriou and got the same null result.

For another test, I constructed a capacitor measuring 10 by 13.5 centimeters and placed it horizontally on a milligram balance that was sensitive to 1 milligram weight changes. When energized with the RC-Norton sawtooth waveform, no weight change was observed. Since

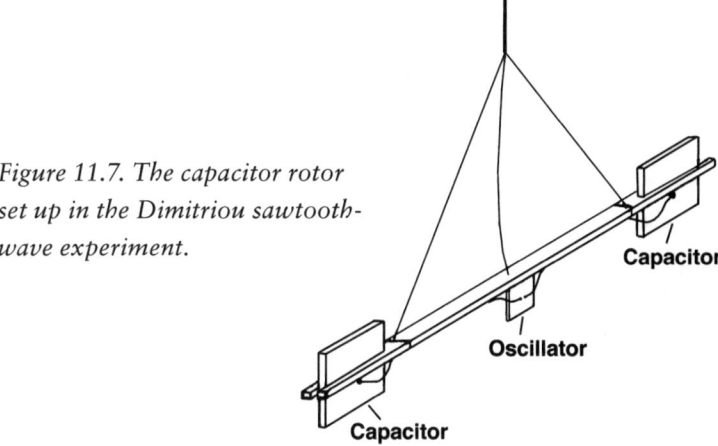

Figure 11.7. The capacitor rotor set up in the Dimitriou sawtooth-wave experiment.

Capacitor

Oscillator

Capacitor

the capacitor itself weighed 85 grams, this indicated that there was no change in gravitational acceleration larger than 0.001 percent *g*.

To check the Dimitriou capacitor rotor experiment, I built a rotor setup similar to that shown in figure 11.7. The capacitors were constructed from two 30-mil rectangular copper slabs measuring 10.5 by 14 centimeters and separated from one another by a thin polyethylene-film layer. The capacitors each weighed about 200 grams and had a capacitance of 655 picofarads. They were mounted at opposite ends of a 90-centimeter-long stick that was suspended from the ceiling at its center point (see figure 11.8). The sawtooth-wave generator and its battery-power supply was attached to the stick. I worked with Professor Panagiotis Pappas and his assistants to carry out tests of the apparatus in his Athens laboratory. We energized both capacitors with the RC-Norton wave, but could see no persistent rotation of the apparatus.

Checking the sawtooth wave with an oscilloscope, we found that an unwanted high-frequency oscillation was present in the waveform, which was due to inductance added by a long lead wire connecting the wave generator to both capacitors. To eliminate this oscillation, we placed the wave generator as close as possible to one of the capacitors and disconnected the wire supplying RF to the other capacitor. The second capacitor, then, was used as an inert counterbalance weight at the opposite end of the rotor arm.

Before carrying out a retest, we constructed a photo-relay circuit

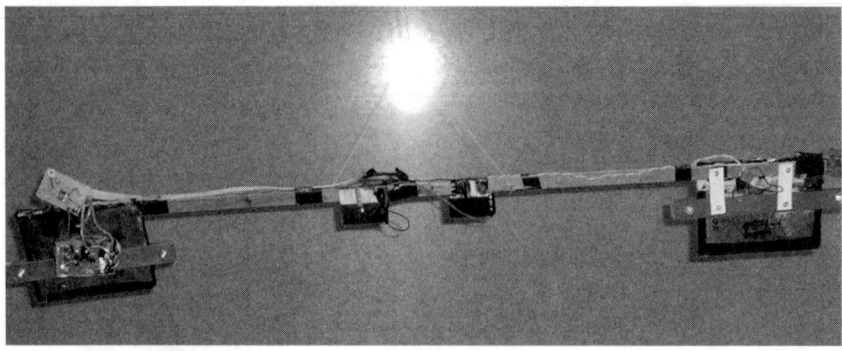

*Figure 11.8. A capacitor rotor constructed by the author. (P. LaViolette, ©
2007)*

that would reverse the sawtooth-wave polarity to the capacitor. Thus,
by operating a switch in the adjoining observation room, we could
momentarily turn on a room light and activate the photo-relay circuit
to reverse the capacitor polarity without actually touching the apparatus. Pappas also set up a video camera on the ceiling that looked down
on the capacitor rotor so we could view any movement of the apparatus remotely from the observation room. This was done to minimize
the chance that air currents produced by movement of people in the
room or by their breathing would disturb the apparatus. We found
that it would take at least an hour for the swinging of the apparatus to
subside, and even then oscillations would still be present due to room
drafts.

We energized the capacitor with a 1.5-megahertz RC-Norton
wave, remotely reversing the waveform polarity on the capacitor, but
we saw no comparable reversal or alteration in the swing of the apparatus. We concluded that no gravitational-thrust effect on the rotor
could be seen other than the rotor's ongoing slow oscillations arising
from room drafts. Using a makeshift tensometer, I determined that
if the capacitor was able to develop a gravitational force of at least
0.04 percent *g*, a persistent rotary movement of the apparatus should
have been observed. The null result, then, calls into question claims
that low-voltage waveforms are able to induce gravitational thrusts on
capacitors.

Later, Pappas modified the rotor assembly so that the sawtooth-wave generator was supplied with 16 volts DC via fine feed wires attached to the rotor's suspension wire, with the power switch being situated in the observation room. In this way, we were able to turn the wave generator on and off from our remote location. We also deactivated the polarity-reversing relay to omit any vibrations that arose when the relay was energized. After allowing several hours for the apparatus to equilibrate, power was turned on and later turned off, but no thrust could be seen other than the swings arising from room drafts.

Naudin reported that he had duplicated the Dimitriou capacitor rotor experiment and had observed a 1-degree rotation of the apparatus. Our rotor was constructed in a fashion similar to that of Naudin's except our capacitor plates were made from sheet copper instead of sheet aluminum and we activated only one of the two rotors. Based on our results, we are left to wonder whether the 1-degree rotation Naudin reported was due to air currents and not to a true detection of gravitational thrust. In his description of the experiment, Naudin acknowledged that air currents could pose a problem to the stability of the capacitor rotor arm.

Dimitriou's original rotor experiment used capacitors that were much more massive than those used in our experiment, since they each incorporated a 1-centimeter-thick copper slab. Thus, the rotor experiment should be repeated using copper slabs of similar thickness to see if a positive result is obtained. Another experiment that should be duplicated is a rotor experiment that Dimitriou demonstrated to professors at the University of Manchester. In that case, two 1.5-centimeter-diameter mica capacitors were placed at opposite ends of a 25-centimeter rod that was suspended at its center. When energized with an RC-Norton waveform, the arm reportedly rotated 20 degrees.

Dimitriou has developed a working theory that has guided his experimental discoveries, details of which are given in his master's thesis. He makes a number of deductions from conventional electrostatic and gravitation theory that have led him to assume an equivalence between the rate of change of a capacitor's charging current and a gravitational acceleration acting on the capacitor. Alternatively, he formulates this as a relationship between an accelerating rate of change of a capacitor's electric field intensity and a consequent gravitational acceleration acting on the capacitor. As shown in the text box below, his electrogravitic acceleration relationship (13) is in agreement with the subquantum kinetics prediction of how virtual charge induces gravitational force given by equation 8 in chapter 4.

Equivalence of the Dimitriou Electrogravitic Theory with That of Subquantum Kinetics

Dimitriou assumes that a constant current, i, producing a constant rate of change of charge, dQ/dt, on a capacitor should be equivalent to effecting a proportionate state of motion in the capacitor and inducing its movement at a constant velocity, v, relative to the charge's frame of reference, mathematically expressed as:

$$i = \frac{dQ}{dt} \propto -v \qquad (10)$$

Alternatively, given that a constantly charging current will produce a constant increase in the rate of change of electric field intensity, $d\mathbf{E}/dt$, he states that this rate of increase should be equivalent to a proportionate state of motion at velocity, v, expressed as:

$$i = \varepsilon S \, \frac{d\mathbf{E}}{dt} \propto -v \qquad (11)$$

in which ε is the dielectric constant of the capacitor and S is its surface area. Note that electric field intensity is the same as the negative of the voltage gradient across the capacitor (i.e., $\mathbf{E} = -\nabla\varphi_E$). Thus he posits that a linear increase in charge on a capacitor plate, or linear increase in voltage gradient across a set of capacitor plates, is equivalent to a virtual velocity vector directed toward the capacitor's positive pole, but that this does not result in any acceleration or motional displacement of the capacitor.

Further, Dimitriou deduces that the rate of change of current, di/dt, producing an accelerating change in the amount of charge on a capacitor, d^2Q/dt^2, or an accelerating change in the electric field intensity across the capacitor plates, $d^2\mathbf{E}/dt^2$, should be equivalent to effecting a state in which the capacitor behaves as though it was subject to a gravitational acceleration, $\mathbf{a}_g$, expressed as:

$$\frac{di}{dt} = \frac{d^2Q}{dt^2} = \varepsilon S \frac{d^2\mathbf{E}}{dt^2} \propto -\mathbf{a}_g \tag{12}$$

Thus, he supposes that an accelerating increase in charge on a capacitor plate, or accelerating increase in voltage gradient across a set of capacitor plates, is equivalent to an acceleration of the capacitor toward its positive pole. He presumes that exertion of a gravitational force and displacement of a capacitor occur only in situations of the second kind, as in equation 12, in which the capacitor's electric field intensity increases or decreases nonlinearly with time.

On the assumption that the electric field intensity is conceived as a wave traveling at the speed of light and that its amplitude changes with distance in the same manner as it changes with time, equation 12 may be expressed in terms of the change of electric potential with respect to distance, r, rather than with respect to time, t, as follows:

$$\varepsilon \frac{d^2\mathbf{E}}{dr^2} = \varepsilon \frac{d^2}{dr^2}\left(\nabla\varphi_E(r)\right) = \varepsilon\nabla\left(\nabla^2\varphi_E(r)\right) \propto \mathbf{a}_g \tag{13}$$

in which $\mathbf{E} = -\nabla\varphi_E$. Given that $\mathbf{a}_g = \mathbf{F}_g/m$, this may be seen to be identical to equation 8, the subquantum kinetic coupling relation derived in chapter 4 that expresses the electrogravitic effects of virtual charge densities.

Dimitriou's theory relates electrogravitic acceleration to charging current, which is advantageous from an electrical engineering standpoint since the output of a wave generator is often described in terms of the current it generates. Subquantum kinetics, however, has the advantage of offering a conceptual model that allows one to see what might be going on to cause there to be this electrogravitic linkage. It allows one to see the connection between electric field potential, virtual-electric-charge density, gravity field potential, and gravitational acceleration. Subquantum kinetics also shows that a first derivative of electric field potential may also be important in the high-frequency regime. It is interesting that Dimitriou arrived independently at an experimentally based formulation that is equivalent to the electrogravitic formulation of subquantum kinetics.

In summary, further research is needed to check out Dimitriou's findings. I believe that an electrogravitic thrust effect should exist, but at wave voltages much higher than the ones Dimitrou and Naudin were using. Additional experimentation with sawtooth waves in the kilovolt range should hopefully bring this electrogravitic thrust phenomenon up to a detectable level. Similarly, additional experiments should be conducted to excite Dimitriou's disc antenna with RF in the kilovolt range to determine with greater certainty whether gravity screening is produced.

12

HIGH-VOLTAGE ELECTROGRAVITICS EXPERIMENTS

Investigations into electrogravitics have continued outside of the classified world as amateur researchers, inspired by Brown's work, have striven to reproduce his results. Experiments conducted by a few researchers are reviewed below. The reader should be aware that one takes a considerable risk when working with high voltages, since high-voltage power supplies can deliver lethal shocks. Thus, it is not recommended that people undertake these experiments unless they are thoroughly familiar with the hazards involved and have taken proper safety precautions.

12.1 ▪ TOM TURMAN'S ELECTROKINETICS EXPERIMENTS

Electrical engineer Tom Turman's initial inspiration to do research in electrogravitics came after reading a 1958 article by Gaston Burridge about Brown's work. In an attempt to duplicate some of Brown's flying-disc experiments, Turman conducted private electrogravitics research between 1965 and 1972, while studying electrical engineering at Texas

Tech University. In 1968, he began corresponding with Brown, both by telephone and by letter. He told Brown about the experiments he was performing and asked if Brown could clarify some aspects of his flying-disc experiments. Brown was impressed with Turman's independent work and at one point was seriously considering hiring him as his assistant. Unfortunately, circumstances did not permit him to follow through.

Turman did not have university funding to help him carry out his research. Most of the equipment he acquired for his task was either given to him or purchased at a low price from electrical surplus dealers and subsequently reconditioned. He had a homemade power supply capable of delivering 300 kilovolts DC at up to 100 milliamperes, an eight-channel oscillograph for use in measuring voltage, current, and force, and a capacitance-type gauge capable of measuring small changes in the weight of a suspended electrogravitic test device.

Turman built several types of lightweight, asymmetrical capacitor devices. One cylindrical device that he built weighed between 3 and 6 grams and achieved maximum thrusts equal to as much as half of its weight.[1] For this design, he used a sheet of insulating plastic film that was a few mils thick and was wrapped around a cylinder 4.75 inches in diameter and 4.4 inches long (figure 12.1). A 2.4-inch-wide aluminum-foil skirt was wrapped around the lower end of the cylinder, with a 0.5-inch overlap onto the plastic film, to serve as the negative electrode. The positive electrode was an aluminum-foil tube measuring 0.25 inch in diameter and 3.6 inches in length that was located at the opposite end of the cylinder and positioned in line with the cylinder's axis so that half the tube extended into the cylinder's interior.

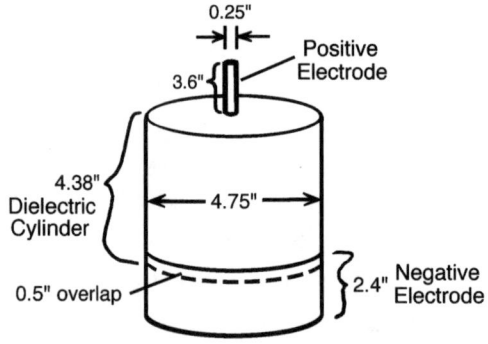

Figure 12.1. A cylindrical-shaped ion-producing device built and tested by Tom Turman. Its construction was based on reports of "lifter" tests carried out by Thomas Townsend Brown. (Based on a sketch by Turman)

Turman found that lift increased exponentially with increasing voltage, V, approximately as V^2 to V^3, confirming similar results found in Brown's earlier work. Turman's cylinders developed thrusts ranging from 0.3 to 3.5 grams (5 to 50 grains) when energized with voltages ranging from 35 to 135 kilovolts with a current draw of a few milliamps. He found that the amount of thrust depended on the type of insulating film he used in making the cylinder. He obtained greater thrusts with materials having greater dielectric constants, observing thrust to increase according to K^2 to $K^{2.3}$ (figure 12.2). He also found that thrust depended on the dimensions of the device, such as the length of the positive electrode and its depth of penetration into the cylinder, the length of the cylinder, and to some extent the width of the aluminum-foil skirt. Data from thrust tests he conducted on a 13.75-inch-diameter cylinder are presented in table 1.

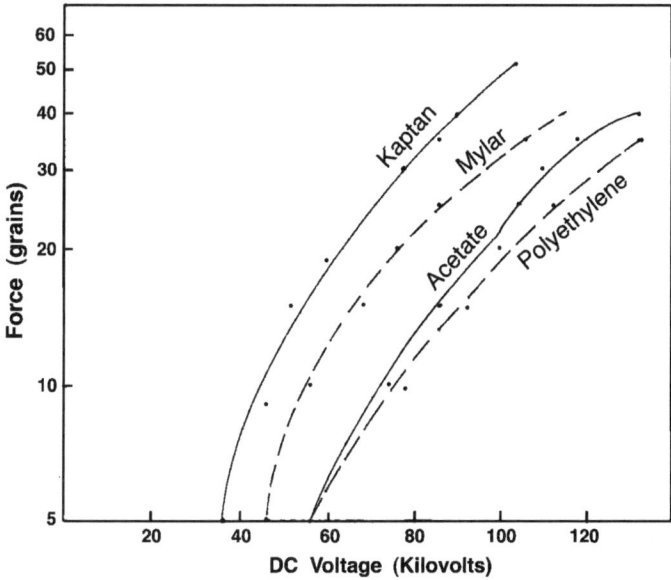

Figure 12.2. Chart of lift produced by 4.75-inch-diameter cylindrical test devices made from various types of plastic film with differing values of dielectric constant (K). Curves are shown for Kaptan (K = 3.7), Mylar (K = 3.1), acetate film (K = 2.9), and high-density polyethylene (K = 2.3). (After T. Turman)

TABLE 1. VARIATION OF THRUST WITH CHANGES IN LENGTH
OF CYLINDER AND DEPTH OF POSITIVE ELECTRODE
(Tested at 250 Kilovolts)

Length (inches)	Depth (inches)	Lift (grams)
10.9	2.5	22
11.4	2.3	20
11.6	1.1	20
12.1	2.4	16

Turman obtained the greatest thrust when applying high-voltage DC pulses, an effect Brown also had noted. When energized in a pulsed fashion, Turman's device achieved thrusts nearly sufficient to self-levitate. He found that his cylinder would also develop a thrust when energized with AC, but not as much as when energized with DC. The unbalanced electrostatic force effect described in chapter 3 accounts for the thrust that would be developed when a reversed voltage polarity was applied to the cylinder. Regardless of whether the smaller upper electrode had a positive or negative polarity, more ions would have been emitted in the vicinity of the small upper electrode, where the electric field density was greater, and this would have exerted a strong upward-repulsive force. This force would have been greater than the downward-repulsive force component produced in the vicinity of the lower-cylindrical electrode, where fewer oppositely charged ions would have been generated and where the consequent repulsive force would primarily have been directed radially outward, away from the cylinder's central axis. Consequently, electrostatic ion-repulsion effects appear to dominate over electrogravitic effects in lightweight devices producing a nonlinear field. The same applies to the lifter experiments carried out by later researchers.

Turman noted that his cylindrical thruster device was a copious producer of ion wind, and hence, he could not rule out ion wind as the principle mode of propulsion. However, observing that the electric field between the two electrodes was highly nonlinear, he predicted that the device should exhibit a discernible Biefeld-Brown effect when tested in a vacuum environment. He had almost finished building the vacuum equipment necessary to conduct these tests when he had to disassemble everything and move his residence due to a job change. Afterward, he never reassembled his equipment.

Turman's cylindrical device did not come close to giving the kind of vertical lift that Brown had obtained from his 100-gram, triarcuate-shaped disc (figure 3.2). Nevertheless, Brown expressed considerable interest in the design. In one of their telephone conversations, when Turman told Brown about getting really good thrust from his cylinder devices, Brown quizzed him extensively about them and said, "[M]ake some drawings and send me those drawings because I am really interested in those cylinders." Turman sent him some drawings and data, and subsequently, on November, 1, 1971, Brown wrote back, saying, "Your sketch shows a point and ring configuration of electrodes with an intermediate dielectric tube. I take it the ends of this tube are open and the airflow is in the direction of the divergent field. This would make the tube assembly move in the opposite direction, that is, toward the small positive electrode. Is this not so? . . . Have you observed any thrust with the positive end of the tube closed?"[2]

Another asymmetrical capacitor design that Turman tested consisted of a flat 8-inch-diameter disc of polyethylene film with aluminum-foil electrodes attached to its upper and lower surfaces.[3] The upper

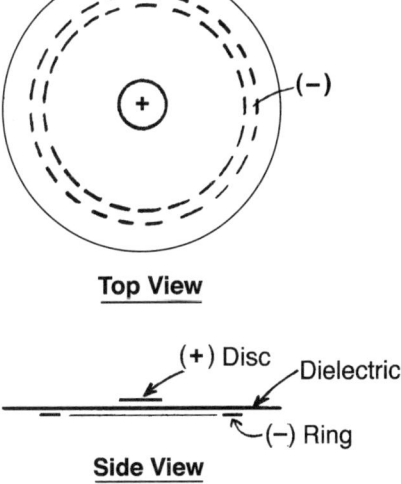

Figure 12.3. A flat-profile electric disc designed and tested by Tom Turman. (Based on a sketch made by T. Turman)

(positive) and lower (negative) electrodes consisted of a 1-inch-diameter foil disc and a 7-inch-diameter foil ring (see figure 12.3). The disc was supported by a balsa-wood structure attached by monofilament lines to an equal arm balance for weight measurement. The device was found to develop a lift of up to 30 percent of its weight when energized. As with the cylindrical device, this flat disc also developed lift when AC was applied to it, but not as much as with DC voltage. Also, the device was found to perform better with DC pulses than with steady DC.

Turman also attempted to duplicate Brown's flying-disc device. He made a 28-inch-diameter saucer out of cardboard and covered it with aluminum foil (figure 12.4).[4] The disc was 2.5 inches thick at its center and tapered to 0.12-inch-diameter blunt edge at its periphery. He curved a 125-mil-diameter brass rod 70 degrees around the disc to serve as the outboard positively charged electrode and spaced it 4.75 inches from the negatively charged disc with a series of Plexiglas insulators. In a static test, the device developed only 4.5 grams of thrust when energized to 80 kilovolts. This yielded a thrust-to-weight ratio of only 1 percent, far lower than even the thrusts observed in the 1952 Office of Naval Research tests in which Brown's discs developed thrusts of 18 grams under a charge of 47 kilovolts. Because of this disappointing performance, Turman wrote to Brown inquiring what might be wrong with the basic design he had used.

In his November 1, 1971, letter (see appendix A), Brown responded by drawing a picture of his electric disc (shown in figure 2.6). This indicated that Brown used a much smaller-gauge wire as his positive electrode, one that had a diameter of 1 mil (0.001 inch). In his electrokinetic apparatus patent, Brown noted that small-diameter wires should be used for discs energized with voltages less than 125 kilovolts. For discs energized at higher voltages, he recommended that the positive electrode consist of a hollow pipe or rod having a diameter of 0.25 to

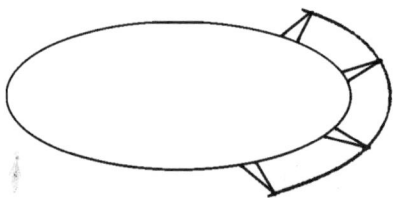

Figure 12.4. A device, similar to Thomas Townsend Brown's electric disc, built and tested by Tom Turman. (Based on a sketch by T. Turman)

0.50 inch. Turman, however, tested his discs in the lower-voltage range with a wire that was more than one hundred times larger in diameter than Brown would have used. This may explain why Turman got lower thrust from his disc. Also, the radius of curvature of the edge of the triarcuate disc in Brown's drawing was eight times larger than in Tom's design (0.50 inch rather than 0.06 inch). In designing his disc, Turman had originally been guided by the diagram Brown gave in his 1960 electrokinetic apparatus patent.

In a subsequent telephone conversation, Turman learned from Brown that the patent did not include a depiction of an optimal design and that the blunt-edge design performed better because it produced a more nonlinear field configuration between its electrodes. Brown emphasized the importance of creating a nonlinear E-field to maximize thrust. Although Brown's patents mentioned nonlinearity, Turman found that Brown placed far more emphasis on this point in his personal conversations. If Turman had decreased the diameter of his positive leading-edge electrode wire by a factor of 120 and had shaped the edge of his negatively charged disc to have a gradual curvature, he would have greatly increased the nonlinearity of the disc's electric field. This would have produced a substantially greater ion emission from the vicinity of his positive electrode, whose forward-acting repulsive forces would have translated into a substantially higher forward thrust for the disc.

Turman asked Brown many other questions as well: Compared with the results of a static test, is the propulsion efficiency of the device increased if it is allowed to run in a circular course? Does the ratio of thrust to weight increase as the size of the disc is increased? What was the largest-size disc that you constructed and what were the problems you encountered? Brown was reticent on the subject.

Turman also built an asymmetrical capacitor device to test the performance of a slatlike device described in Brown's 1960 electrokinetic apparatus patent (see figure 12.5).[5] It bore a close resemblance to the lifter devices that later became popular among electrogravitics hobbyists, but it was of much heavier construction. Turman used a stack of four brass slats as the negative electrode, each slat measuring 1 inch by 12 inches, and a 12-gauge (110-mil-diameter) copper wire as the

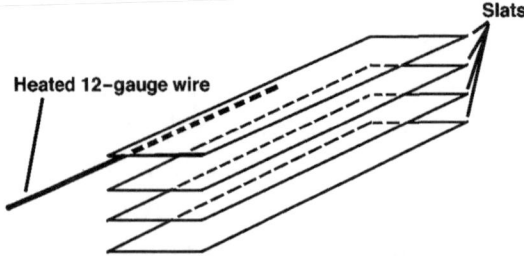

Figure 12.5. A slat-style asymmetrical capacitor device built by Tom Turman that used a heated positive electrode. (Based on a sketch by T. Turman)

outboard positive electrode.* Turman said that his device produced a tremendous ion wind when energized with 30 kilovolts. In addition to applying high-voltage DC to the electrodes, he electrically heated the positive electrode using a modified 12-volt X-ray-tube filament transformer. He found that by heating the positive electrode, he was able to get a greatly increased thrust. He noted that luminous ionization beads would form at regularly spaced intervals along the wire, forming sites where ions were discharged at a higher rate. As the wire was heated to a higher temperature, an increasing number of beads would form along the length of the wire. In his 1928 patent, Brown similarly proposed electrode heating as a means of improving the electrogravitic thrust of his vacuum tube gravitator cells (recall figure 1.6).

Turman noted that after a black oxide coating had formed on the wire, he could still get a lot of thrust, even when he stopped heating the wire. The oxide-coated wire apparently produced a lot more thrust than did a clean copper wire. Turman felt that Brown's flying discs may have used a positive electrode wire that was coated with some exotic material to enhance ion emission. Such a film may have formed on the wire's surface as the result of heating and oxidation. For example, rare-earth oxides are used in radio tubes to enhance the emission of electrons from their filaments. The same type of coating might also enhance the formation of positive ions at the surface of a positively charged electrode. Brown did not mention using heated or coated wires in his flying-disc experiments, and Turman never brought up the subject with him since he performed his tests on heating wires years after he had talked with Brown.

More recently, after reading my paper on the B-2 bomber, Turman

---

*Turman did not state the thickness of his slats.

speculated that the leading edge of the B-2's wing may have an oxide or chemical coating to enhance ionization. Another way of inducing ionization might be to use RF excitation. According to one source, a major aerospace company had taken out a patent in the late 1950s on a method of using high-frequency voltage on the skin of planes to reduce air drag. A similar technique might be employed in electrifying the B-2.

12.2 ▪ LARRY DEAVENPORT'S ELECTROKINETIC DISC TEST

In 1995, independent researcher Larry Deavenport carried out high-voltage tests designed to investigate Brown's electric disc experiment. He constructed a 16-inch-long armature made from shellacked balsa wood and suspended two aluminum discs 5.5 inches below each end of the arm (figure 12.6).[6] Each disc measured about 2.6 inches in diameter

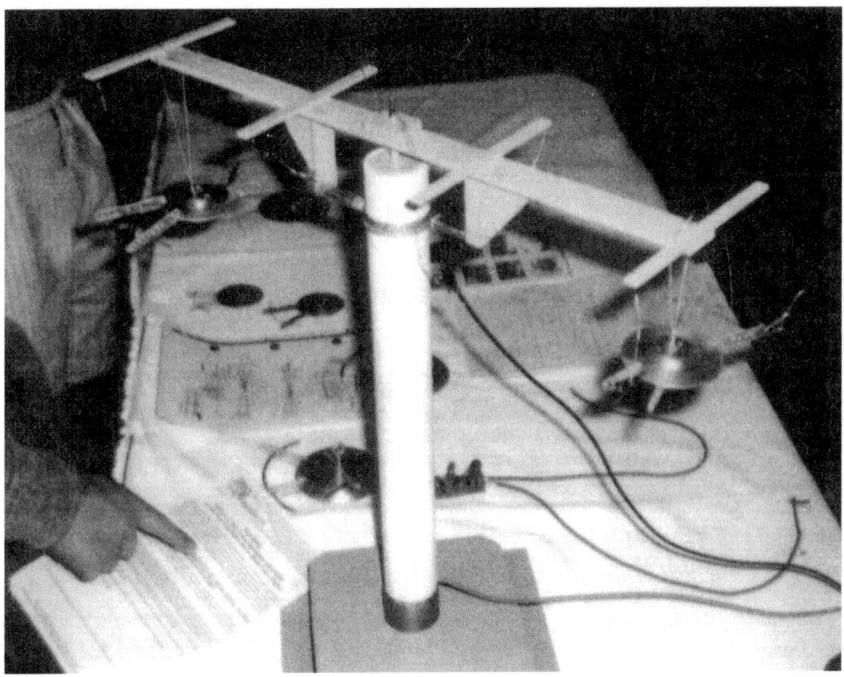

Figure 12.6. A small-size rotary electric disc setup built by Larry Deavenport to duplicate Thomas Townsend Brown's electrokinetic disc experiment. (Photo courtesy of L. Deavenport)

and was one-eighth of an inch thick at the center, tapering to 20 mils (0.02 inch) at the edge. A curved piece of brass wire measuring about 50 mils in diameter and held 1.8 inches from the disc by shellacked balsa wood fingers served as the positive leading-edge electrode. Each disc weighed approximately 33.5 grams. The entire carousel rig was pivoted at its center of gravity on a needle bearing.

When the discs were energized with 0.8 milliamp at 30 kilovolts DC, the apparatus revolved at a speed averaging three-quarters of a revolution per second and reaching as high as one revolution per second (4 feet per second). Ballistic pendulum measurements determined that the discs produced a thrust of 0.58 gram when energized at 25 kilovolts and 1.7 grams when energized at 50 kilovolts.

Deavenport had used a 50-mil-diameter wire, much finer than the 125-mil-diameter wire that Turman had used. However, Deavenport's wire still was about fifty times thicker than what Brown recommended in his letter to Turman. According to Brown, using a smaller-diameter wire would have increased the field nonlinearity around the leading electrode and that would have boosted the thrust developed by the discs.

Deavenport also conducted carousel tests of a cylindrical electro-kinetic device made from aluminum bottles.[7] He was able to get the apparatus to revolve at up to one revolution per second by applying high voltage between 50-mil-diameter curved emitter wires secured at the bow and stern of the cylinder and separated from the cylinder by about 2 inches. The rear wire was connected to the cylinder body. He found that the apparatus revolved slightly faster when a negative potential was applied to the lead wire, indicating that the propulsion he was seeing was primarily electrostatic and not gravitic. Deavenport's disc electrodes instead performed better with their lead wire made positive, as in Brown's experiments. Nevertheless, this suggests that Brown's electrokinetic discs most likely would also have revolved if charged with a reverse polarity and that a large fraction of their thrust may have been due to electrostatic force effects.

12.3 ■ ROBERT TALLEY'S ELECTROGRAVITIC ROTOR TEST

Between 1988 and 1991, Robert Talley conducted research at Veritay Technology Inc. to investigate Brown's electrogravitic rotor experiment.[8] The project was financed by a Small Business Innovation Research grant under sponsorship of Phillips Laboratory at Edwards Air Force Base. Talley's experiment was similar to the vacuum chamber experiment that Brown conducted in Paris (figure 3.1), but with two exceptions. Tally used DC voltages ranging up to 19 kilovolts, rather than up to 200 kilovolts as Brown had done. Sparking between Talley's electrodes prevented accurate measurements from being made at higher voltages. Also, unlike Brown's rotor, which was free to revolve, Talley's was restrained by fibers that allowed the rotor's thrust to be assessed through the amount of twist it generated. This arrangement was sensitive to thrusts as small as 0.2 microgram.

Talley's rotor consisted of two capacitors mounted in pinwheel fashion (figure 12.7). Each consisted of an 8-centimeter-diameter brass disc separated by 4 centimeters from a 1-centimeter-diameter aluminum ball electrode. In some cases, a quarter-inch-diameter rod of high-K dielectric such as titanium-lead zirconate (K = 1,750) was placed between the electrodes. The rotor was mounted inside of a chamber that was evacuated to a pressure of 10^{-6} torr (10^{-6} millimeters of mercury, or about a billionth of an atmosphere). Talley found no evidence of thrust when his rotor was powered with steady potentials of up to 19 kilovolts. However,

Figure 12.7. A schematic of the test rotor Robert Talley used in his vacuum chamber experiment.

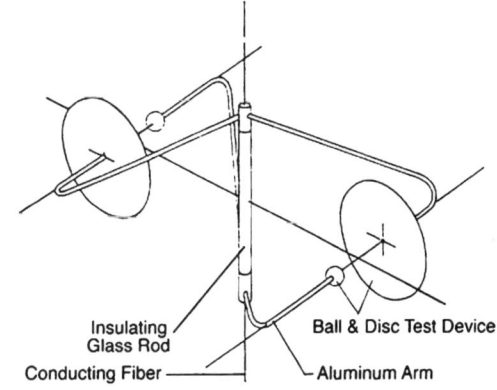

Insulating
Glass Rod

Conducting Fiber

Ball & Disc Test Device

Aluminum Arm

he found that the rotor developed substantially large rotational thrusts when sparks jumped between its electrodes. Since this spark-induced thrust was observed only when he used a high-K dielectric between the rotor's capacitor plates, he concluded that the dielectric material must somehow be directly involved and that this thrust phenomenon could not easily be attributed to ion propulsion or to other known electrodynamic effects. Talley's experiment provides support for the thrust effect that Brown observed when his electrogravitic rotor sparked during tests in a high vacuum.

In 2003, American inventor Hector Serrano duplicated Talley's vacuum chamber rotor experiment.[9] Unlike Talley, Serrano was able to get a 70-degree rotational deflection of the rotor element in the absence of sparking in a vacuum of 10^{-7} torr. Serrano's success may possibly be due to his use of a greater voltage potential, 41 kilovolts instead of 19 kilovolts, so his tests appear to confirm Brown's findings that an electrogravitic force is propelling the rotor in the absence of any ionic discharge. Talley's finding that his rotor did not develop any torque at 19 kilovolts is consistent with Brown's findings that a certain voltage threshold must be exceeded in order to observe a thrust effect. For example, in testing his highly efficient vertical-lift electrokinetic apparatus, Brown observed that he had to apply in excess of 10 kilovolts before any noticeable thrust effect was observed. Also, if we extrapolate the voltage-speed trend line for Brown's electrokinetic disc (figure 2.4), we find that saucer speed drops precipitously, projecting just 9 centimeters per second at 30 kilovolts and 1 centimeter per second at 20 kilovolts. Brown has no data points at such low voltages probably because he found the thrust to be so low that it was unable to overcome his carousel's bearing resistance.

Talley's observation that the spark-induced thrust was greater when a high-K dielectric was placed between the rotor electrodes confirms Brown's statement that the thrust on his electrokinetic apparatus was proportional to the dielectric constant of the support rod placed between its electrodes. For a given voltage differential across the rotor element, a material with a higher dielectric constant would cause more negative charges to accumulate on the negative electrode. Hence, the negative ion cloud formed at the time of spark discharge would have repelled these accumulated charges with greater force to produce a greater thrust in the direction of the positive electrode.

12.4 ▪ THE CORNILLE-NAUDIN PENDULUM EXPERIMENTS

In 1996, physicist Patrick Cornille constructed a double-ball pendulum similar to the one Brown had tested in his 1920s experiments.[10-12] He suspended a pair of aluminum spheres, each weighing 500 grams, from two nylon lines and applied between 30 and 50 kilovolts DC to the spheres through two wires secured to these lines (figure 12.8). Each high-voltage feed wire measured half a millimeter (20 mils) in diameter. Each time he turned on his power supply, the pendulum would swing in the direction of its positively charged sphere, in apparent violation of Newton's third law of motion. That is, Brown's classic pendulum experiment apparently violates the law of conservation of momentum. At 50 kilovolts, the pendulum was acted on by a force of 3.5 grams. Curiously, Cornille found that the effect occurred only when he used bare feed wires, as opposed to insulated feed wires. He found that the

Figure 12.8. The electrogravitic pendulum experiment carried out by Patrick Cornille in July 1996. High-voltage DC is fed to the spheres via the suspension wires. (Photo courtesy of P. Cornille)

bare feed wires were able to emit a 1.5-milliamp ion-leakage current through the air. This demonstrates that the emission of charges into the atmosphere plays an important role. He showed that the conventional ion wind theory, however, failed to explain the pendulum's movement, since ions attracted to the opposite electrode would impact with a force that was two orders of magnitude too small. Also, such a mechanism would not explain his finding that the developed force increased as a moderate function of the pendulum mass, ~$m^{0.5}$.

Cornille theorized that the leakage current was somehow related to the thrust effect, but offered no clear explanation. As we shall see below, the observed thrust effect is most likely electrogravitic, arising from the ion space charge established in the air. For example, Cornille estimates that his feed wires would have been emitting ions at the rate of about 10^{16} ions per second, since their ion-leakage current amounted to 1.5 milliamperes.[13] Consequently, they would have been generating about 5×10^{13} ions per second per centimeter of wire length. Let us suppose that this built up space charges of the order of 10^{13} ions per square centimeter along the wires. Here, we make a very rough estimate, adopting a value similar to that given in Supplement B of the 1960 "Electrohydrodynamics" report for the ion-space charge developed around Brown's vertical-thrust electrokinetic apparatus.[14] In the case of Cornille's experiment, the volume of air lying within 5 centimeters of each wire would have contained on the order of 10^{17} ions. This would be more than 10,000 times the surface charge that would have accumulated on the surface of the feed wires and the pendulum spheres, which according to Cornille's estimate would have been around 3.5×10^{12} ions.[15] So we see that the electrogravitic force would be substantially enhanced by allowing a leakage current to create ions in the vicinity of the feed wires.

In the case in which the wires were insulated, a small amount of ion leakage would have been present between the pendulum's spheres, but according to subquantum kinetics, the gravity gradient created between the resulting positive and negative-ion space charge would have existed between the spheres but would not have intercepted the spheres themselves (see figure 12.9). Only the lightweight plastic spacer between the spheres would have been affected by this field, so the electrogravitic

Figure 12.9. Chart of the gravity gradient in the Patrick Cornille pendulum experiment when the high-voltage feed wires are insulated. (P. LaViolette, © 2006)

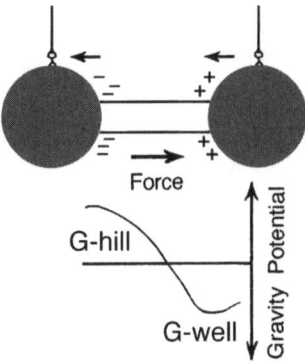

thrust on the pendulum would have been very slight. Also, since the spheres that make up the bulk of the pendulum mass would have lain on the outer sides of the ion clouds, they would have experienced a thrust in a direction opposite to the central thrust vector acting on the insulating spacer. Furthermore, since the ion-leakage current between the spheres was relatively low at 3 microamps, or about five hundred times less than the ion leakage produced when the pendulum was tested with bare feed wires, the spheres would have emitted only a small number of ions. Hence, the induced gravity gradients would have been quite minimal. In summary, it is not surprising that Cornille observed no pendulum movement.

However, in the case in which the pendulum was tested with bare feed wires, most of the ions would have been released along the length of the wires at a considerable distance from the pendulum masses. The ions released from the lower extremity of the feed wires, where the wires attach to the spheres, would have had the greatest influence on the pendulum. Their ion clouds would have been separated by a sufficient distance so that their induced gravity field gradient would have intercepted part of the pendulum masses (see figure 12.10). The G-on fluxes coming from more remote ion space charges located farther up the feed wire may also have had some effect by enhancing the magnitude of the gravity potential hills and wells being generated in the vicinity of the pendulum.

I would like to emphasize, again, that the electrostatically charged pendulum experiences an applied gravitational thrust in the absence of any so-called space-time warping. The general relativistic concept

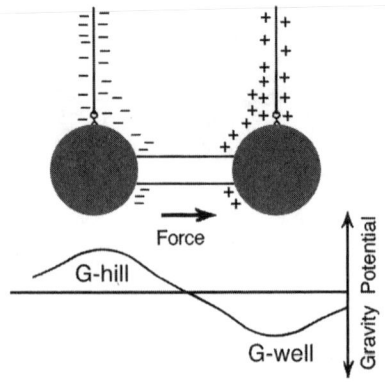

Figure 12.10. Chart of the gravity gradient in the Patrick Cornille pendulum experiment when the high-voltage feed wires are bare. (P. LaViolette, © 2006)

of gravitational space-time warping is a fiction. Understanding electrogravitics requires that we dispense with such outdated ideas and adopt new concepts such as those proposed in subquantum kinetics. Namely, gravity potential gradients are understood to be concentration gradients created in an ether that occupies a Euclidean space. Such gradients cause movement by altering the etheric reactions that are continually regenerating the fields of the propelled object's constituent subatomic particles.

French researcher Jean-Louis Naudin duplicated Cornille's pendulum experiment and found that the pendulum moved even when the vertical ion-emitting feed wires were detached from the metal spheres and supported at the distal ends of the spheres by means of insulating polystyrene blocks,[16] so he demonstrated that charging of the spheres was not crucial to the effect. Like Cornille, he found that the magnitude of the force increased as the mass of the spheres increased, indicating the presence of an unconventional gravitational effect.

These results support the electrogravitic theory suggested above as an interpretation of Cornille's pendulum experiment and also suggested earlier in the analysis of Brown's electric discs. Naudin was generating positive and negative ion clouds on either side of the test mass spheres and, according to the subquantum kinetic theory, a gravity potential gradient would have been generated between these charged ion poles. The test masses, which were situated in the midst of this electric and gravity potential field gradient, then moved in the direction of the outlying gravity potential well, that is, the outlying positive ion cloud (see figure 12.11).

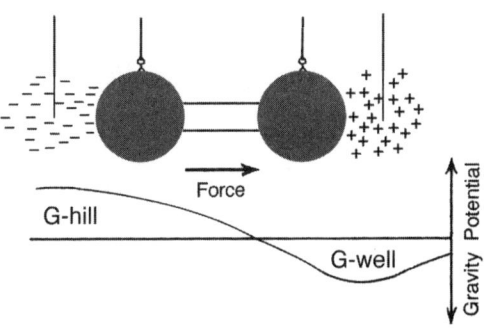

Figure 12.11. Chart of the proposed gravity gradient generated in Jean-Louis Naudin's modification of Patrick Cornille's pendulum experiment. (P. LaViolette, © 2006)

Naudin also tested the pendulum by energizing it through feed wires that were insulated and through which a 0.5-milliamp ionization current was allowed to flow between the two balls. As in Cornille's experiment with insulated feed wires, Naudin's experiment provided no observable pendulum movement. This resolved the question raised by Naudin as to whether current flow might play a crucial role. It showed that it is not the current flow itself that is important, but rather the location of ion discharge in relation to the spherical masses.

12.5 ■ LIFTER RESEARCH

During the 1970s, Jeff Cameron, an engineer working in Huntsville, Alabama, was researching a laser preionizer, a triangular high-voltage filament used to ionize the lasing medium in a gas laser, when he observed unusually strong forces deforming the preionizer element. This led him later to build and test a large-scale replica. It was similar to the parallel slat thruster described in Brown's 1960 electrokinetic apparatus patent and to the thruster built by Turman, but was much lighter in weight and had its slats arranged to form a triangle. This electrostatic thrust device, which has since come to be called a *lifter*, consisted of three vertical aluminum-foil fins connected to form a larger triangular structure with a thin wire supported horizontally above the fins. The upper edge of each fin facing the wire was made to have a rounded contour. When charged with 30 kilovolts, the device was observed to levitate.

After Cameron posted his findings on the Internet, Naudin tested a modified version of Cameron's lifter that used a 2-mil-diameter

(0.05-millimeter) emitter wire and a central baffle to duct the airflow (figure 12.12).[17] Interestingly, this wire diameter comes close to the 1-mil diameter that Brown used in his electrokinetic disc experiments. The lower-fin electrodes in Naudin's device were 1.5 centimeters wide and 30 centimeters long, and had a rounded upper edge. The wire was suspended 7 centimeters above the fin. His device was able to lift its own weight of 4 grams plus an added 2-gram weight when energized with 37 kilovolts.

After Naudin posted the construction plans for building a lifter on his website, hundreds of hobbyists began duplicating the experiment and testing their own versions. The ensuing frenzy even attracted media attention. Lifter researcher Tim Ventura played a central role in catalyzing this widespread activity that even developed into competitions to see who could build and levitate the heaviest lifter. Ventura's website (www.americanantigravity.com) is a good resource for those interested in ongoing lifter research. Experimenters found that they could obtain even more spectacular results when they combined many triangular lifters into a single structure. Some have been made that weigh as much as 250 grams, which includes the weight of a 60-gram payload. An image of one such multielement lifter in flight is shown in figure 12.13.

Saviour, a French researcher, found that he could improve the thrust

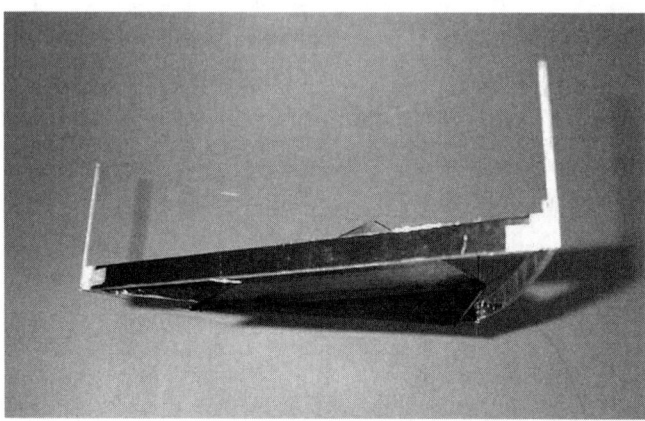

Figure 12.12. A lifter built and tested by Jean-Louis Naudin. (Photo courtesy of J.-L. Naudin, from his website, http://jnaudin.free.fr/)

Figure 12.13. A multielement lifter in flight. (Photo courtesy of Tim Ventura)

of the lifter by using a nichrome emitter wire heated with a current from a 12-volt power supply. Thus, he rediscovered the phenomenon Turman had discovered in the early 1970s in carrying out his brass slat lifter experiment. The heated wire was able to emit more ions at a given voltage potential. Again, these findings confirm ideas hinted at in Brown's writings, which indicate that methods of encouraging ion emission from the wire electrode would increase the resulting thrust. Although ions are important in producing the lifter's thrust, its thrust is not due to ion wind effects, that is, forces arising from the recoil or impact of ions on the electrodes. Such mechanical forces have been shown to fall short by several orders of magnitude in accounting for the observed lift.

As Turman discovered in his cylindrical, asymmetrical capacitor experiments, so, too, lifter researchers found that they could get lifters to work by charging the wire to either a positive or a negative potential. Since the lifters are made of extremely lightweight materials, such as aluminum foil and plastic soda straws, electrogravitic forces would not play an important role in producing their thrust. Rather, the thrust is most likely attributable to unbalanced electrostatic forces (see text box on page 368). Others, such as Naudin, have given similar explanations.[18]

Unbalanced Ionic Electrostatic Forces Acting on a Lifter

Consider a case in which the lifter's emitter wire is positively charged. The positive-ion cloud generated around the wire would be displaced below the wire since the field would be stronger on the lower side of the wire facing the underlying negative electrode. This downward-displaced cloud would produce a net-upward electrostatic repulsive force on the adjacent wire. Since the field is very nonlinear and concentrated near the wire, this upward-repulsive force would be comparably strong. Also, the upper edge of the negative electrode would experience an upward force because it is electrostatically attracted to the positive-ion cloud.

The lower negative-ion cloud would not be as extensive as the upper positive-ion cloud for a number of reasons. First, negative ions would be emitted from the negative electrode at a lower rate because of its lower electric field gradient. Second, positive ions brought downward by the ion wind would neutralize many of the negative ions in the air and would also impinge on the negative electrode to neutralize negative charges on the electrode. Furthermore, the mixture of positive and negative ions in the vicinity of the negative electrode would tend to screen the negative-ion space charge. Any net-negative-ion charge present in the vicinity of the negative electrode would direct its force nearly horizontal to the electrode fin, pushing toward the fin from either side. In addition, the downward ion-wind flow would cause a modest negative-ion space charge to build up below the negative electrode, and this would tend to produce an upward-directed repulsion force on that electrode. All of these forces together would cause the entire lifter structure to levitate. If the wire was instead negatively charged and the fins were positively charged, unbalanced electrostatic forces would again produce lift.

Although an ion wind continually rushes downward, that is, in the direction of lower electric field intensity, these ions are continually replaced by newly emitted ions, so these ion space charges will always be present to exert their upward forces on the wire and fin. Any means of encouraging greater ion emission from the upper electrode would increase the ion-space charge in the vicinity of the wire as well as the upward repulsive electrostatic forces, thereby improving lift.

Such electrostatic forces would produce an upward thrust regardless of the applied field polarity. Similar unbalanced electrostatic forces would account for the thrust developed by Turman's cylinders. Electrogravitic forces become more significant in the electrokinetic pendulum experiments of Brown and Cornille, which involve the propulsion of a heavy mass. Accordingly, we find in such cases that the apparatus always moves toward its positive pole.

However, the explanation of electrostatic forces produced by ion-space charges does not tell the full story of what is happening in the lifter. For example, Purdue University researcher William Stein carried out a lifter test in which a 12-centimeter-long lifter was energized with 17 kilovolts in a high vacuum.[19] Although the lifter was unable to support its weight, it reportedly produced a levitating thrust of 0.3 millinewtons. Stein's test would indicate that with ion emission essentially eliminated, a lifter is still able to generate a measurable thrust, which would be about 12 percent of what it would generate if allowed to operate in air.* Hungarian researcher Zoltan Losonc has done a computer analysis of the electrostatic forces that a lifter's charged electrodes would generate when electrified in a vacuum. He has concluded that no lift force should be produced and, hence, that some exotic principle must be operating to explain the results of the Purdue vacuum test.[20] The question remains as to whether Stein's measurements may have been influenced by electrostatic forces developed between his lifter element and the vacuum chamber walls. His 12-centimeter-long lifter should have weighed less than 2 grams, as compared with the arcuate discs that Brown vacuum tested, which ranged in weight from 17 to 125 grams.

Naudin has also performed tests that, like Stein's, suggest the lifter may be generating a force in the direction of its smaller electrode in the absence of ion electrostatic forces. He placed plastic soda straws over a lifter's emitter wires to prevent its emission of ions and found that the lifter still produced a measurable lifting force.[21] He also found

---

*To compare this with Naudin's lifter model, whose three fins had a combined length of 90 centimeters and which was energized at a voltage about twice as high, we must scale up this thrust by a factor of 16, giving a thrust of 5 millinewtons, or 0.5 gram. By comparison, Naudin's lifter tested in air achieved a thrust of 4 grams, eightfold greater.

that his lifter produced a measurable thrust even when enclosed in a plastic bag, thereby containing its ion wind. Naudin has posted only general information about this on the Internet, leaving some questions unanswered, such as whether the force he measured may have been due to electrostatic attraction between his lifter and the beam balance on which it was placed. So, as with Stein's results, it may be premature to draw any conclusions. If the thrust in a vacuum is observed only when the wire is positively charged, then perhaps the force could be explained as a manifestation of the Biefeld-Brown effect, that is, an electrogravitic force acting on the lifter's mass. However, because of the absence of any massive dielectric element between the lifter electrodes, it is unlikely that an appreciable electrogravitic effect would be present.

Unclassified public research on electrogravitics research is, for the most part, being conducted by independent researchers, some of whom have been mentioned above. With few exceptions, no similar research is being conducted at universities or government research institutions. Clearly, the science and engineering establishment needs to take a serious interest in conducting additional electrogravitics research before the secrets of field-effect propulsion, currently locked away in black R&D programs, becomes openly applied to make mankind's dream of antigravity a reality.

12.6 ■ THE LAFFORGUE THRUSTER

French inventor Jean-Claude Lafforgue has patented an asymmetrical capacitor field propulsion thruster having a shape similar to that shown in figure 12.14.[22] Like Brown's asymmetrical capacitor, Lafforgue's device develops a net thrust through unbalanced electrostatic forces, with the thrust acting in the same direction regardless of plate polarity.

To determine the thrust acting on his capacitor, Lafforgue calculated the magnitude and direction of the force per unit surface area acting on the capacitor's plates at various plate locations, this quantity alternately being referred to as the *force density* or the *electrostatic pressure*, P. It is mathematically expressed as: $P = F/A = E \bullet q/A = E \bullet \sigma$, in which F is electrostatic force, A is surface area, E is electric force-field intensity, q is charge, and σ is surface-charge density. Thus, he relies on

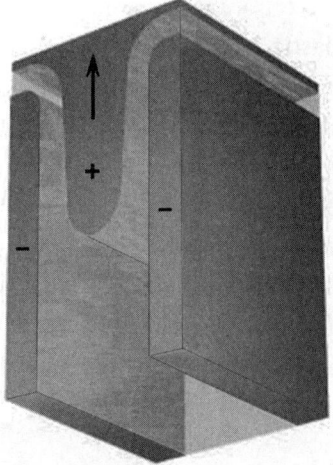

Figure 12.14. The Jean-Claude Lafforgue field propulsion thruster. (Adapted from Lafforgue's 1991 patent)

the conventional practice of calculating electrostatic force as the product of the electric field intensity and the electric charge.

Subquantum kinetics achieves the same outcome, except that it works with the negative electric-potential gradient, $-\nabla\varphi$, instead of electric force-field intensity, with the two being equivalent (i.e., $E = -\nabla\varphi$). As mentioned earlier, subquantum kinetics prefers to work with energy potential (ether concentration), since it regards this as the real existent rather than the force-field intensity. Lafforgue's approach of summing the electrostatic pressures acting over a particular plate surface area to get a resultant force vector is equivalent to the subquantum kinetics approach of multiplying the field-potential gradient present on a given electrode sector by the surface-charge density present in that sector and summing the resulting force vectors. This approach was described earlier in analyzing the electrostatic forces acting on Brown's electrokinetic apparatus (see chapter 3, section 3.3).

Also, Lafforgue proposed that the electric field intensity is seated in the local space-time continuum and, hence, exerts its force on the plate surface charges from a reference frame that is not attached to the capacitor. Thus, any resulting force imbalance would be able to displace the capacitor as a whole. Lafforgue's approach, which presumably was arrived at through experience gained from experimental observation, is in accord with the theoretical approach of subquantum kinetics, which views the electric-field potential as being seated in the ether and able to act on a capacitor independent of the capacitor's reference frame and thereby cause it to be displaced. Subquantum kinetics, though, goes into much greater detail to explain how the electrostatic potential field is generated and how it exerts its force on a charge without any countering reaction force. Since the approaches of Lafforgue and of subquantum kinetics were developed independently and produced similar conclusions, we are reassured from both observation and theory that it is correct to conclude that unbalanced electrostatic forces can propel an asymmetrical capacitor if it is properly designed. An analysis of how unbalanced forces form on the Lafforgue capacitor is presented in the text box below.

Analysis of Electrostatic Forces
on the Lafforgue Capacitor

The unbalanced thrust acting on the Lafforgue asymmetrical capacitor may be understood to arise as follows. Referring to figure 12.14, the field lines coming from the upper ends of vertical, negatively charged outer plates diverge as they approach the central, positively charged plate, which curves to a horizontal T-shape at the upper end of the capacitor. As a result, the field lines and surface charge are more concentrated on the negative electrode than on the positive, which causes the attractive force, or electrostatic pressure, that is directed from the negative electrode out to the positive to be greater than the attractive force that is directed from the positive electrode in toward the negative.

The opposing horizontal components directed in toward the positive electrode cancel one another, but the upward-directed component is unopposed, leaving a net-upward thrust. In addition, the field lines emerging downward from the lower tip of the central, positively charged electrode diverge toward the horizontal as they approach the two flanking negatively charged electrodes, resulting in a net force, or pressure, directed downward away from the positive electrode. The forces, or pressures, attracting the two negative electrodes toward the central positive electrode, being for the most part horizontal and opposed to one another, will cancel each other out, leaving the downward residual force on the positive electrode unopposed.

However, since the field gradient at the upper end of the negative electrode is much greater in magnitude than that formed around the lower portion of this central electrode, the upward thrust from the negative electrode will be anywhere from 3 to 18 percent greater than the downward thrust generated at the lower end of the positive electrode. As a result, a net-upward thrust will act on the capacitor as a whole. The result will come out the same even, if the positive and negative plate polarities are reversed.

Naudin offers a considerable amount of information about the Lafforgue thruster on his website. He has taken the force equations given in Lafforgue's patent, made a few minor corrections, and used them to create a calculator for computing a capacitor's thrust.[23] Visitors can enter values for a capacitor's dimensions, charge voltage, and dielectric constant and then compute what thrust would be expected. For example, a 50-kilogram thruster measuring 38.5 centimeters high, 8.3 centimeters wide, and 33 centimeters long, using a 4,000-K dielectric and charged to 100 kilovolts, is computed to develop a phenomenally high thrust of 0.68 ton, a force that measures almost fourteen times the capacitor's normal weight! Thirty of these thrusters would be capable lifting a 20-ton vehicle. Forward movement could be obtained simply by vectoring the direction of one of the thrusters. However, experimental data on a high-K Lafforgue thruster that might substantiate these projections is currently not available.

Since barium titanate has a volume resistance of about 10^{11} ohm-meters, a capacitor of this size would have a total resistance of about 10^{10} ohms, provided that its electrodes are properly insulated from contact with the outside air. This would amount to a current leakage of 10 microamps, or a power dissipation of 1 watt. So, theoretically, all thirty thrusters could be powered with a 100-watt power supply. A propulsion device yielding 20 metric tons of force for a power dissipation of 100 watts would have a thrust-to-power ratio of 2 million newtons per kilowatt, about 130,000 times that of a jet engine.

However, it is likely that due to the opposing thrust vector developed by its polarized dielectric, the Lafforgue thruster loses its thrust once it becomes fully charged. Hence, like Brown's electrokinetic apparatus, it may need to be repeatedly charged and discharged to create a continuing thrust effect. If 30 of these Lafforgue thrusters, then, were to have a combined capacitance of about 30 microfarads and were to be charged to 100 kilovolts once every second, they would draw 300 kilowatts of power. This would project a lower thrust-to-power ratio of about 670 newtons per kilowatt or about 45 times that of a jet engine.

In his patent, Lafforgue notes that in addition to its use for air transport, his thruster could be used for power generation by mounting a number of thrusters around the circumference of an axis and connecting the axis to a generator. For example, four thrusters of the size estimated above, each producing 680 kilograms of force and each mounted at the end of a rotor arm extending 2 feet out from the axis, would collectively generate 12,000 foot-pounds of torque. Spinning at 5,254 rpm, this electrostatic motor would be generating 12,000 horsepower, or 8.95 megawatts, of power. Accounting for efficiency losses in the electric generator and due to bearing resistance, a motor-generator combination should be capable of producing 5 megawatts of power, but the thrusters would require only 40 kilowatts of power if the capacitors were being charged and discharged once per second. Hence their output power would exceed their input power by a factor of 125.

Lafforgue's patent was issued in 1991. If we are even somewhat close in our thrust projections, the question that arises is, What has everyone been waiting for? Why aren't these being offered for sale to power our homes or electric cars? Is it perhaps that people just don't believe that something this simple might solve the energy problem? Indeed, some people's belief in and subservience to the law of energy conservation (and to Newton's third law of motion) are so ingrained that they would rather continue to burn oil and gas and ultimately create ecological disaster on our planet than give up their cherished misconceived belief.

Naudin has built and tested some small-size Lafforgue thrusters measuring just 0.5 millimeter thick and has demonstrated that they produce a net thrust just as Lafforgue claims. Naudin used a low-K epoxy dielectric (K = 3.7) and operated his thruster at a much lower voltage of 9,500 volts, using the test setup shown in figure 12.15.[24] When energized, his Lafforgue thruster generated about 0.03 gram of force, as indicated by the upward swing of the armature. Naudin commented

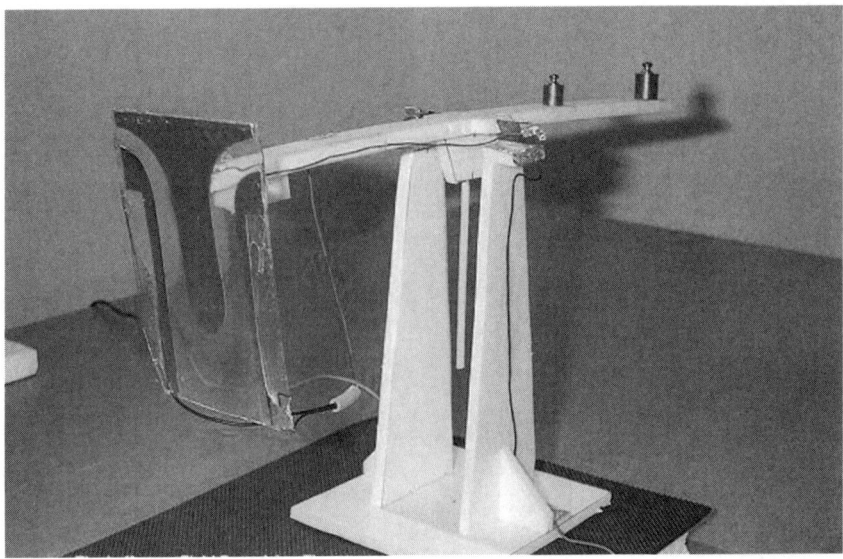

Figure 12.15. A test of the Lafforgue thruster carried out by Jean-Louis Naudin. (Photo courtesy of J.-L. Naudin, from his website, http://jnaudin.free.fr/)

TABLE 2. THRUST-TO-POWER RATIO COMPARISON
Field Propulsion versus Conventional Technologies

Propulsion Technology	newtons/kilowatts
T. T. Brown's electrokinetic apparatus (barium titanate)	70,000
T. T. Brown's electrokinetic apparatus (pyrex dielectric)	2,200
T. T. Brown's gravitator	2,000
Lafforgue thruster (pulsed barium titanate dielectric)*	approx. 700
Lafforgue thruster (epoxy dielectric) tested by Naudin	approx. 30
Jet engine	15
Podkletnov gravity impulse beam (improved version)	0.5
Space Shuttle Main Engine (NASA)	0.22
NASA Lewis Research Center ion thruster	0.23
Phoebus nuclear thruster	0.20
SERT II mercury-propellant ion thruster (NASA)	0.03
Micro-Pulsed Plasma Thruster (Air Force)	0.01

*Based on Naudin's theoretical thrust projection

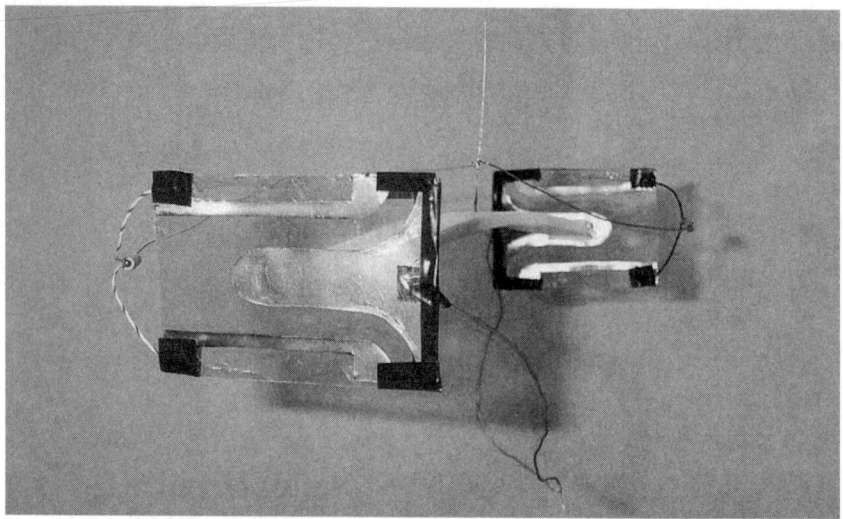

Figure 12.16. A test of the rotating Lafforgue thruster carried out by Jean-Louis Naudin. (Photo courtesy of J.-L. Naudin, from his website, http://jnaudin.free.fr/)

that he was able to reduce his leakage current to "near zero" by properly insulating his capacitor so that the electrodes were not in contact with the ambient air environment. By this, he meant that the current was "not measurable in a microampere range." Supposing his thruster was drawing less than 1 microamp at 10 kilovolts, this would be a power consumption of less than 10 milliwatts, which amounts to a thrust-to-power ratio of about 30 newtons per kilowatt.

Table 2 on page 375 compares the thrust-to-power ratios of various field propulsion technologies to those of conventional jet and rocket propulsion techniques.

Lafforgue's equations indicate that thrust should scale in direct proportion to the capacitor's dielectric constant, in direct proportion to the capacitor's length (e.g., plate area), and according to the square of the applied voltage. This K- and V-dependence is essentially the same as what Brown found in testing the performance of his asymmetrical capacitor, as disclosed in the "Electrohydrodynamics Report," discussed in chapter 3. Scaling Naudin's experiment up to a K = 6,000 capacitor measuring 33 centimeters long with slightly different electrode curvatures and energized at 100 kilovolts, we find that thrust increases by a factor of more than 100 million, projecting a thrust of about 3 metric tons!

In another experiment, Naudin placed two 0.5-millimeter-thick epoxy dielectric Lafforgue thrusters at opposite ends of a rotor arm, as shown in figure 12.16.[25] When energized with 15 kilovolts, the apparatus began rotating, reaching a top speed of 40 rpm. It would continue to rotate as long as 0.18 watt of power was supplied to maintain the capacitor's charge. The thrust was not attributable to an ion-wind effect, since the electrodes were shielded to minimize any ion emission. Hence, Naudin's experiment confirms that the rotation arises from the creation of action without reaction, a clear violation of Newton's third law of motion.

There is no indication that anyone has conducted high-voltage tests of a Lafforgue thruster made with a high-K dielectric such as barium titanate. One electrogravitics researcher, Anthony Colacchio, reported having constructed a Lafforgue thruster made with a low-to-medium-K dielectric consisting of barium-titanate powder mixed into an epoxy matrix.[26] Such a mixture would typically have a K-value of about 30. He tested his thruster at a potential of 100 kilovolts but says he found no indication that any thrust was produced. Given that his thruster was forty times thicker than Naudin's, used a dielectric having an eightfold-higher K-value, and was tested at a voltage about tenfold higher than Naudin's, one would expect a thrust about 35,000 times greater. Hence, this experiment should have produced about 1 kilogram of force if the scaling relations are correct. Perhaps Colacchio observed no force because he applied DC to his capacitor. The dielectric's tendency to create an electric dipole moment directed in opposition to the applied electric field may then have negated the thrust effect. Recall that such was the case in Brown's gravitator experiment. Thus, perhaps better results might be achieved if the voltage is applied to the capacitor as a pulse rather than as steady DC. Naudin, for example, was pulsing his epoxy dielectric Lafforgue thruster with a 5 percent duty cycle. Clearly, more research needs to be done on this design before it can be said to be ready for commercial application. Again, a word of caution: Experimenting with high-voltage capacitors can be lethal.

13

BLACK HOLE DISCOVERED IN NASA

13.1 ■ THE SPACE EXPLORATION OUTREACH PROGRAM

On July 20, 1989, President George H. W. Bush proposed that the United States undertake an ambitious mission of manned and robotic exploration of the solar system that would include building a permanent base on the moon and landing humans on Mars beginning around the year 2014. This was known as the Space Exploration Initiative. Shortly thereafter, Vice President Dan Quayle, who was chairman of the National Space Council, requested that NASA "cast a net widely to find the most innovative ideas in the country" for carrying out the initiative. Thus was born the Space Exploration Outreach Program (SEOP).

To initiate the program, NASA administrator Richard Truly made a public request, inviting anyone who was interested to submit new technology ideas that might help NASA execute the space exploration mission it was undertaking. I was one of about 45,300 individuals who, early in 1990, received a flyer describing SEOP and inviting us to contribute our ideas. All suggestions were to be sent to the RAND Corporation, which was responsible for their review. Ideas were solicited in the following categories:

1. Mission concepts and architectures
2. System design and analysis
3. Space transportation, launch vehicles, and propulsion
4. Space and surface power
5. Life-support systems, space medicine and biology, and human factors
6. Space processing, manufacturing, and construction
7. Structures, materials, and mechanisms
8. Communications, telemetry, and sensing
9. Automation, robotics, and teleoperators
10. Information systems
11. Ground support, simulation, and testing
12. None of the above (specify category)

Seeing that this might be a good opportunity to inform NASA of the benefits of electrogravitic propulsion technology, I decided to make a concept submission under category 3, "Space transportation, launch vehicles, and propulsion." Certainly a means of transporting people to Mars in five days rather than 224 should be of some interest to NASA.

13.2 ■ IDEA CENSORSHIP

The submission I made to SEOP, cataloged by NASA as idea number 100159, is reproduced in appendix G.[1] My submission pointed out that electrogravitics could make an important contribution toward helping NASA meet its space exploration challenge. I noted that development work on electrogravitic propulsion is currently in progress at major aerospace companies, but that the work is restricted by military classification. Furthermore, I explained that application of electrogravitic technology to NASA's space program to replace outmoded rocket propulsion technology would entail a minimal amount of R&D if aircraft designs already perfected in the military aerospace sector could be declassified. Hence, the issue would not be one of technological feasibility, but rather one of political decision—the decision to declassify an advanced technology already in existence. I suggested that NASA make a serious lobbying effort to convince military authorities to declassify the technology for more open use in space exploration. In addition to citing Brown's electrogravitics work, I included several quotes from the February 1956 *Aviation Studies* report.

A total of 1,697 people responded to NASA's submission request (about 4 percent of the people originally solicited). The ideas were initially screened by Peat Marwick Main & Co. to remove submissions that were deemed to contain classified or proprietary ideas. About 149 were removed as a result of this screening. The remaining 1,548 ideas were sent on to RAND, which divided the workload among five review panels. Each panel reviewed idea submissions concerned with a particular aspect of NASA's activities and each wrote up its own summary report. The review panels carried out an additional screening of the ideas, with the result that only 215 ideas (13 percent of the total number submitted to RAND) were passed on for final synthesis. A synthesis group summarized the RAND panels' reports along with ideas obtained from other sources. These other sources were the American Institute of Aeronautics and Astronautics, the Department of Defense, the Department of Energy, the Department of Interior, the Aerospace Industries Association, several aerospace contractors, and NASA. The overall organization of the outreach synthesis process is illustrated in figure 13.1.

The synthesis group summarized this information in a document titled *America at the Threshold,* which was publicly distributed in the fall of 1991.[2] This appeared to be more on the level of a NASA public relations document than a report with any kind of technical substance. It was replete with pictures of planets and astronauts constructing space stations and attractive artwork of spacecraft. The text did not go into much technical depth. It appeared to be directed primarily to a general audience.

Upon receiving this synthesis group report, I discovered that nowhere did it mention anything about electrogravitics. Puzzled as to the report's silence on the subject, I called up the SEOP synthesis group

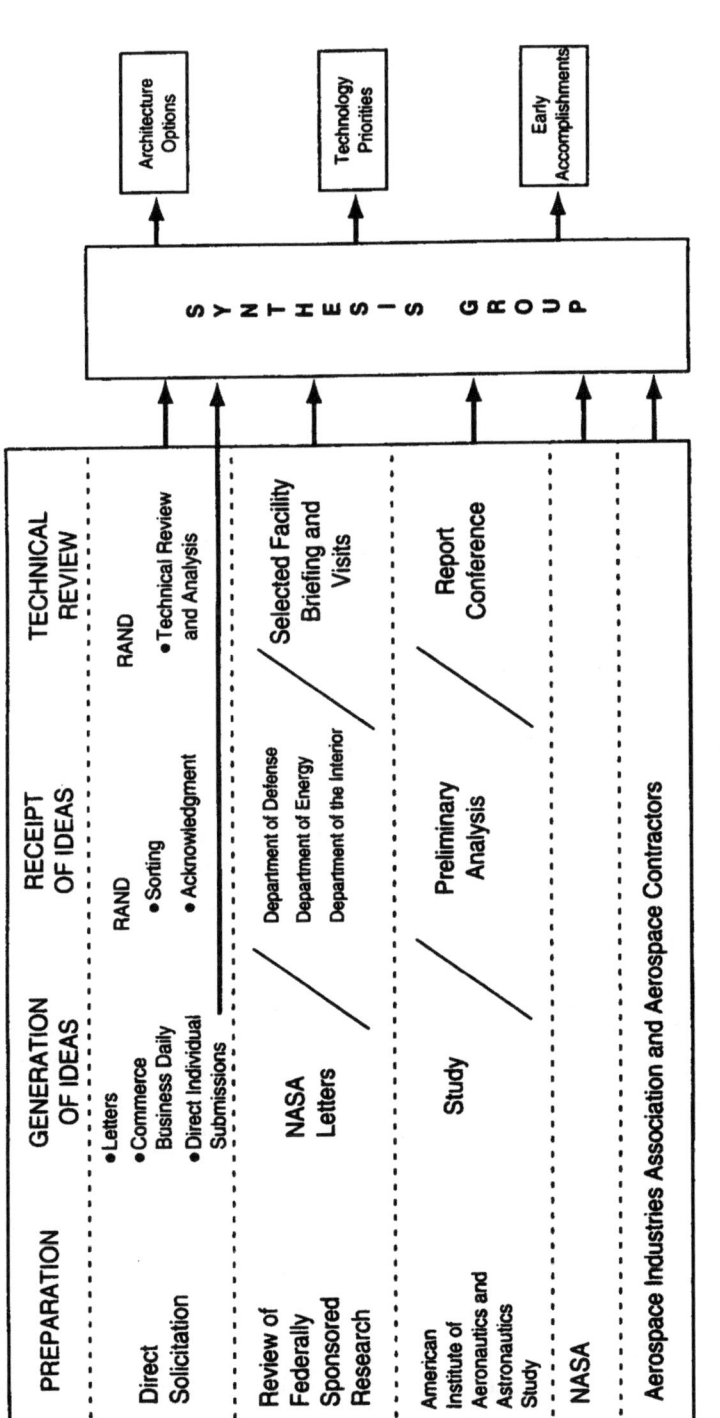

Figure 13.1. Chart showing the outreach and synthesis process followed in NASA's Space Exploration Outreach Project. (From America at the Threshold, A-49)

office, but I was dismayed to find that the project had been disbanded. Late in 1990, after the review process had been completed, the Space Exploration Initiative office went through a dramatic change. Aerospace engineer Dr. Michael Griffin took over as its director and replaced almost all of its personnel, leaving only one person who had some knowledge of the preceding activities. I contacted the office at the end of June 1991, but Lieutenant General Thomas Stafford (U.S. Air Force, retired), who had been responsible for chairing the project, was no longer there, and George Abbey, who had coordinated the synthesis group, had been moved to the White House, where he was working on another project. Personnel from NASA who had worked on SEOP later became scattered between two NASA offices—the Office of Exploration in Washington, D.C., and the Exploration Program Office at the Johnson Space Center near Houston. Neither office was able to give me a reasonable explanation as to why electrogravitics had been excluded from the synthesis group report. They suggested I talk with the people at RAND who had administered the project.

However, personnel at RAND were of little help. As far as they were concerned, their contract was over, and they wanted nothing more to do with SEOP. Any telephone inquiries were directed to a spokesperson who would not allow me to speak directly with RAND employees who had been involved in the project. However, they did send me a copy of their technical report summarizing the findings of the panel that reviewed the ideas in category 3.[3] This contained considerably more information on submissions in this particular category and had an appendix listing the titles of the 348 submissions that had been reviewed in this category. However, the main body of the report remained curiously silent on the subject of electrogravitics.

A review of the titles in the report's appendix indicated there were several other submissions beside my own that also had suggested NASA look into nonconventional propulsion technologies. The titles of some of those are in table 3.

I wanted to get copies of this subset of submissions along with the names and addresses of their respective authors for the purpose of correspondence, but was stonewalled. The RAND representative told me all material processed for SEOP had been turned over to NASA.

TABLE 3. A SAMPLE OF NONCONVENTIONAL SUBMISSIONS MADE TO SEOP

ID No.	Name	Title
100105	William D. Taylor	Whirley-go
100136	(unknown)	Inertial drive unit
100153	Joe Hughes	Beyond electric propulsion
100159	Paul LaViolette	Electrogravitics: An energy-efficient means of spacecraft propulsion
100174	Fred R. Nehen	Gyro propulsion
101456	William W. Few	The Searl levity disc
101570	Roger Fritz	Inertial engine
200453	Gordon C. Campbell	How to build a flying saucer

However, people at NASA's Office of Exploration and Exploration Programs Office were not of much help either. One person at NASA thought that the submissions were being stored temporarily in someone's office but did not know whose. An individual at the Office of Exploration seemed to become nervous when I asked him about the whereabouts of the submissions. I got the impression that he actually knew where they were being kept but was trying to avoid telling me. He instead directed my request to the Johnson Space Center office. People at that office, in turn, directed my request to the Washington office. Thus, I very quickly got the impression that I was being sent in circles. This was supposed to be an open, unclassified solicitation of ideas. Why should they be trying to avoid public inquiries into the ideas that had been presented? Was there something about this project they were trying to hide?

After about four months of calling one office or another and getting nowhere, finally at the end of May 1992, I instituted a request through NASA's Freedom of Information Act Office. After some difficulty, they eventually located the archived documents and in September sent me copies of most of the requested submissions. NASA would not divulge the addresses of the submitters, only their names. A review of these SEOP submissions confirmed what I had suspected, that there were others who also had attempted to make NASA aware of nonconventional propulsion technologies and that their ideas had also been omitted from the final report.

Two submissions had informed NASA about electrogravitic propulsion: my own (no. 100159) and that of Joe Hughes, on electric propulsion (no. 100153), which is reproduced in appendix H. Hughes referred to Brown's flying disc experiments as well as to Brown's proposed design for a spacecraft powered by a plasma-jet, high-voltage ion generator. He included a copy of Brown's electrokinetic generator patent. Hughes also cited Dr. George McDonough, director of science and engineering at NASA's Marshall Space Flight Center in Alabama, as saying that electric propulsion "is an interesting alternative to nuclear propulsion which is the only one being considered by the agency" and that the Soviets "consider it a viable way to do the job."

In the case of the submission titled "The Searle levity disc," the search could locate only the abstract. Quite mysteriously, the backup paper that was supposed to contain an explanation of the disc's operation could not be found. As discussed in chapter 10, this is a device developed by the British engineer John Searl that nullifies gravity by rotating a set of roller magnets. It was foolish of NASA not to take a serious look at this concept, since a few years after the SEOP report was issued, Roshchin and Godin, working in Russia on a shoestring budget, built and successfully tested a version of the Searl disc in which the rotor and its test platform were observed to lose 35 percent of their weight with the rotor spinning at the modest rate of 600 rpm.[4]

Several individuals had also sent SEOP submissions suggesting that NASA look into gyroscopic inertial drive as a feasible method of spacecraft propulsion (submission nos. 100105, 100136, 100174, and 101570). Inertial drive technology is entirely mechanical in nature. It generally involves various methods of either rotating or repeatedly jerking back and forth the bearing supports of a massive spinning gyroscope wheel so as to produce a reactionless vectored thrust of the entire apparatus. Several devices immediately come to mind. One prototype inertial propulsion engine developed by American inventor Robert Cook (U.S. patent 4,238,968) has been shown to develop a thrust of 1 pound.[5] Another device, built by Canadian inventor Roy Thornson, has demonstrated a thrust of 8 pounds. Yet another inertial propulsion prototype, developed by Scottish engineer Sandy Kidd, has produced about 0.50 pound of thrust.[6] With the financial backing of an Australian oil drilling equipment company, Kidd has subsequently begun work on a much larger, passenger-carrying prototype. All of these inertial devices blatantly violate Newton's third law of motion, which states that every action must have an equal and opposite reaction. That is, unlike a rocket, an inertial drive unit moves forward without ejecting mass in the opposite direction. Nevertheless, like electrogravitics, such devices have had a long history of development and a proven track record.

The inertial drive idea submissions to NASA were generally quite well written, and in particular, the authors of two of these (nos. 100105 and 100136) indicated that they either had working devices or had done considerable computer simulation work establishing concept feasibility. Yet nowhere in RAND's space transportation/propulsion panel report or in the synthesis group report was there any mention of the inertial propulsion concept. By all reasonable standards, NASA should have looked into these ideas, yet like the other nonconventional propulsion concepts, the RAND and synthesis group reports totally ignored them. If NASA was asked to "cast a net widely to find the most innovative ideas in the country," why had these ideas not been considered? Had RAND selected panelists who were grossly inept, scientists with tunnel vision who callously weeded out some of the best ideas of the bunch just because they did not fit standard textbook theories, or was there a concerted effort to exclude such ideas in the name of national security? The latter seems more likely since the RAND Corporation, which has a history of being involved in intelligence projects and weapons development, is said to be a front organization for the Central Intelligence Agency.

It is doubtful that this screening operation was put in place to avoid criticisms that might have been leveled by academics skeptical of nonconventional ideas. More likely its intent was to discourage NASA from considering technologies that were already being worked on in defense-sponsored black projects. The censorship of the SEOP ideas submitted to RAND can be traced directly to the screening procedures that RAND had adopted. Some revealing information in this regard may be found in the transportation, launch vehicles, and propulsion panel's technical report. It states that of the 350 submissions reviewed by that particular panel, "approximately 30 percent were judged infeasible" either because they "violated known physical laws or the performance claimed for a concept would be impossible to achieve."[7] Actually, the total number rejected on this pretense was closer to 39 percent, since, at the end of the screening process, only 213 submissions had been passed on for more-formal analysis.

The report states that of the submissions that made it through this screening, none contained "any new scientific laws or principles, or wholly new areas of technology . . . nothing was presented that is truly new and revolutionary." Moreover, it states that most of the submissions had proposed "concepts or ideas that are currently being considered [by NASA] or have been examined in the past."[8] These observations about the outcome of the SEOP project should not be surprising; it is obvious the screening process was set up so that any ideas that were truly new and revolutionary were omitted. As the expression goes, Quayle's team of appointed panelists and hired think-tank consultants "threw out the baby with the bathwater." It is apparent that this was not just one other instance of Murphy's Law at work. These people knew what they were doing, as one panelist privately admitted to me that SEOP had received quite a few "advanced technology" suggestions and that none of these had been included in the final summary report.[9]

The bias against considering innovative ideas is evident in the procedure used to rank the submissions. Each submission was ranked on a scale of 1 to 5 (5 being the best) in each of five attribute areas: utility, feasibility, safety, innovativeness, and cost. However, for reasons not stated, these areas were not given an equal weighting. For example, the transportation, launch vehicles, and propulsion panel weighted these attributes respectively as 25 percent, 25 percent, 25 percent, 5 percent, and 15 percent. Thus, feasibility was considered to be five times more important than innovativeness, with such feasibility being judged according to whether the ideas violated "known" physical laws. It is no wonder that after being asked to "cast a net widely" for new technologies, the panelists had come up with nothing, with American taxpayers footing the bill.

What is even more shocking, the panelists usually ranked the submissions by considering their scores on just the first two attributes—utility and feasibility. Thus, safety, innovativeness, and cost did not seem to count much in the final outcome. This explains why the transportation, launch vehicles, and propulsion panel report seriously discusses concepts such as explosion-driven spacecraft and antimatter annihilation propulsion that, by most standards, fail miserably to meet safety or cost-effectiveness criteria. In the case of explosion-driven propulsion, a nuclear bomb is detonated behind the spacecraft and the shock wave is made to impact a pusher plate that propels the craft forward. Nowhere did the report express any concern about matters such as passenger safety and dangers associated with the proliferation of nuclear weapons in space. The report's discussion of antimatter annihilation propulsion is another case in point. It would take a million years using the current multibillion-dollar CERN and Fermilab facilities just to accumulate 1 milligram of antiprotons, enough to propel a 1-ton payload to escape velocity from Earth. Accumulating and storing such large quantities of antimatter is an even more formidable problem, from both the technical and the cost standpoints. If cost was really of any concern, the report should not have even bothered discussing the antimatter subject. Perhaps they were trying to please the fans of *Star Trek*.

In view of the above, it is quite disappointing that no space at all was given to reviewing propulsion technologies like electrogravitics, magnetic drive, and inertial drive that not only are feasible, but are safe and cost-effective as well. One wonders whether the Space Exploration Outreach Project was worth the millions of dollars that taxpayers spent.

It seemed that I had exposed a very large black hole, one that happened to be centered right smack in the middle of NASA and was swallowing up a lot of money and with it a lot of good ideas. We are told that one of the characteristics of a black hole is that whatever goes into it can never emerge again. This definition very much fits NASA's Space Exploration Outreach Program, for in 1991, after the synthesis group report was completed and copies mailed out, participants found it nearly impossible to learn anything about the fate of their submitted ideas or to get any information about ideas submitted by other participants. As soon as the report was completed, the arm of this "outreach program" was immediately retracted, with no plans in place for carrying out follow-up activities. It seemed the program organizers planned it that way from the start. By the end of 1991, NASA had dismantled the project's office and transferred its personnel to other jobs. The original idea submissions were diverted either to someone's office closet or to an obscure archive repository. The rapidity with which the SEOP office was dissolved and its personnel and raw data scattered is reminiscent of the sudden dissolution of an FBI front operation after it has caught its crooks. It is clear that SEOP had been planned to be a one-way information-gathering intelligence operation.

13.3 ■ THE MISSING DISCS

The transportation, launch vehicles, and propulsion panel report states that information about the evaluation of each SEOP submission was logged into a Macintosh computer database (Fourth Dimension by ACIUS). This included "the unique ID number of the submission, the reviewer, the date of review, the name of the panel performing the review, and the title or subject of the review."[10] Also, the database included the score assigned to the proposal (ranked on a scale of 1 to 5) and a written justification for that score. The report states, "Each reviewer was required to briefly explain the reasons for scoring a submission as he or she did."[11]

Information stored in this computer database should have been made available to anyone requesting information about the fate of his or her submission, but when attempts were made to locate the computer disks, they were nowhere to be found. Neither RAND nor NASA personnel claimed to know their whereabouts. One NASA employee went as far as to claim that there was no computer database. However, one of the transportation, launch vehicles, and propulsion panelists had previously told me he had used the database in his proposal review. Thus, the database definitely existed at one time. I initiated a NASA Freedom of Information Act request to obtain a copy of the information on this disc, but the officials failed to locate this information, either in the magnetic media archives or in the archived boxes containing the written submissions. An appeal also failed to turn up anything. Later, I came in contact with Debra Ladwig, a NASA employee who had worked as a computer support person during the synthesis group phase of SEOP. She told me she had initially received a copy of the discs from RAND and that at the close of the project she had turned the discs over to Dr. Brenda Ward of Johnson Space Center. However, in May 1992, I asked Ward if there had been a computer disc summarizing the submission reviews, and strangely, she maintained she did not know of any. At present, then, the magnetic disc records of RAND's evaluations of the SEOP submissions remain missing. Was their disappearance just an accident, or did someone not want the public to know why certain idea submissions were not included in RAND's final report?

The disappearance of the proposal evaluation database is particularly disturbing. The closed-door nature of the whole evaluation process sounds more like a classified, black R&D project than a NASA program. It is reminiscent of what reportedly was going on in the UFO study conducted by the Air Force under Project Blue Book. According to informants connected with that project, the more-unusual UFO reports submitted to the project were routinely siphoned off to a highly classified intelligence group, never to be seen again by the public. The missing reports not only were absent from Project Blue Book's database but also were omitted from its final report. It is not surprising to find that the same operating procedures are being practiced at NASA to screen out information relating to antigravity and field-propulsion technologies.

13.4 ▪ THE NATIONAL AERO-SPACE PLANE

At the time the SEOP report was published, NASA had plans in the works to develop the X-30 National Aero-Space Plane, also called the space plane, which was to be the eventual replacement for the space shuttle. The plane was to use three different propulsion systems. The "low-speed" propulsion system, whose technology was then classified, was designed to take the craft up to a speed of about Mach 3 and to altitudes of greater than 50,000 feet. At Mach 3, a liquid-hydrogen-burning ramjet would take over. Unlike the space shuttle, which uses a liquid-oxygen oxidizer, the X-30 ramjet would be air breathing. A ramjet is a jet engine that has no moving parts and that depends on the high pressures created in front of it to force air through its combustion chamber. Since liquid oxygen accounts for 89 percent of the fuel weight in a standard liquid-oxygen/liquid-hydrogen rocket, this ramjet system allows a substantial reduction in the rocket's overall weight. At Mach 6, the ramjet would convert into a scramjet—a supersonic ramjet—as the airflow in its combustion chamber transitioned from subsonic to supersonic flow. The scramjet then would propel the X-30 into speeds of Mach 12 and greater. Ground tests achieved Mach 12 speeds, and much higher Mach numbers were expected to be forthcoming when actual flight tests would be conducted. For comparison, at Mach 12, a trip from New York to Tokyo would take just one hour. As the space plane would attain an increasingly high altitude, it increasingly would rely on the addition of liquid oxygen to sustain combustion in its scramjet. Outside the atmosphere, it was to rely entirely on liquid-oxygen/liquid-hydrogen rocket combustion.

One problem NASA anticipated with the space plane was that its wing leading edges would experience excessive frictional heating during the plane's high-velocity flight through the atmosphere (at Mach 3 and above). Just in this one area, the project could have substantially benefited from knowledge about Brown's electrogravitics work, which was discussed in two of the SEOP idea submissions, ideas that were weeded out from SEOP's final report. In particular, high-voltage electrification of the leading edge of the aircraft's body, in the manner Brown had suggested, would have assisted in deflecting the approaching airstream so it would not directly contact the vehicle's surface, thereby reducing the air frictional drag and softening the vehicle's transition through the sonic barrier. However, personnel working on the space plane project indicated that they had not heard of electrogravitics or about the potential of high-voltage charge to alleviate air drag, one of the project's most pressing technical problems.

Determined to circumvent SEOP's idea censorship, in May 1992 I sent copies of my electrogravitic propulsion SEOP submission and a copy of the 1956 Aviation Studies report, "Electrogravitics Systems," to Charles Morris, director of the space plane project. He said he would circulate this material among the project's engineers. One month later, he sent a letter stating that "the concept is not appropriate for consideration within the National Aero-Space Plane (NASP) program" (see appendix I). A year later, I convinced him to reconsider the idea. Since he had not kept any of the material I had sent earlier, in September 1993 I sent him a new packet of information (see appendix J). Later that year, when I inquired whether he thought the space plane might benefit by using a high-voltage charge, the director commented that he found the ideas very interesting, but was not optimistic that NASA would adopt such a technology in the immediate future.

Subsequently, a scientist at NASA's Marshall Space Flight Center commented to me:

> Electrogravitics is one of those things that certainly is worth looking at because we're running up against boundaries, and nuclear propulsion isn't going to happen in our time, as far as I can tell . . . We don't have a program. That's the problem. We don't have anything on the horizon where there's support at headquarters for really futuristic things . . . There are some real interesting things out there like this. NASA used to be a lot better at forward thinking than we have gotten to be, and if we are going to survive in this age, we are going to have to take off our "things-as-usual hat" and think about some of these things.[12]

Work on the space plane proceeded for several years but was discontinued in 1994 due to budget cuts and because the program could not deliver the kind of results Congress was expecting. In 2003, there was a strong lobbying effort to resurrect the project, but none of the proposed ideas made any mention of the idea of applying electrostatic charge to a wing's leading edge to solve the hull-heating problem.

13.5 ▪ THE *COLUMBIA* DISASTER

On February 1, 2003, the *Columbia* space shuttle crashed to Earth in flames as the result of damage its wing had suffered earlier in the mission. During takeoff, a suitcase-size chunk of insulating foam had broken off from its main propellant tank and impacted the leading edge of the shuttle's left wing, causing damage to one of the wing's thermal protection tiles. The damage went undetected and later caused a major problem when the vehicle attempted to reenter the atmosphere during landing. Air friction normally heats the wing surface to incandescent temperatures during the high-velocity atmospheric reentry. As a result of the earlier damage, superheated gases were able to penetrate a gap in the wing's thermal protection tiles and cause damage to the shuttle's internal wing structure. Exposed to these hot gases, the wing structure ultimately failed, the vehicle became uncontrollable, and it was eventually destroyed by the extreme heat of reentry. Seven crew members perished in the crash.

Had Brown's aerospace technology been implemented on the space shuttle, the Columbia disaster would never have happened. Thirteen years earlier, two SEOP submissions had pointed out the advantages of electrogravitics, but the SEOP review panel discarded the ideas and did not include them in its final report to NASA. As mentioned earlier, in 1992 I had also contacted space plane project director Charles Morris of NASA to suggest a solution to the hull-heating problem foreseen to plague the project. I had pointed out to him how air-friction heating of the leading edge of a shuttlecraft's wing could be prevented simply by applying a high-voltage charge to the wing. Again, in my 1993 letter to him (reproduced in appendix J), I wrote ". . . electrostatic charging of the plane's leading edge would also have the added benefit of reducing air friction heating of the hull surface." However, NASA personnel did not employ the idea.

Somewhat later I spoke with Jonathan Campbell, an engineer who works on electrical propulsion systems at NASA's Marshall Space Flight Center. I learned that he had been trying for years to convince NASA to look into electrokinetics as a means for spacecraft propulsion, but his requests for money were routinely turned down by management. He took out two patents on a thrust-producing apparatus (see figure 13.2) that is very similar to one of Brown's electrohydrodynamic devices. Although he has acknowledged that Brown's work inspired him to develop his cylindrical thruster, curiously, Campbell's patent did not cite or discuss Brown's prior work.[13]* Campbell has a more conventional view on the

---

*Interestingly, Turman's cylindrical thruster experiment (see chapter 12) bears a close resemblance to Campbell's asymmetrical capacitor; compare figure 13.2 to figure 12.1. Turman's research predates Campbell's patent application by at least twenty-six years.

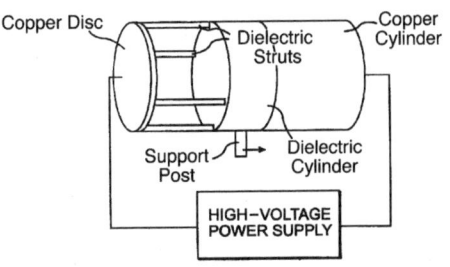

Figure 13.2. A symmetrical capacitor apparatus for generating thrust. A high-voltage DC potential causes air to flow from the copper cylinder toward the copper disc. (Campbell, 2001)

operation of asymmetrical capacitors than Brown and others, denying that any exotic principle such as electrogravitics operates. His patent did not discuss the idea of charging the leading edge of a spacecraft's wing, so even if NASA had funded Campbell's research, there is no guarantee that Brown's airframe-charging idea would have been employed.

In March 2003, I submitted a suggestion to the Columbia Accident Investigation Board (reproduced in appendix J). I pointed out again the benefits of charging the wing leading edge, stating, "One technology that could prevent a Columbia-type hazard from happening in the future would be to apply a high voltage charge to the Space Shuttle hull during reentry, in particular to the leading edge of its wing. The ion sheath so formed would create a buffer zone around the craft, ionizing, repelling and deflecting oncoming air molecules and thereby preventing them from directly impacting and heating the hull." I also noted that Northrop had researched this technique thirty-five years earlier and named some references they could consult. I also summarized Brown's work and the use of electrogravitics on the B-2 bomber. I noted how I had earlier attempted to make NASA aware of the technology, both through my submission to SEOP and through my contacts with Morris. I named Campbell as someone at NASA whom they could contact and also offered my own assistance to point them in the right direction, but nothing ever came of my suggestions. All they sent me was a form letter thanking me for my input. Efforts made to resurrect the space plane as a future space shuttle replacement made no mention of my suggestion for wing electrification.

The aerospace industry has not shown the same bureaucratic dinocracy. In 1994, one year after NASA turned down the idea for its space plane, BAE Corporation (formerly British Aerospace) became seriously interested in Russian research into plasma air-drag reduction. Together with the Defence Evaluation and Research Agency (DERA), Britain's military research organization, it began researching the idea of generating plasmas upwind of an aircraft as a means of reducing air drag.[14]

In 1996, Terry Cain, a research engineer with DERA, traveled to Russia to meet Anatoly Klimov and his colleagues and to repeat the plasma drag reduction experiments they had performed. At the Central Aerohydrodynamics Institute, near Moscow, they carried out supersonic wind tunnel tests on 10-centimeter, conical-shaped bodies that used plasma generators to create upstream plasmas. One method they used to generate the plasmas was to energize the cones with a Tesla coil that created voltages high enough to cause air to ionize over large distances. The high-voltage fields generated "little streamers of lightning" that propagated into the airflow ahead of the test model.[15] They measured drag reductions of 10 percent.

In the United States, the Arnold Engineering Development Center, at Arnold Air Force Base in Tennessee, has modified its wind tunnels to conduct airflow tests of plasma-assisted models. The long list of aerospace planes that have undergone aerodynamic testing at Arnold includes the B-2 Stealth Bomber and the X-30 National Aero-Space Plane. In 2000, the head of its applied technology directorate was quoted as saying that a number of organizations have shown interest in plasma air-drag reduction, although he would not give their names. Might we expect NASA to be among those showing interest?[16]

13.6 ■ NASA: A MILITARY FRONT ORGANIZATION?

Donna Hare, a former employee of a NASA contractor, has disclosed evidence that implicates NASA in covering up evidence of the presence of advanced-technology spacecraft. During the 1970s, Hare worked for NASA contractor Philco Ford in its photo lab at the Johnson Space Center. Since she had a secret clearance, she was able, one day, to walk into the NASA photo lab where a friend of hers worked. The lab was involved in developing satellite pictures and pictures taken during NASA's various missions. Her friend directed her attention to an area of a photo mosaic he had been working on. Then, smiling, he suggested she look at a particular area of one of these photo panels. There she saw a round white dot with a very crisp outline. She asked if it was a dot on the emulsion. Grinning, he said, "Dots on the emulsion don't leave shadows on the ground." Sure enough, there was a round shadow on the trees at the correct angle from where the sun would have been shining. She asked, "Is this a UFO?" He answered, "I can't tell you that," meaning it was a UFO, but he wasn't allowed to tell her it was. He said, "We always have to airbrush them out before we sell them to the public." Hare's astounding testimony can be found in Stephen Greer's book *Disclosure*.[17]

Hare also disclosed stories she had heard from NASA employees about some astronauts having seen extraterrestrial craft. One gentleman whom she knew very well said just about every one of the astronauts who had gone to the moon had seen things. One said there were three craft on the moon at the time the Apollo 11 mission had landed. He said that as a precaution, the astronauts were put in quarantine for a while after they had returned and that some of the astronauts who wanted to talk were threatened.

Hare also related a story told to her by someone who used to work at Johnson Space Center as a security guard.[18] He said one day soldiers came in fatigues and ordered him to burn photographs. At one point, he stole a glance at one of the pictures and could see that it was a UFO on the ground. One of the guards apparently caught him doing this and hit him in the head with a gun butt. She said she could see that as he told his story, he was very frightened.

So we see that NASA and its employees are being threatened and manipulated into silence to maintain the status quo of a cover-up about the presence of alien craft and the existence of advanced aerospace propulsion technologies. Seeing that the NASA administration seem determined to steer clear of electrogravitics technology despite repeated attempts by several people to interest them, one is led to sympathize with Tom, my Project Skyvault contact, who said in the note he passed me at the 1994 Tesla conference that NASA is essentially a public relations organization or a front that obscures Air Force space research.

Over the phone, Tom later told me an astounding story of the scope of the Air Force's involvement in space. He said he had been in the

Civil Air Patrol and had been given a Mitchelson Award, the highest award that one could get. As a result, in 1963 he was chosen to represent the state of Idaho and go to Chanute Air Force Base along with Civil Air Patrol representatives from the other forty-nine states. One day they were all gathered in an auditorium, and onstage there were about eight generals who were available for a "no bars" question-and-answer period. One person popped up and asked them about Air Force Major Donald E. Keyhoe, who at the time was writing about UFOs and had been severely censored. One of the generals responded that they had a way of taking care of people who gave out a little too much information. He said they would use physical injury or whatever was necessary to make them shut up, indicating they would kill a person (extreme prejudice, if you will). Someone else started to ask more about UFOs and one of the generals said the United States had a defense system in place at the time that consisted of a number of satellites, in orbit not only around the Earth, but also around Mercury, Venus, Mars, and a few other, more distant planets they couldn't talk about. He said the satellites together functioned as an early warning system, that they were afraid of the "people out there" because they didn't know very much about them. This satellite defense system was built to observe three possible sources: missiles that might come from the Soviet Union, missiles that might come from China, and intrusions of aliens coming in toward Earth. Someone asked why the generals were being so candid with them. According to Tom, one responded by saying, "If you want, you can go ahead and tell people what we told you, but they're not going to believe you. Besides, if you did get anyone to believe you and they came back to ask us, we would just deny it. So we have nothing to lose by telling you this."[19]

Russia had put the Sputnik satellite in orbit around the Earth in 1957, and the United States followed by putting the Explorer in orbit the next year. In 1959, the Soviets photographed the far side of the moon with Luna 3. In 1962, the Mariner 2 probe of Venus sent back close-range information about Venus, and that same year, the Russians launched the first probe to Mars, but contact was lost. So in 1963, Tom was told that the United States at that time had a network of sophisticated warning satellites scattered throughout the solar system, orbiting planets farther out than Mars! The well-funded military space program was apparently several decades ahead of what was being publicly acknowledged! Tom said he had heard rumors that the first satellite ever was launched by the United States in 1948 using a modified V-2 rocket. He said the Soviets never really were ahead of the United States in the space race. The military used the publicity of the Soviet effort to their own advantage to get more money from Congress.

North American Rockwell delivered its first space shuttle to NASA in March 1979. This was the Columbia shuttle, which made its maiden voyage two years later, in 1981, but Tom said that Rockwell had been delivering space shuttles to the U.S. Air Force as early as 1976. He also said the Air Force has its own shuttle system and that its shuttles were being launched from a highly secured island in the Pacific known as Johnston Island. He said he had been working for the Air Force between 1976 and 1978 and that during this time he met a captain who was an engineer with the Air Force and who had returned from Johnston Island after being there for a year or two. He said this captain told him he had heard rumors that the United States had a base on the moon. The captain said that from looking at the cargo manifest for one of these shuttle launchings, one could conclude that provisions were routinely being shipped out. This was several years after the Apollo program had been terminated, the last Apollo mission to the moon having been completed in December 1972.

Thus, it is apparent that there has been an effort to keep secret the military's capabilities in space. While NASA was mesmerizing the public with its rocket flights, aerospace companies were carrying out secret research on electrogravitics and microwave beam propulsion technologies. A good guess is that the U.S. military currently has large fleets of craft capable of hypersonic flight in space that use nonconventional means of propulsion.

This educated guess may be fact. In 2002, a forty-year-old British computer buff named Gary McKinnon succeeded in using his home computer to hack into the computer network of several U.S. military organizations. Although not part of any terrorist organization and only snooping to satisfy his own curiosity, he now faces up to seventy years' imprisonment in a U.S. jail, but what he found on one of his Internet forays was quite astounding. In a U.S. Space Command database, McKinnon found a list of officer's names under the heading "Non-Terrestrial Officers." He also found a list of "fleet-to-fleet transfers" and a list of ship names. He tried to look them up, but found they were not Navy ships. He came to conclude that these were off-planet vessels.

The U.S. Space Command is headquartered at Peterson Air Force Base in Colorado Springs. Its website states that its "mission is to conduct joint space operations in accordance with the Unified Command Plan assigned missions." These include "Space Force Support, Space Force Enhancement, Space Force Application, and Space Force Control."[20] With records of nonterrestrial officers carrying out fleet-to-fleet transfers, its missions appear to be far bolder than the average U.S. citizen might have guessed. It seems the United States has ongoing, manned space operations that go beyond anyone's wildest dreams, all taking place under a cloak of secrecy and all made possible by advanced field-effect propulsion technologies that were under intense development in the middle of the twentieth century.

The Russians may also have a substantial presence in space. In his book *The Awakening of the Red Bear,* Dimosthenes Liakopoulos wrote that the Russians have large electrogravitic-propulsion craft called cosmospheres that are equipped with particle beam weapons.[21] He maintains that these are used to ferry supplies to ten bases on the moon.

As for the application of field-propulsion technology to civilian aerospace flight, it is apparent that NASA, with its rocket-oriented approach, will not be the one to take the initiative. That, instead, will likely be undertaken by farsighted aerospace corporations such as the Spaceship Company and Virgin Galactic. The ability of nonmilitary private enterprise to compete in the space arena became evident on October 4, 2004, when Brian Binnie piloted SpaceShipOne to an altitude of 114 kilometers to win the $10-million X-Prize. This prize, which was offered by the X-Prize Foundation, was available to anyone who succeeded in reaching an altitude of at least 100 kilometers twice within a two-week period. That was SpaceShipOne's second voyage, its maiden voyage having been made just five days earlier. The winning team, led by aerospace engineer Burt Rutan, showed the world that, with a little ingenuity, space flight is possible even on a shoestring budget.

Figure 13.3. SpaceShipOne landing in the Mohave Desert on October 4, 2004. It was the first privately owned plane to achieve suborbital flight. (Photo courtesy of Mike Massee)

In July 2005, Rutan's company, Scaled Composites, signed an agreement with Virgin Galactic, a spin-off of the Virgin Group of Companies, founded by Sir Richard Branson, to form a new aerospace production company that both companies will jointly own. This new company, called the Spaceship Company, has plans to build a fleet of commercial suborbital spaceships and launch aircraft and to market them to spaceline operators, one of which will be Virgin Galactic. Rutan, who will head up the company's technical development team, said, "[T]his will truly herald an era of personal space flight first described by the visionary science fiction writers of the 1940s and 1950s. Richard Branson and I share a vision that commercially viable and safe space tourism will provide the foundation for the human colonization of space."[21] It is more likely that entrepreneurs such as these, who are accustomed to thinking out of the box, will ultimately be the ones who will develop field-effect propulsion for aerospace flight.

14

A TECHNOLOGY THAT COULD CHANGE THE WORLD

Clearly, gravity control would be a boon to society. So why is work on this important technology being kept so highly classified? One obvious reason is that the military sector wants to make sure that its defense technologies are always one step ahead of everyone else's. According to one estimate, black-program technologies are at present at least fifty to one hundred years ahead of those used in the commercial world. A second motivation for secrecy is the concern over whether society is able to monitor and control the public use of this new science effectively. A case in point is the advent of nuclear technology toward the middle of the twentieth century. While ways were later found to harness atomic energy for peaceful uses, its initial development was for use as a weapon, the atomic bomb. This brought with it the accompanying threat of nuclear holocaust, and today, even though the cold war has ended, the threat still lingers that terrorists might detonate a dirty bomb. Similarly, the same physics that gives us a proper understanding of gravity control and that could be put to many peaceful uses could also be used to build very destructive weapons.

According to Ray, the black-project scientist I had spoken to, the curators of this technology do not feel that society as a whole has matured emotionally to an extent that this knowledge could safely be made available to the public. Apparently the behind-the-scenes individuals who monitor black-world research and decide whether or not it stays classified are following a program of "controlled evolution." If these "powers that be" determine that the world has advanced to a point where it can handle a new technology, they will allow it to be introduced. This leaves us with the question of whether this self-chosen group is itself sufficiently qualified to be making these kinds of decisions. With problems like global warming, global deforestation, acid rain, widespread pollution of the oceans, radioactive waste, overpopulation, hunger, and the AIDS epidemic looming ever larger, shouldn't at least some of this knowledge be declassified to help the world? Some members of the black-programs community feel that it should, and as a result, they are making efforts to push things in a more liberal direction.

When the internal combustion engine was developed at the turn of the twentieth century, should it have been classified because of its military potential to create tanks and war vehicles? If it had, we may still be driving a horse and buggy and would undoubtedly be living in a world that had a much lower standard of living. Today, it seems that our government is intent on keeping our current technology status quo. On January 15, 2008, four group directors that head the U.S. Patent and Trademark Office (USPTO) sent out a memo to all USPTO technology center patent examiners that is just as reactionary as if they had outlawed the automobile. The memo reminded the examiners about the USPTO Sensitive Application Warning System (SAWS) program and required that they "flag" any patent application that contains subject matter of "special interest," specifically those containing the following topics: "1) perpetual motion machines [i.e., over unity energy generators], 2) antigravity devices, 3) room temperature superconductivity, 4) free energy—tachyons, etc., 5) gain assisted superluminal light propagation (faster than the speed of light), 6) other matters that violate the general laws of physics . . ."[1]

Further, the directive required the examiners to, among other things, flag: "applications with pioneering scope" and "applications dealing with inventions that, if issued, would potentially generate extensive publicity." It stated that the SAWS program "is intended to ensure that the [USPTO] examination standards and guidelines are applied properly to such applications." Such guidelines instruct examiners to reject any applications that violate the "known laws of physics." Obviously, the laws the Patent Office are referring to is the catechism taught in university physics courses around the country. By those laws, patent applications for any invention using over-unity energy generators, electrokinetics technology, or superluminal beam generators (such as that developed by Podkletnov) should be promptly rejected.

But let us imagine for a moment that the prevailing bureaucratic suppression has faded, ushering in a host of possibilities. Once field-effect propulsion technologies and energy generators are commercialized, they could dramatically improve people's lives. For example, Earth-based transportation would be revolutionized, commuters would be able to travel vertically as well as horizontally, roads and bridges would no longer be needed, and ground-level rush-hour traffic jams would be a thing of the two-dimensional past. However, in populated areas, special navigation computers would be required in order to prevent midair collisions. Transport speeds would be vastly increased, and there would be few limits as to where such vehicles could go. Antigravity vehicles would revolutionize farming, mining, building construction, and shipping, stimulating the world economy beyond our wildest dreams.

Space flight would be made practical. Travel from one's Earth-based home to an orbital space station would become as easy as making flights from one town or city to another. Flight from the Earth to bases on the moon or Mars, or even journeys to the more-remote planets of the solar system, would be accomplished as easily as present-day intercontinental flights. Such journeys could be completed with a minimal expenditure of fuel. Just think of all the billions of dollars that would be saved if nations used antigravity propulsion instead of rocket propulsion to accomplish their space missions. Electrogravitic spaceships could theoretically attain velocities exceeding the speed of light, making it practical also to travel to nearby star systems.

World peace would be aided. Antigravity transportation would erase the distance barriers separating the nations of the world. Imagine, traveling anywhere in the world in one hour! International air traffic would skyrocket. Geographical space would shrink, bringing people from different parts of the world closer together. People would become more internationally oriented and would see one another as close neighbors. As people grew more tolerant of other cultures and as poverty declined, a new planetary world order would arise. With luck, war might even become a thing of the past.

Ecologically safe methods of energy production would become available. Electrogravitics could be usefully applied as a means for generating pollution-free electric power. One method would use permanent magnet generators such as the Searl effect generator or the MEC, discussed in chapter 10. Another method might be to use phase-conjugating parametric amplifiers that had overunity outputs. Yet another technique might be to develop rotary generators such as the electrostatic motors suggested by Lafforgue (see chapter 12) and Brown (see figure 1.10).

Electrogravitic free-energy machines, as well as other types of scientifically advanced energy generators, would provide society with clean energy. Besides producing affordable power, such technology would be environmentally and socially safe because it would not produce dangerous waste products that could pose health hazards. Nor would it release carbon dioxide or waste heat into the atmosphere that could threaten global warming. The energy output would be almost entirely in the form of mechanical motion. In the case of the Searl effect generator, it could actually have the side benefit of refrigerating the environment as it generates power. Although electrogravitic devices use high voltage, they may be rendered safe by being properly enclosed. Moreover, power production would become decentralized. Each home, factory, or vehicle would have its own power unit. The miles of unsightly power lines that presently clutter our landscape would become a thing of the past as each person became his own energy czar.

Declassifying the black-world ether physics would substantially benefit society from an intellectual standpoint. It would galvanize a whole new era of expansion in fundamental physical theory, which today has largely stagnated as a result of dogged perpetuation of outmoded ideas. It would also benefit society from a humanistic standpoint. One could argue that the spiritual vacuum and overemphasis on materialism that characterize modern society stem in part from the teaching of positivistic science that recognizes only experimentally observable quantities as having a real existence. The new ether physics, on the other hand, acknowledges that the physical world is only a manifestation of a much more fundamental, subtle realm that is not directly accessible to our physical senses, but whose operation to some extent may be elucidated with the aid of reaction-kinetic models. It leads to a worldview in which science becomes united with mystical teachings, rather than separated from them. Widespread knowledge of this new conceptual paradigm could bring humanity back toward a more ethical track, to a global mind-set better prepared to receive the advanced technologies that are now kept from us.

LETTERS WRITTEN BY T. TOWNSEND BROWN TO THOMAS TURMAN

The following letters (on pages 405 to 410) from T. Townsend Brown to Thomas Turman were written in 1968 and 1971.

THE TOWNSEND BROWN FOUNDATION
LIMITED
SASSOON BLDG. (P.O.BOX 272)
NASSAU, BAHAMAS

Cables:
TOBROFO, NASSAU

Kindly Address Reply to:

1001 Third Street
Santa Monica, Calif. 90403

January 23, 1968

Mr. Thomas M. Turman
1506-A Ave. T.
Lubbock, Texas 79401

Dear Mr. Turman:

Again I must apologize for my delay in responding to your requests. Yesterday I came home after a stay in the hospital which included an operation. I am not feeling very well and I am not able to think too clearly. The company to which I serve as consultant has not released some of the information you have requested and I am very much at a loss to know what to say.

The Patent No. 3,187,206 contains the essential teaching in electro-gravitics. A definition of the electrogravitic force might be " the ponderomotive force developed within a high-K dielectric under electrical strain". The patent teaches the use of non-linear electric fields such as those internally developed in truncated cones of dielectric material.

In short, at voltages from 75 KV to 125 KV or more, this thrust appears pronounced in vacuum. It persists as long as the voltage is maintained and is independent of the orientation with respect to the Earth's gravity vector.

There are a number of mysteries concerning the nature of the force, largely the variations which it undergoes. There appear to be at least three semi-diurnal cycles:

1) relating to mean solar time (with maxima at 4 AM and 4 PM)
2) relating to lunar hour angle with maxima approximately 2 hours after the upper and lower meridian transit of the moon, and
3) relating to sidereal time with a sharp peak at 16^h S.T. and a minor maximum at 4^h S.T.

The reasons for these variations as well as for the reasons for the almost continuous secular variations is completely unknown.

The belief that the phenomenon is gravitic in nature is based almost entirely upon the appearance of the effects of mass (in the dielectric material) on the force exhibited.

Mr. Thomas M. Turman   - 2 -   Jan. 23, 1968

Your best bet in undertaking experimentation in this field would be to make up conical sections of hi-K, hi-density dielectric (barium titanate and lead monoxide) and suspend the same with thrust sensors in high vacuum. For the gravitic effects every effort must be made to eliminate thrust contamination by ambient ions or electrostatic fields.

You ask if I have ever been able to get an anti-gravitic device to lift from the ground, its weight being entirely removed. The answer is no. The relatively small forces are detectable, usually only in vacuum and then only at exceedingly high electric gradients. Contamination of the forces make it necessary to be extremely careful in the design of the apparatus and this, needless to say is expensive. For example the equipment used to observe the cyclic effects cost us over $25,000.

I am sorry that I am unable at this time to give you more detailed information for the reasons I have explained above, maybe by next summer these problems will be sufficiently resolved so that we can go further.

Sincerely yours,

T. Townsend Brown

TTB:jb

P. O. Box 1321
Avalon, California 90704

Nov. 1, 1971

Mr. Thomas M. Turman
1611 16th Street
Lubbock, Texas 79401

Dear Mr. Turman:

I have reviewed your letter and also tried to remember what we talked about in our recent telephone conversations. I will probably not be able to answer all of your questions and I shall leave something for future correspondence.

It was a pleasure to look over the photographs of the equipment you have assembled for it indicates some rather thorough planning on your part and the expenditure of quite a sum of money. It appears you have put together the essential pieces of apparatus to perform high quality experimental work and this, I know, will be a great satisfaction to you.

Someday I would like for you to send me more detailed information and a circuit diagram of your high voltage supply. I thought I understood it was a voltage doubler but I would like to know in greater detail the capacitors and rectifiers you are using. This is of some concern, particularly as to the ripple present in the dc output.

Your sketch shows a point and ring configuration of electrodes with an intermediate dielectric tube. I take it the ends of this tube are open and the air flow is in the direction of the divergent field. This would make the tube assembly move in the opposite direction, that is, toward the small positive electrode. Is this not so?

I am also interested in the way you measure forces. You mentioned a capacitor type pick up, which would feed, thru an amplifier, to the recorder. Have you observed any for thrust with the positive end of the tube closed?

Going to the disc airfoils, you ask about the spacers used to mount the electrodes from the disc body. I was not interested then in lift, only thrust and the top speed around a 50' turnstile. The attached diagram will give you some idea of the setup.

Experiments involving lift were of a different nature. We used a triarcuate ballistic electrode as the anode and a small electrode underneath as the cathode, substantially as follows:

The large electrode was made of a balsa umbrella-like frame with aluminum foil covering. A thin glass stand-off insulator mounted the cathode as shown in the drawing. The lift of this unit at 170 kv was about 125 grams. The electrode structure itself weighed only about 100 grams, so it was actually self-levitating.

In my next letter I will send you some pressure profiles across the ballistic electrode where we attempted to analyze this lift, but I shall make it the subject of another letter.

I shall try to get this other letter off to you within a few days.

Sincerely,

T. Townsend Brown

TTB:jb
Encl.

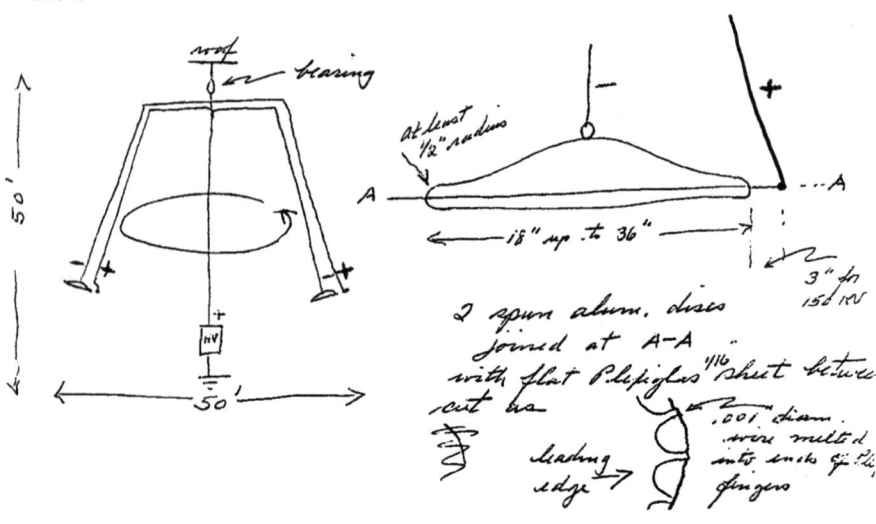

P. O. Box 1321
Avalon, California 90704

November 12, 1971

Mr. Thomas M. Turman
1611 16th Street
Lubbock, Texas 79401

Dear Mr. Turman:

I am sorry I have been so long in getting this letter off to you but, during my recuperation, I seldom get to the office.

In my letter of November 1st I promised to send you some pressure profiles, such as I remember them, which caused the movement of the triarcuate ballistic electrode. A sketch of the electrode and the approximate pressure profile is as follows:

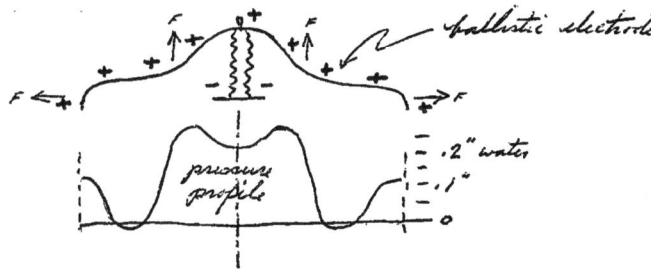

It was found that, by canting the center electrode, the pressures could be unbalanced so that one side or the other could be lifted. This could provide horizontal stability in a large prototype. An alternate way of doing this is to provide three independent electrodes in triangular configuration instead of one center electrode. These electrodes can be differently charged in order to change the electric field configuration under the ballistic canopy and this did away with the necessity for a mechanical moving part. Horizontal stability could be maintained entirely electrically. The structure is substantially as follows:

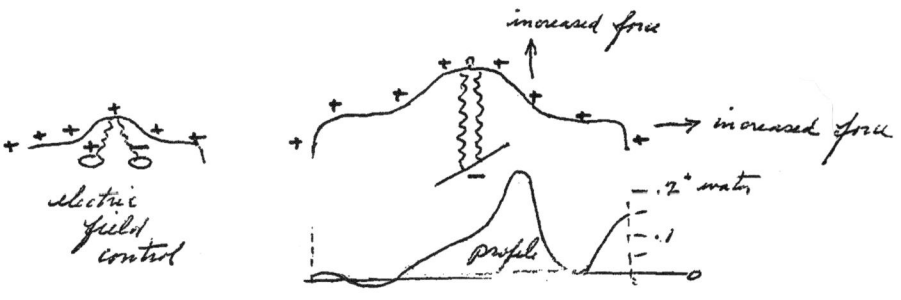

If you are able to perform any of these experiments, I would certainly like to hear about them.

Kindest personal regards,

Sincerely yours,

T. Townsend Brown

TTB:jb

APPENDIX B

AVIATION STUDIES INTERNATIONAL LTD. PUBLICATIONS
Publications Printed in 1960

AVIATION STUDIES (INTERNATIONAL) LIMITED
Aviation House 66 Sloane St. London S W 1. Tel. SLO 0657/8. Cables AVIAREP LONDON

A MAINLY CIVIL SERVICES   (fee quoted is annual rate)
1   The Aviation Reports that appear on Tuesdays $88 or £30 Including airmail
2   Finances of Increased Airline Profitability 3-part treatise on airline finance policy.$27.or.£9
3   Financial Discipline for cost computation. for major turbine aircraft $20
4   Airline Fleet Record:Quarterly statistical survey of the industry Price $18. or £6
5   Civil Transport Data Sheets:quarterly weight and performance figures Price $9 or £3 inc airmail
6   About half the Aviation Report Supplements are purely civil Price each $3 or £1
   or   No 79 & 84 onwards $6 or £2

B MAINLY MILITARY SERVICES
1   Aviation Reports issued on Fridays Price $88 or £30 Including airmail
2   Military Record of Atomic Happenings Quarterly $75 or £25 (1956 7 8 and 9 available.
3   Armament Data Sheets:Quarterly Price $30 or £10
4   Army Airforce and Naval Air Statistical Record - a quarterly military user survey $18 or £6
5   Army Vehicle and Military Aircraft Data Sheets:quarterly, performance summary $12 or £4
6   About half the Aviation Report Supplements are military Price the same as A 6
7   Weapons Compendium:a combination of No 3 4 5 and the Engine Data Sheets Price $60 or £20

C CIVIL AND MILITARY INTEREST
1   Aviation Reports (New Dimensions) issued on Wednesdays Price $105 or £35, including airmail
2   Engine Data Sheets:400 Variants of engines listed quarterly.Price $12 or £4
3   Official Price List:600 current quotations Quarterly $12 or £4

AVIATION REPORT SUPPLEMENTS

1957
17. Jan. An Appraisal of the Wright Brothers Lecture
18. Jan. The Turboprop and Propeller
19. Feb. Statistics and Records
20. Feb. Comparison of World Air and Surface Fares
21. Mar. Missiles
22. Mar. 1950 56 Cost Rise Examination
23. Apr. Diversification Analysis
24. Apr. Scheduled Airlines Freight Traffic Growth
25. May. Survival in the Sixties
26. May. North Atlantic Traffic Analysis
27. Jun. World Production Survey
28. Jun. Airmail Analysis
29. Jul. Costs and Prices
30. Jul. Accident Trends of Individual Airlines
31. Aug. Manufacturing Financial Analysis 1956
32. Aug. Traffic Directional Trends
33. Sep. Abbreviations
34. Sep. Workforce and Plant Size (Revised)
35. Oct. Production and Price Statistics
36. Oct. Helicopter Market Revised.
37. Nov. Fuel Capacity Revised
38. Nov. Traffic Growth
39. Dec. Manufacturing Ownership
40. Dec. Airline Ownership

1958
41. Jan. Quarterly Production & Price Statistics
42. Jan. Capital Gains Case
43. Feb. Military Air Transport
44. Feb. New Propulsion Sources
45. Mar. Airline Statistics Survey
46. Mar. Accident Analysis
47. Apr. Industrial Statistics Survey
48. Apr. Quarterly Production & Price Statistics
49. May. Military Statistics Survey
50. May. Electronics Statistics Survey
51. Jun. Accident Trends of Individual Airlines
52. Jun. Scheduled Airlines Freight Traffic Growth
53. Jul. Manufacturing Ownership
54. Jul. U.K. Civil/Military Policy Portfolio.

55. Aug. Comparison of World Air and Surface Fares
56. Aug. Air Mail Statistics
57. Sep. Wrecks - 1
58. Sep. All-Cargo Freight Finances
59. Oct U.S. Policy Portfolio
60. Oct Airline Financial Appreciation
61. Nov. Costs Statistics
62. Nov. Spares Statistics and Policies
63. Dec Nuclear Propulsion 1950-1960
64. Dec Airlines Ownership

1959
65. Jan Civil/Military Airfields & Runways
66. Jan Subsidy Statistics
67. Feb Research & Development Cost Statistics
68. Feb. Accident Chronology 1958
69. Mar Transport Aircraft Prices & Related Costs
70. Mar. Traffic Statistics
71. Apr. Civil/Military Procurement Practices
72. Apr Operations/Maintenance Statistics
73. May Price Statistics
74. May Electronics Statistics
75. Jun World Distribution of Military Aircraft
76. Jun Aircraft Production & Procurement
77. Jul Engine Production & Procurement
78. Jul. Helicopter Statistical Analysis
79. Aug Bilateral/Multilateral Agreements (2nd edition)
80. Aug Civil/Military Transport Fleets
81. Sep Dimensions of World Traffic-Mail
82. Sep Civil/Military Transport Specifications
83. Oct. Civil/Military Aviation Manpower
84. Nov. Dimensions of World Air Traffic - 2
85. Dec. Company ownership   /Passengers

1960
86. Jan. Horizons in New Propulsion
87. Feb. Dimensions of World Air Traffic - 3 Freight
88. Mar. Aircraft & Missile Production.
89. Apr. Operating cost comparison I - Helicopters.
90. May Operating cost comparison II - fixed wing

To: Aviation Studies (International) Ltd
     Aviation House, 66 Sloane Street,
     LONDON. S.W.1. England

Please send us ring ones required;. A.1, 2, 3, 4, 5, 6   B 1, 2, 3, 4, 5, 6, 7 .   C 1, 2, 3.

Send us Supplement No. 17 18, 19, 20, 21, 22, 23, 24, 25, 26 27 28, 29, 30, 31 32 33 34 35, 36 37, 38, 39 40 41,
42, 43, 44, 45 46 47 48 49, 50, 51, 52, 53 54, 55, 56, 57 58 59 60 61 62 63 64 65 66 67, 68 69 70, 71 72 73, 74
75. 76. 77, 78, 79 80, 81 82, 83. 84. 85. 86 87 88, 89 90

Publications Printed in 1993

AVIATION STUDIES

43rd YEAR - 1993

TITLES FOR 1992/1993
Rates annually including sourcebook, updates, binder, index, titles, taxes & postage.

SUSSEX HOUSE. PARKSIDE LONDON S.W.19 ENGLAND. TEL 081-946-5082

* Update services also available on preformatted 5.25" DEC Computer diskettes at 50% surcharge.
* AVAILABLE IN U.S. $ OR ANY CONVERTIBLE CURRENCY. ALL SERVICES IN 1992 OFFERED FOR 1993.

1. <u>Aviation Report</u> Aerospace policy containing equipment and intelligence data, economic/technical acquisition needs, government news, analysis and trends. Weekly $1,100 per year.
2. <u>Aircraft Industry Record</u> Indispensable planning historical record of thousands of first flight/delivery dates, selected chronologies, milestones, production runs. Periodic amendments $245 per year.
3. <u>Official Price List</u> More than 1,000 prices of aircraft, ships, avionics, radio, weapons, missiles, helicopters, engines, ACVs, new and historical prices/costs/comparisons. Quarterly $395 per year.
4. <u>Engine Data Sheets</u> Design features/quantities of 200 piston, 1,000 turbojet/fan, liquid rocket and 450 turboprop, rocket, hybrid and freefans, engine designs/versions. Three updates $245 per year.
5. <u>Military Record of CBR/Atomic Happenings</u> - 1993 Theme: "Post Soviet Challenge" - Analyses of CW, BW and nuclear infrastructure, costs, plans. Quarterly $540, with Armament Data Sheets $735 per year.
6. <u>Airline Fleet Record</u> Lists by continent of the fleets of 1,500 airlines.Three updates $245 per year.
7. <u>Nuclear Weapon Data File</u> Lists features of 50,000 weapons of more than 100 designs - weights, dimensions, physics, PALs, manpower, infrastructure, quantities, costs, updated quarterly $726.
8. <u>Civil Transport Data Sheets</u> Payload, fuel, weight, dimensions, English/metric characteristics of 500 passenger and cargo aircraft current/projected designs and derivatives. Three updates $245 per year.
9. <u>Armament Data Sheets</u> Rockets, missiles, mines, LGBs, guns, gas-fills, clusters, depth charges, comparisons, fire bombs, warheads, ECM, torpedoes, RPVs. readiness, sonobuoys. Quarterly $320 per year.
10. <u>Missile and Military Aircraft Data Sheets</u> 2,500 designs of fighters, bombers, transports, helos, trainers: dimensions, engines, weights, fuel, speed, quantities. Quarterly updates $245 per year.
11. <u>Army, AF & Naval Air Statistical Record</u> Aircraft, helos, AFVs, nuclears, missiles, guns, ships, personnel, funds: AF, Navy, Army, Marine, Coast Guard, Customs, Police, DOT, CRAF/airline assets, for 200 territories, six updates $675 per year or with weekly Report $1,400 or with Weekly Profile $1,200.
12. <u>Weapons Compendium</u> Nos. 9, 10, 11 with amendments $830 or with weekly Military Profile $1,080.
13. <u>Civil Compendium</u> Combined items 2, 3, 4, 6 & 8 listed above with amendments, $1,200 per year.
14. <u>Forecast Cumulative Sheets</u> Civil/Military aircraft, helos and general aviation forecasts listed individually by type in 200 territories, 4 actual (1975/80/85/90) and 9-10 forecast years. Six updates. Aircraft $910 per year, with Weekly Profile $1,250 and/or missiles, vehicles and weapons $495 per year.
15. <u>Directed Energy/Avionics Data Sheets</u> Thermal weapon types, prospects, roles, R&D, also avionics suites for tactical, strategic, patrol, support aircraft/helos: periodic updates $495 per year.
16. <u>Air Cyclopaedia</u> Unduplicated copy of all listed services including 150 updatings per year $6,635.
17. <u>Military Intelligence Critical Attributes</u> AF Leadership/Intelligence Advisories, Chiefs, organization, roles, missions, costs, manpower, NIE planning, kinds of intel., periodic updatings $665 per year.
18. Nos. 5, 7 and 17 listed above combined: 3 for less than the price of 2, special rate of $1,200.
19. Nos. 1 and 14 listed above combined special rate $1,400 per year.

To: Aviation Studies Atlantic,		(1992/93)		
Sussex House, Parkside,	We wish to:	1	7	13
LONDON, SW19 5NB. England.	* Procure (ring items)	2	8	14
	* See free sample sheets (arrow items)	3	9	15
Name/address to be sent:		4	10	16
		5	11	17
	We enclose U.S. $	6	12	18
	Invoice us later			19

APPENDIX C

ELECTROGRAVITICS SYSTEMS

An Examination of Electrostatic Motion, Dynamic Counterbary, and Barycentric Control

The following excerpt is a reproduction of the Aviation Studies (International) Ltd.'s 1956 Gravity Research Group report entitled "Electrogravitics Systems: An examination of electrostatic motion, dynamic counterbary and barycentric control," also known as "Report GRG-013/56—Electrogravitics Systems."

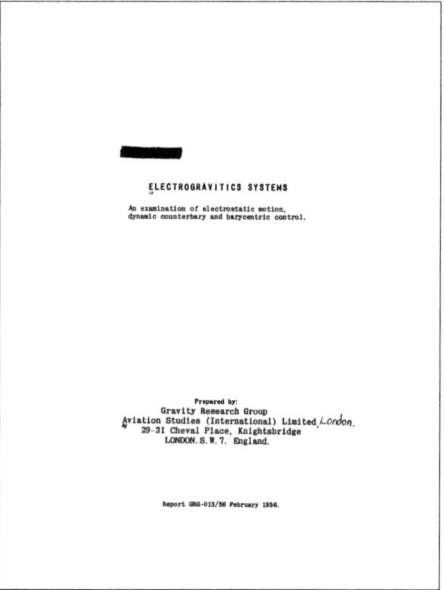

CONTENTS

It has been accepted as axiomatic that the way to offset the effects of gravity is to use a lifting surface and considerable molecular energy to produce a continuously applied force that, for a limited period of time, can remain greater than the effects of gravitational attraction. The original invention of the glider, and evolution of the briefly self-sustaining glider, at the turn of the century led to progressive advances in power and knowledge. This has been directed to refining the classic Wright Brothers' approach. Aircraft design is still fundamentally as the Wrights adumbrated it, with wings, body, tails, moving or flapping controls, landing gear and so forth. The Wright biplane was a powered glider, and all subsequent aircraft, including the supersonic jets of the nineteen-fifties are also powered gliders. Only one fundamentally different flying principle has so far been adopted with varying degrees of success. It is the rotating wing aircraft that has led to the jet lifters and vertical pushers, coleopters, ducted fans and lift induction turbine propulsion systems.

But during these decades there was always the possibility of making efforts to discover the nature of gravity from cosmic or quantum theory, investigation and observation, with a view to discerning the physical properties of aviation's enemy.

It has seemed to Aviation Studies that for some time insufficient attention has been directed to this kind of research. If it were successful such developments would change the concept of sustentation, and confer upon a vehicle qualities that would now be regarded as the ultimate in aviation.

This report summarizes in simple form the work that has been done and is being done in the new field of electrogravitics. It also outlines the various possible lines of research into the nature and constituent matter of gravity, and how it has changed from Newton to Einstein to the modern Hlavaty concept of gravity as an electromagnetic force that may be controlled like a light wave.

The report also contains an outline of opinions on the feasibility of different electrogravitics systems, and there is reference to some of the barycentric control and electrostatic rigs in operation.

Also included is a list of references to electrogravitics in successive *Aviation Reports* since a drive was started by Aviation Studies International Limited to suggest to aviation business eighteen months ago that the rewards of success are too far-reaching to be overlooked, especially in view of the hopeful judgment of the most authoritative voices in microphysics. Also listed are some relevant patents on electrostatics and electrostatic generators in the United States, United Kingdom and France.

Gravity Research Group
25 February 1956

DISCUSSION

Electrogravitics might be described as a synthesis of electrostatic energy used for propulsion—either vertical propulsion or horizontal or both—and gravitics, or dynamic counterbary, in which energy is also used to set up a local gravitational force independent of the earth's.

Electrostatic energy for propulsion has been predicted as a possible means of propulsion in space when the thrust from a neutron motor or ion motor would be sufficient in a dragless environment to produce astronomical velocities. But the ion motor is not strictly a part of the science of electrogravitics, since barycentric control in an electrogravitics system is envisaged for a vehicle operating within the earth's environment and it is not seen initially for space application. Probably large scale space operations would have to await the full development of electrogravitics to enable large pieces of equipment to be moved out of the region of the earth' s strongest gravity effects. So, though electrostatic motors were thought of in 1925, electrogravitics had its birth after the War, when Townsend Brown sought to improve on the various proposals that then existed for electrostatic motors sufficiently to produce some visible manifestation of sustained motion. Whereas earlier electrostatic tests were essentially pure research, Brown's rigs were aimed from the outset at producing a flying vehicle. As a private venture he produced evidence of motion using condensers in a couple of saucers suspended by arms rotating round a central tower with input running down the arms. The massive-k situation was summarized subsequently in a report, Project Winterhaven, in 1952. Using the data some conclusions were arrived at that might be expected from ten or more years of intensive development—similar to that, for instance, applied to the turbine engine. Using a number of assumptions as to the nature of gravity, the report postulated a saucer as the basis of a possible interceptor with Mach 3 capability. Creation of a local gravitational system would confer upon the fighter the sharp-edged changes of direction typical of motion in space.

The essence of electrogravitics thrust is the use of a very strong positive charge on one side of the vehicle and a negative on the other. The core of the motor is a condenser and the ability of the condenser to hold its charge (the k-number) is the yardstick of performance. With air as 1, current dielectrical materials can yield 6 and use of barium aluminate can raise this considerably, barium titanium oxide (a baked ceramic) can offer 6,000 and there is promise of 30,000, which would be sufficient for supersonic speed.

The original Brown rig produced 30 fps on a voltage of around 50,000 and a small amount of current in the milliamp range. There was no detailed explanation of gravity in Project Winterhaven, but it was assumed that particle dualism in the subatomic structure of gravity would coincide in its effect with the issuing stream of electrons from the electrostatic energy source to produce counterbary. The Brown work probably remains a realistic approach to the practical realization of electrostatic propulsion and sustentation. Whatever may be discovered by the Gravity Research Foundation of New Boston, a complete understanding and synthetic reproduction of gravity is not essential for limited success. The electrogravitics saucer can perform the function of a classic lifting surface—it produces a pushing effect on the under surface and a suction effect on the upper, but, unlike the airfoil, it does not require a flow of air to produce the effect.

First attempts at electrogravitics are unlikely to produce counterbary, but may lead to development of an electrostatic VTOL vehicle. Even in its developed form this might be an advance on the molecular heat engine in its capabilities. But hopes in the new science depend on an understanding of the close identity of electrostatic motivating forces with the source and matter of gravity. It is fortuitous that lift can be produced in the traditional fashion and if an understanding of gravity remains beyond full practical control, electrostatic lift might be an adjunct of some significance to modern thrust producers. Research into electrostatics could prove beneficial to turbine development, and heat engines in general, in view of the usable electron potential round the periphery of any flame. Materials for electrogravitics and especially the development of commercial quantities of high-k material is another dividend to be obtained from electrostatic research even if it produces no counterbary. This is a line of development that Aviation Studies' Gravity Research Group is following.

One of the interesting aspects of electrogravitics is that a breakthrough in almost any part of the broad front of general research on the intranuclear processes may be translated into a meaningful advance towards the feasibility of electrogravitics systems. This demands constant monitoring in the most likely areas of the physics of high-energy sub-nuclear particles. It is difficult to be overoptimistic about the prospects of gaining so complete a grasp of gravity while the world's physicists are still engaged in a study of fundamental particles—that is to say those that cannot be broken down any more. Fundamental particles are still being discovered—the most recent was the Segre-Chamberlain-Wiegand attachment to the bevatron, which was used to isolate the missing anti-proton, which must—or should be presumed to—exist according to Dirac's theory of the electron. Much of the accepted mathematics of particles would be wrong if the anti-proton was proved to be non-existent. Earlier Eddington has listed the fundamental particles as:

e. The charge of an electron.

m. The mass of an electron.

M. The mass of a proton.

h. Planck' s constant

c. The velocity of light.

G. The constant of gravitation, and

λ. The cosmical constant.

It is generally held that no one of these can be inferred from the others. But electrons may well disappear from among the fundamental particles, though, as Russell says, it is likely that e and m will survive. The constants are much more established than the interpretation of them and are among the most solid of achievements in modern physics.

Gravity may be defined as a small scale departure from Euclidean space in the general theory of relativity. The gravitational constant is one of four dimensionless constants: first, the mass relation of the nucleon and electron. Second is e^2/hc; third, the Compton wavelength of the proton; and fourth is the gravitational constant, which is the ratio of the electrostatic to the gravitational attraction between the electron and the proton.

One of the stumbling blocks in electrogravitics is the absence of any satisfactory theory linking these four dimensionless quantities. Of the four, moreover, gravity is decidedly the most complex, since any explanation would have to satisfy both cosmic and quantum relations more acceptably and intelligibly even than in the unified field theory. A gravitational constant of around 10^{-39} has emerged from quantum research and this has been used as a tool for finding theories that could link the two relations. This work is now in full progress, and developments have to be watched for the aviation angle. Hitherto Dirac, Eddington, Jordan and others have produced differences in theory that are too wide to be accepted as consistent. It means therefore that (i) without a cosmical basis, and (ii) with an imprecise quantum basis and (iii) a vague hypothesis on the interaction, much remains still to be discovered. Indeed some say that a single interacting theory to link up the dimensionless constants is one of three major unresolved basic problems of physics. The other two main problems are the extension of quantum theory and a more detailed knowledge of the fundamental particles.

All this is some distance from Newton, who saw gravity as a force acting on a body from a distance, leading to the tendency of bodies to accelerate towards each other. He allied this assumption with Euclidean geometry, and time was assumed as uniform and acted independently of space. Bodies and particles in space normally moved uniformly in straight lines according to Newton, and to account for the way they sometimes do not do so, he used the idea of a force of gravity acting at a distance, in which particles of matter cause in others an acceleration proportional to their mass, and inversely proportional to the square of the distance between them.

But Einstein showed how the principle of least action, or the so-called cosmic laziness means that particles, on the contrary, follow the easiest path among geodesic lines and as a result they get readily absorbed into space-time. So was born non-linear physics. The classic example of non-linear physics is the experiment in bombarding a screen with two slits. When both slits are open particles going through are not the sum of the two individually but follows a non-linear equation. This leads on to wave-particle dualism and that, in turn, to the Heisenberg uncertainty principle in which an increase in accuracy in measurement of one physical quantity means decreasing accuracy in measuring the other. If time is measured accurately energy calculations will be in error; the more accurate the position of a particle is established the less certain the velocity will be, and so on. This basic principle of the acausality of microphysics affects the study of gravity in the special and general theories of relativity. Lack of pictorial image in the quantum physics of this interrelationship is a difficulty at the outset for those whose minds remain obstinately Euclidean.

In the special theory of relativity, space-time is seen only as an undefined interval which can be defined in any way that is convenient and the Newtonian idea of persistent particles in motion to explain gravity cannot be accepted. It must be seen rather as a synthesis of forces in a four dimensional continuum, three to establish the position and one the time. The general theory of relativity that followed a decade later was a geometrical explanation of gravitation in which bodies take the geodesic path through space-time. In turn this means that instead of the idea of force acting at a distance it is assumed that space, time, radiation and particles are linked and variations in them from gravity are due rather to the nature of space.

Thus gravity of a body such as the earth, instead of pulling objects toward it as Newton postulated, is adjusting the characteristics of space and, it may be inferred, the quantum mechanics of space in the vicinity of the gravitational force. Electrogravitics aims at correcting this adjustment to put matter, so to speak, "at rest."

One of the difficulties in 1954 and 1955 was to get aviation to take electrogravitics seriously. The name alone was enough to put people off. However, in the trade much progress has been made and now most major companies in the United States are interested in counterbary. Groups are being organised to study electrostatic and electromagnetic phenomena. Most of industry's leaders have made some reference to it. Douglas has now stated that it has counterbary on its work agenda but does not expect results yet awhile. Hiller has referred to new forms of flying platform, Glenn Martin say gravity control could be achieved in six years, but they add that it would entail a Manhattan District type of effort to bring it about. Sikorsky, one of the pioneers, more or less agrees with the Douglas verdict and says that gravity is tangible and formidable, but there must be a physical carrier for this immense trans-spatial force. This implies that where a physical manifestation exists, a physical device can be developed for creating a similar force moving in the opposite direction to cancel it. Clarke Electronics state they have a rig, and add that in their view the source of gravity's force will be understood sooner than some people think. General Electric is working on the use of electronic rigs designed to make adjustments to gravity—this line of attack has the advantage of using rigs already in existence for other defence work. Bell also has an experimental rig intended, as the company puts it, to cancel out gravity, and Lawrence Bell has said he is convinced that practical hardware will emerge from current programs. Grover Leoning is certain that what he referred to as an electro-magnetic contra-gravity mechanism will be developed for practical use. Convair is extensively committed to the work with several rigs. Lear Inc., auto-pilot and electronic engineers have a division of the company working on gravity research and so also has the Sperry division of Sperry-Rand. This list embraces most of the U.S. aircraft industry. The remainder, Curtiss-Wright, Lockheed, Boeing, and North American, have not yet declared themselves, but all these four are known to be in various stages of study with and without rigs.

In addition, the Massachusetts Institute of Technology is working on gravity, the Gravity Research Foundation of New Boston, the Institute for Advanced Study at Princeton, the CalTech Radiation Laboratory, Princeton University and the University of North Carolina are all active in gravity. Glenn L. Martin is setting up a research Institute for Advanced Study which has a small staff working on gravity research with the unified field theory and this group is committed to extensive programs of applied research. Many others are also known to be studying gravity, some are known also to be planning a general expansion in this field, such as in the proposed Institute for Pure Physics at the University of North Carolina.

A certain amount of work is also going on in Europe. One of the French nationalized constructors and one company outside the nationalized elements have been making preliminary studies, and a little company money has in one case actually been committed. Some work is also going on in Britain where rigs are now in existence. Most of it is private venture work, such as that being done by Ed Hull, a colleague of Townsend Brown who, as much as anybody, introduced Europe to electrogravitics. Aviation Studies' Gravity Research Group is doing some work, mainly on k studies, and is sponsoring dielectric investigations.

One Swedish company and two Canadian companies have been making studies, and quite recently the Germans have woken up to the possibilities. Several of the companies have started digging out some of the early German papers on wave physics. They are almost certain to plan a gravitics program. Curiously enough the Germans during the war paid no attention to electrogravitics. This is one line of advance that they did not pioneer in any way and it was basically a U.S. creation. Townsend Brown in electrogravitics is the equivalent of Frank Whittle in gas turbines. This German overlooking of electrostatics is even more surprising when it is remembered how astonishingly advanced and prescient the Germans were in nuclear research. The modern theory of making thermonuclear weapons without plutonium fission initiators returns to the original German idea that was dismissed, even ridiculed. The Germans never went very far with fission, indeed they doubted that this chain would ever be made to work. The German air industry, still in the embryo stage, has included electrogravitics among the subjects it intends to examine when establishing the policy that the individual companies will adopt after the present early stage of foreign licence has enabled industry to get abreast of the other countries in aircraft development.

It is impossible to read thorough this summary of the widening efforts being made to understand the nature of matter of gravity without sharing the hope that many groups now have, of major theoretical breakthroughs occurring before very long. Experience in nucleonics has shown that, when attempts to win knowledge on this scale are made, advances are soon seen. There are a number of elements in industry, and some managements, who see gravity as a problem for later generations. Many see nothing in it all and they may be right. But as said earlier, if Dr. Vaclav Hlavaty thinks gravity is potentially controllable that surely should be justification enough, and indeed inspiration, for physicists to apply their minds and for management to take a risk. Hlavaty is the only man who thinks he can see a way of doing the mathematics to demonstrate Einstein's unified field theory—something that Einstein himself said was beyond him. Relativity and the unified field theory go to the root of electrogravitics and the shifts in thinking, the hopes and fears, and a measure of progress is to be obtained only in the last resort from men of this stature.

Major theoretical breakthroughs to discover the sources of gravity will be made by the most advanced intellects using the most advanced research tools. Aviation's role is therefore to impress upon physicists of this calibre with the urgency of the matter and to aid them with statistical and peripheral investigations that will help to clarify the background to the central mathematical and physical puzzles. Aviation could also assist by recruiting some of these men as advisers. Convair has taken the initiative with its recently established panel of advisers on nuclear projects, which include Dr. Edward Teller of the University of California. At the same time much can be done in development of laboratory rigs, condenser research and dielectric development, which do not require anything like the same cerebral capacity to get results and make a practical contribution.

As gravity is likely to be linked with the new particles, only the highest powered particle accelerators are likely to be of use in further fundamental knowledge. The country with the biggest tools of this kind is in the best position to examine the characteristics of the particles and from those countries the greatest advances seem most likely.

Though the United States has the biggest of the bevatrons—the Berkeley bevatron is 6.2 bev—the Russians have a 10 bev accelerator in construction which, when it is completed, will be the world's largest. At Brookhaven a 25 bev instrument is in development which, in turn, will be the biggest. Other countries without comparable facilities are of course at a great disadvantage from the outset in the contest to discover the explanation of gravity. Electrogravitics, moreover, unfortunately, competes with nuclear studies for its facilities. The clearest thinking brains are bound to be attracted to localities where the most extensive laboratory equipment exists. So, one way and another, results are most likely to come from the major countries with the biggest undertakings. Thus the nuclear facilities have a direct bearing on the scope for electrogravitics work.

The OEEC report in January made the following points:

> The U.S. has six to eight entirely different types of reactor in operation and many more under construction. Europe has now two different types in service.
>
> The U.S. has about 30 research reactors plus four in Britain, two in France.
>
> The U.S. has two nuclear-powered marine engines. Europe has none, but the U.K. is building one. Isotope separation plants for the enrichment of uranium in the U.S. are roughly 11 times larger than the European plant in Britain.
>
> Europe's only heavy water plant (in Norway) produces somewhat less than one-twentieth of American output.

In 1955 the number of technicians employed in nuclear energy work in the U.S. was about 15,000; there are about 5,000 in Britain, 1,800 in France, and about 1,000 in the rest of Europe. But the working party says that pessimistic conclusions should not be drawn from these comparisons. European nuclear energy effort is unevenly divided at the moment, but some countries have notable achievements to their credit and important developments in prospect. The main reason for optimism is that, taken as a whole, "Europe's present nuclear effort falls very far short of its industrial potential."

Though gravity research, such as there has been of it, has been unclassified, new principles and information gained from the nuclear research facilities that have a vehicle application is expected to be withheld.

The heart of the problem to understanding gravity is likely to prove to be the way in which the very high energy sub-nuclear particles convert something, whatever it is, continuously and automatically into the tremendous nuclear and electromagnetic forces. Once this key is understood, attention can later be directed to finding laboratory means of duplicating the process and reversing its force lines in some local environment and returning the energy to itself to produce counterbary. Looking beyond it seems possible that gravitation will be shown to be a part of the universal electro-magnetic processes and controlled in the same way as a light wave or radio wave. This is a synthesis of the Einstein and Hlavaty concepts. Hence it follows that though in its initial form the mechanical processes for countering gravity may initially be massive to deal with the massive forces involved, eventually this could be expected to form some central power generation unit. Barycentric control in some required quantity could be passed over a distance by a form of radio wave. The prime energy source to energise the waves would of course be nuclear in its origins.

It is difficult to say which lines of detailed development being processed in the immediate future is more likely to yield significant results. Perhaps the three most promising are: first, the new attempt by the team of men led by Chamberlain working with the Berkeley bevatron to find, the anti-neutron, and to identify more of the characteristics of the anti-proton* and each of the string of high energy particles that have been discovered during recent operation at 6.2 bev.

A second line of approach is the United States National Bureau of Standards program to pin down with greater accuracy the acceleration values of gravity. The presently accepted figure of 32.174 feet per

---

*The reaction is as follows: protons are accelerated to 6.2 bev, and directed at a target of copper. When the proton projectile hits a neutron in one of the copper atoms the following emerge: the two original particles (the projectile and the struck neutron) and a new pair of particles, a proton and anti-proton. The anti-proton continues briefly until it hits another proton, then both disappear and decay into mesons.

second per second is known to be not comprehensive, though it has been sufficiently accurate for the limited needs of industry hitherto. The NBS program aims at re-determining the strength of gravity to within one part of a million. The present method has been to hold a ball 16 feet up and chart the elapsed time of descent with electronic measuring equipment. The new program is based on the old, but with this exceptional degree of accuracy it is naturally immensely more difficult and is expected to take 3 years.

A third promising line is the new technique of measuring high-energy particles in motion that was started by the University of California last year. This involves passing cosmic rays through a chamber containing a mixture of gas, alcohol and water vapour. This creates charged atoms, or positive ions, by knocking electrons off the gas molecules. A sudden expansion of the chamber results in a condensation of water droplets along the track that can be plotted on a photographic plate. This method makes it easier to assess the energy of particles and to distinguish one from the other. It also helps to establish the characteristics of the different types of particle. The relationship between these high-energy particles, and their origin, and characteristics, have a bearing on electrogravitics in general.

So much of what has to be discovered as a necessary preliminary to gravity is of no practical use by itself. There is no conceivable use, for instance, for the anti-proton, yet its discovery even at a cost of $9-million is essential to check the mathematics of the fundamental components of matter. Similarly it is necessary to check that all the nuclear ghosts that have been postulated theoretically do in fact exist. It is not, moreover, sufficient, as in the past only to observe the particles by radiation counters. In each instance a mechanical maze has to be devised and attached to a particle accelerator to trap only the particle concerned. Each discovery becomes a wedge for a deeper probe of the nucleus. Many of the particles of very high energy have only a fleeting existence and collisions that give rise to them from bevatron bombardment is a necessary prerequisite to an understanding of gravity. There are no shortcuts to this process.

Most of the major programs for extending human knowledge on gravity are being conducted with instruments already in use for nuclear research and to this extent the cost of work exclusively on gravitational examinations is still not of major proportions. This has made it difficult for aviation to gauge the extent of the work in progress on gravity research.

CONCLUSIONS

1. No attempts to control the magnitude or direction of the earth's gravitational force have yet been successful. But if the explanation of gravity is to be found in the as yet undetermined characteristics of the very high energy particles it is becoming increasingly possible with the bevatron to work with the constituent matter of gravity. It is therefore reasonable to expect that the new bevatron may, before long, be used to demonstrate limited gravitational control.

2. An understanding and identification of these particles is on the frontiers of human knowledge, and a full assessment of them is one of the major unresolved puzzles of the nucleus. An associated problem is to discover a theory to account for the cosmic and quantum relations of gravity, and a theory to link the gravitational constant with the other three dimensionless constants.

3. Though the obstacles to an adequate grasp of microphysics still seem formidable, the transportation rewards that could follow from electrogravitics are as high as can be envisaged. In a weightless environment, movement with sharp-edged changes of direction could offer unique maneuverability.

4. Determination of the environment of the anti-proton, discovery of the anti-neutron and closer examination of the other high energy particles are preliminaries to the hypothesis that gravity is one aspect of electromagnetism that may eventually be controlled like a wave. When the structure of the nucleus becomes clearer, the influence of the gravitational force upon the nucleus and the nature of its behaviour in space will be more readily understood. This is a great advance on the Newtonian concept of gravity acting at a distance.

5. Aviation's role appears to be to establish facilities to handle many of the peripheral and statistical investigations to help fill in the background on electrostatics.

6. A distinction has to be made between electrostatic energy for propulsion and counterbary. Counterbary is the manipulator, of gravitational force lines; barycentric control is the adjustment to such manipulative capability to produce a stable type of motion suitable for transportation.

7. Electrostatic energy sufficient to produce low speeds (a few thousand dynes) has already been demonstrated. Generation of a region of positive electrostatic energy on one side of a plate and negative on the other sets up the same lift or propulsion effect as the pressure and suction below and above a wing, except that in the case of electrostatic application no airflow is necessary.

8. Electrostatic energy sufficient to produce a Mach 3 fighter is possible with megavolt energies and a k of over 10,000.

9. k figures of 6,000 have been obtained from some ceramic materials and there are prospects of 30,000.

10. Apart from electrogravitics there are other rewards from investment in electrostatic equipment. Automation, autonetics and even turbine development use similar laboratory facilities.

11. Progress in electrogravitics probably awaits a new genius in physics who can find a single equation to tie up all the conflicting observations and theory on the structure and arrangement of forces and the part the high energy particles play in the nucleus. This can occur any time, and the chances are improved now that bev. energies are being obtained in controlled laboratory conditions.

ADDENDUM I
Extracts from Aviation Report

ANTI-GRAVITATION RESEARCH

The basic research and technology behind electro-anti-gravitation is so much in its infancy that this is perhaps one field of development where not only the methods but the ideas are secret. Nothing therefore can be discussed freely at the moment. Very few papers on the subject have been prepared so far, and the only schemes that have seen the light of day are for pure research into rigs designed to make objects float around freely in a box. There are various radio applications, and aviation medicine departments have been looking for something that will enable them to study the physiological effects on the digestion and organs of an environment without gravity. There are however long-term aims of a more revolutionary nature that envisage equipment that can defeat gravity.

Aviation Report, 20 August 1954

Managerial Policy for Anti-Gravitics

The prospect of engineers devising gravity-defeating equipment—or perhaps it should be described as the creation of pockets of weightless environments—does suggest that as a long term policy aircraft constructors will be required to place even more emphasis on electromechanical industrial plant, than is now required for the transition from manned to unmanned weapons. Anti-gravitics work is therefore likely to go to companies with the biggest electrical laboratories and facilities. It is also apparent that anti-gravitics, like other advanced sciences, will be initially sponsored for its weapon capabilities. There are perhaps two broad ways of using the science—one is to postulate the design of advanced type projectiles on their best inherent capabilities. And the more critical parameters (that now constitutes design limitation) can be eliminated by anti-gravitics. The other, which is a longer term plan, is to create an entirely new environment with devices operating entirely under an anti-gravitic envelope.

Aviation Report, 24 August 1954

The Greater the Easier

Propulsion and atomic energy trends are similar in one respect: the more incredible the long term capabilities are, the easier it is to attain them. It is strange that the greatest of nature's secrets can be harnessed with decreasing industrial effort, but greatly increasing mental effort. The Americans went through the industrial torture to produce tritium for the first thermonuclear experiment, but later both they and the Russians were able to achieve much greater results with the help of lithium 6 hydride. The same thing is happening in aviation propulsion: the nuclear fuels are promising to be tremendously powerful in their effect, but excessively complicated in their application, unless there can be some means of direct conversion as in the strontium 90 cell, But lying behind and beyond the nuclear fuels is the linking of electricity to gravity which is an incomparably more powerful way of harnessing energy than the only method known to human intellect at present—electricity and magnetism. Perhaps the magic of barium aluminum oxide will perform the miracle in propulsion that lithium 6 hydride has done in the fusion weapon. Certainly it is a well-known material in dielectrics, but when one talks of massive-k, one means of course five figures. At this early stage it is difficult to relate k to Mach numbers with any certainty but realizable k can, with some kinds of arithmetic, produce astounding velocities. They are achievable, moreover, with decreasing complexity, indeed the ultimate becomes the easiest in terms of engineering, but the most hideous in terms of theory. Einstein's general theory of relativity is, naturally, an important factor, but some of the postulates appear to depend on the unified field theory, which cannot yet be physically checked because nobody knows how to do it. Einstein hopes to find a way of doing so before he dies.

Aviation Report, 31 August 1954

Gravitics Formulations

All indications are that there has still been little cognizance of the potentialities of electrostatic propulsion and it will be a major undertaking to re-arrange aircraft plants to conduct large-scale research and development into novel forms of dielectric and to improve condenser efficiencies and to develop the novel type of materials used for fabrication of the primary structure. Some extremely ambitious theoretical programs have been submitted and work towards realization of a manned vehicle has begun. On the evidence, there are far more definite indications that the incredible claims are realizable than there was for instance, in supposing that uranium fission would result in a bomb. At least it is known, proof positive, that motion, using surprisingly low k, is possible. The fantastic control that again is feasible, has not yet been demonstrated, but there is no reason to suppose the arithmetic is faulty, especially as it has already led to a quite brisk example of actual propulsion. That first movement was indeed an historic occasion, reminiscent of the momentous day at Chicago when the first pile went critical, and the phenomenon was scarcely less weird. It is difficult to imagine just where a well-organized examination into long-term gravitics prospects would end. Though a circular planform is electrostatically convenient, it does not necessarily follow that the requirements of control by differential changes would be the same. Perhaps the strangest part of this whole chapter is how the public managed to foresee this concept, though not of course the theoretical principles that gave rise to it, before physical tests confirmed that the mathematics was right. It is interesting also that there is no point of contact between the conventional science of aviation and the New: it is a radical offshoot with no common principles. Aerodynamics, structures heat engines, flapping controls, and all the rest of aviation is part of what might be called the Wright Brothers era—even the Mach 2.5 thermal barrier piercers are still Wright Brothers concepts, in the sense that they fly, and they stall, and they run out of fuel after a short while, and they defy the earth's pull for a short while. Thus this century will be divided in two parts—almost to the day. The first half belonged to the Wright Brothers who foresaw nearly all the basic issues in which gravity was the bitter foe. In part of the second half, gravity will be the great provider. Electrical energy, rather irrelevant for propulsion in the first half becomes a kind of catalyst to motion in the second half of the century.

Aviation Report, 7 September 1954

Electro-Gravitics Paradox

Realization of electro-static propulsion seems to depend on two theoretical twists and two practical ones. The two theoretical puzzles are: first, how to make a condenser the centre of a propulsion system, and second is how to link the condenser system with the gravitational field. There is a third problem, but it is some way off yet, which is how to manipulate kva for control in all three axes as well as for propulsion and lift. The two practical tricks are first how, with say a Mach 3 weapon in mind, to handle 50,000 kva within the envelope of a thin pancake of 35 feet in diameter and second how to generate such power from within so small a space. The electrical power in a small aircraft is more than in a fair sized community the analogy being that a single rocketjet can provide as much power as can be obtained from the Hoover Dam. It will naturally take as long to develop electro-gravitic propulsion as it has taken to coax the enormous power outputs from heat engines. True there might be a flame in the electro-gravitic propulsion system, but it would not be a heat engine—the temperature of the flame would be incidental to the function of the chemical burning process.

The curious thing is that though electrostatic propulsion is the antithesis of magnetism,* Einstein's unified field theory is an attempt to link gravitation with electro-magnetism. This all-embracing theory goes on logically from the general theory of relativity, that gives an ingenious geometrical interpretation of the concept of force that is mathematically consistent with gravitation but fails in the case of electro-magnetism, while the special theory of relativity is concerned with the relationship between mass and energy. The general theory of relativity fails to account for electro-magnetism because the forces are proportional to the charge and not to the mass. The unified field theory is one of a number of attempts that have been made to bridge this gap, but it is baffling to imagine how it could ever be observed. Einstein himself thinks it is virtually impossible. However Hlavaty claims now to have solved the equations by assuming that gravitation is a manifestation of electro-magnetism.

This being so it is all the more incredible that electro-static propulsion (with kva for convenience fed into the system and not self-generated) has actually been demonstrated. It may be that to apply all this very abstruse physics to aviation it will be necessary to accept that the theory is more important than this or that interpretation of it. This is how the physical constants, which are now regarded as among the most solid of achievements in modern physics, have become workable, and accepted. Certainly all normal instincts would support the Einstein series of postulations, and if this is so it is a matter of conjecture where it will lead in the long term future of the electro-gravitic science.

Aviation Report, 10 September 1954

---

*Though in a sense this is true, it is better expressed in this report than it was here in 1954.

Electro-Gravitic Propulsion Situation

Under the terms of Project Winterhaven the proposals to develop electro-gravitics to the point of realizing a mach 3 combat type disc were not far short of the extensive effort that was planned for the Manhattan District.* Indeed the drive to develop the new prime mover is in some respects rather similar to the experiments that led to the release of nuclear energy in the sense that both involve fantastic mathematical capacity and both are sciences so new that other allied sciences cannot be of very much guide. In the past two years since the principle of motion by means of massive-k was first demonstrated on a test rig, progress has been slow. But the indications are now that the Pentagon is ready to sponsor a range of devices to help further knowledge, In effect the new family of TVs would be on the same tremendous scope that was envisaged by the X-1, 2, 3, 4 and 5 and D-558s that were all created for the purpose of destroying the sound barrier—which they effectively did, but it is a process that is taking ten solid years of hard work to complete. (Now after 7 years the X-2 has yet to start its tests and the X-3 is still in performance testing stage). Tentative targets now being set anticipate that the first disc should be complete before 1960 and it would take the whole of the sixties to develop it properly, even though some combat things might be available ten years from now.

One thing seems certain at this stage, that the companies likely to dominate the science will be those with the biggest computors to work out the ramifications of the basic theory. Douglas is easily the world's leader in computor capacity, followed by Lockheed and Convair. The frame incidentally is indivisible from the engine. If there is to be any division of responsibility it would be that the engine industry might become responsible for providing the electrostatic energy (by, it is thought, a kind of flame) and the frame maker for the condenser assembly which is the core of the main structure.

Aviation Report, 12 October 1954

Gravitics Study Widening

The French are now understood to be pondering the most effective way of entering the field of electro-gravitic propulsion systems. But

---

*The proposals, it should be added, were not accepted.

not least of the difficulties is to know just where to begin. There are practically no patents so far that throw very much light on the mathematics of the relation between electricity and gravity. There is, of course, a large number of patents on the general subject of motion and force, and some of these may prove to have some application. There is, however a series of working postulations embodied in the original Project Winterhaven, but no real attempt has been made in the working papers to go into the detailed engineering. All that had actually been achieved up to just under a year ago was a series of fairly accurate extrapolations from the sketchy data that has so far been actually observed. The extrapolation of 50 mph to 1,800 mph, however, (which is what the present hopes and aspirations amount to) is bound to be a rather vague exercise. This explains American private views that nothing can be reasonably expected from the science yet awhile. Meanwhile, the NACA is active, and nearly all the Universities are doing work that borders close to what is involved here, and something fruitful is likely to turn up before very long.

Aviation Report, 19 October 1954

Gravitics Steps

Specification writers seem to be still rather stumped to know what to ask for in the very hazy science of electro-gravitic propelled vehicles. They are at present faced with having to plan the first family of things—first of these is the most realistic type of operational test rig, and second the first type of test vehicle. In turn this would lead to sponsoring of a combat disk. The preliminary test rigs which gave only feeble propulsion have been somewhat improved, but of course the speeds reached so far are only those more associated with what is attained on the roads rather than in the air. But propulsion is now known to be possible, so it is a matter of feeding enough KVA into condensers with better k figures. 50,000 is a magic figure for the combat saucer—it is this amount of KVA and this amount of k that can be translated into Mach 3 speeds.

Meanwhile Glenn Martin now feels ready to say in public that they are examining the unified field theory to see what can be done. It would probably be truer to say that Martin and other companies are now looking for men who can make some kind of sense out of Einstein's equations. There's nobody in the air industry at present with the faintest idea of what it is all about. Also, just as necessary, companies have somehow to find administrators who know enough of the mathematics to be able to guess what kind of industrial investment is likely to be necessary for the company to secure the most rewarding prime contracts in the new science. This again is not so easy since much of the mathematics just cannot be translated into words. You either understand the figures, or you cannot ever have it explained to you. This is rather new because even things like indeterminacy in quantum mechanics can be more or less put into words.

Perhaps the main thing for management to bear in mind in recruiting men is that essentially electro-gravitics is a branch of wave technology and much of it starts with Planck's dimensions of action, energy and time, and some of this is among the most firm and least controversial sections of modern atomic physics.

Aviation Report, 19 November 1954

ELECTRO-GRAVITICS PUZZLE

Back in 48 and 49, the public in the U.S. had a surprisingly clear idea of what a flying saucer should, or could, do. There has never been any realistic explanation of what propulsion agency could make it do those things, but its ability to move within its own gravitation field was presupposed from its maneuverability. Yet all this was at least two years before electro-static energy was shown to produce propulsion. It is curious that the public were so ahead of the empiricists on this occasion, and there are two possible explanations. One is that optical illusions or atmospheric phenomena offered a preconceived idea of how the ultimate aviation device ought to work. The other explanation might be that this was a recrudescence of Jung's theory of the Universal Mind which moves up and down in relation to the capabilities of the highest intellects and this may be a case of it reaching a very high peak of perception.

But for the air industries to realize an electro-gravitic aircraft means a return to basic principles in nuclear physics, and a re-examination of much in wage technology that has hitherto been taken for granted. Anything that goes any way towards proving the unified field theory will have as great a bearing on electro-gravitics efforts on the furtherance of nuclear power generally. But the aircraft industry might as well face up to the fact that priorities will in the end be competing with the existing nuclear science commitments. The fact that electro-gravitics has important applications other than for a weapon will however strengthen the case for governments to get in on the work going on.

Aviation Report, 28 January 1955

MANAGEMENT NOTE FOR ELECTRO-GRAVITICS

The gas turbine engine produced two new companies in the U.S. engine field and they have, between them, at various times offered the traditional primes rather formidable competition. Indeed General Electric at this moment has, in the view of some, taken the Number two position. In Britain no new firms managed to get a footing, but one, Metro-Vick, might have done if it had put its whole energies into the business. It is on the whole unfortunate for Britain that no bright newcomer has been able to screw up competition in the engine field as English Electric have done in the airframe business.

Unlike the turbine engine, electro-gravitics is not just a new propulsion system, it is a new mode of thought in aviation and communications, and it is something that may become all-embracing. Theoretical studies of the science unfortunately have to extend right down to the mathematics of the meson and there is no escape from that. But the relevant facts wrung from the nature of the nuclear structure will have their impact on the propulsion system, the airframe and also its guidance. The airframe, as such, would not exist, and what is now a complicated stressed structure becomes some convenient form of hard envelope. New companies therefore who would like to see themselves as major defence prime contractors in ten or fifteen years' time are the ones most likely to stimulate development. Several typical companies in Britain and the U.S. come to mind—outfits like AiResearch, Raytheon, Plessey in England, Rotax and others. But the companies have to face a decade of costly research into theoretical physics and it means a great deal of trust. Companies are mostly overloaded already and they cannot afford to, but when they sit down and think about the matter they can scarcely avoid the conclusion that they cannot afford not to be in at the beginning.

Aviation Report, 8 February 1955

ELECTRO-GRAVITICS BREAKTHROUGHS

Lawrence Bell said last week he though that the tempo of development leading to the use of nuclear fuels and antigravitational vehicles (he meant presumably ones that create their own gravitational field independently of the earth's) would accelerate. He added that the breakthroughs now feasible will advance their introduction ahead of the time it has taken to develop the turbojet to its present pitch. Beyond the thermal barrier was a radiation barrier, and he might have added ozone poisoning and meteorite hazards, and beyond that again a time barrier. Time however is not a single calculable entity and Einstein has taught that an absolute barrier to aviation is the environmental barrier in which there are physical limits to any kind of movement from one point in space-time continuum to another. Bell (the company not the man) have a reputation as experimentalists and are not so earthy as some of the other U.S. companies; so while this first judgment on progress with electrogravitics is interesting, further word is awaited from the other major elements of the air business. Most of the companies are now studying several forms of propulsion without heat engines though it is early days yet to determine which method will see the light of day first. Procurement will open out because the capabilities of such aircraft are immeasurably greater than those envisaged with any known form of engine.

Aviation Report, 15 July 1955

THERMONUCLEAR-ELECTROGRAVITICS INTERACTION

The point has been made that the most likely way of achieving the comparatively low fusion heat needed—1,000,000 degrees provided it can be sustained (which it cannot be in fission for more than a microsecond or two of time)—is by use of a linear accelerator. The concentration of energy that may be obtained when accelerators are rigged in certain ways make the production of very high temperatures feasible but whether they could be concentrated enough to avoid a thermal heat problem remains to be seen. It has also been suggested that linear accelerators would be the way to develop the high electrical energies needed for creation of local gravitational systems. It is possible therefore to imagine that the central core of a future air vehicle might be a linear accelerator which would create a local weightless state by use of electrostatic processes and turn heat into energy without chemical processes for propulsion. Eventually—towards the end of this century—the linear accelerator itself would not be required and a ground generating plant would transmit the necessary energy for both purposes by wave propagation.

Aviation Report, 30 August 1955

POINT ABOUT THERMONUCLEAR REACTION REACTORS

The 20-year estimate by the AEC last week that lies between present research frontiers and the fusion reactor probably refers to the time it will take to tap fusion heat. But it may be thought that rather than use the molecular and chemical processes of twisting heat into thrust it would be more appropriate to use the new heat source in conjunction with some form of nuclear thrust producer which would be in the form of electrostatic energy. The first two Boeing nuclearjet prototypes now under way are being designed to take either molecular jets, or nuclear jets in case the latter are held up for one reason, or another. But the change from molecular to direct nuclear thrust production in conjunction with the thermonuclear reactor is likely to make the aircraft designed around the latter a totally different breed of cat. It is also expected to take longer than two decades, though younger executives in trade might expect to live to see a prototype.

Aviation Report, 14 October 1955

ELECTROGRAVITICS FEASIBILITY

Opinion on the prospects of using electrostatic energy for propulsion, and eventually for creation of a local gravitational field isolated from the earth's has naturally polarized into the two opposite extremes. There are those who say it is nonsense from start to finish, and those who are satisfied from performance already physically manifest that it is possible and will produce air vehicles with absolute capabilities and no moving parts. The feasibility of a mach 3 fighter (the present aim in studies) is dependent on a rather large k extrapolation, considering the pair of saucers that have physically demonstrated the principle only achieved a speed of some 30 fps. But, and this is very important, they have attained a working velocity using a very inefficient (even by to-day's knowledge) form of condenser complex. These humble beginnings are surely as hopeful as Whittle's early postulations.

It was, by the way, largely due to the early references in *Aviation Report* that work is gathering momentum in the U.S. Similar studies are beginning in France, and in England, some men are on the job full time.

Aviation Report, 15 November 1955

Electro-Gravitics Effort Widening

Companies studying the implications of gravitics are said, in a new statement, to include Glenn Martin, Convair, Sperry-Rand, Sikorsky, Bell, Lear Inc. and Clark Electronics. Other companies who have previously evinced interest include Lockheed Douglas and Hiller. The remainder are not disinterested, but have not given public support to the new science—which is widening all the time. The approach in the U.S. is in a sense more ambitious than might have been expected. The logical approach, which has been suggested by Aviation Studies, is to concentrate on improving the output of electrostatic rigs in existence that are known to be able to produce thrust. The aim would be to concentrate on electrostatics for propulsion first and widen the practical engineering to include establishment of local gravity forcelines, independent of those of the earth's to provide unfettered vertical movement as and when the mathematics develops.

However, the U.S. approach is rather to put money into fundamental theoretical physics of gravitation in an effort first to create the local gravitational field. Working rigs would follow in the wake of the basic discoveries. Probably the correct course would be to sponsor both approaches, and it is now time that the military stepped in with big funds. The trouble about the idealistic approach to gravity is that the aircraft companies do not have the men to conduct such work.

There is every expectation in any case that the companies likely to find the answers lie outside the aviation field. These would emerge as the masters of aviation in its broadest sense.

The feeling is therefore that a company like A.T. & T. is most likely to be first in this field. This giant company (unknown in the air and weapons field) has already revolutionized modern warfare with the development of the junction transistor and is expected to find the final answers to absolute vehicle levitation. This therefore is where the bulk of the sponsoring money should go.

Aviation Report, 9 December 1955

ADDENDUM II
Electrostatic Patents

ELECTROSTATIC MOTOR

(a) American patents still in force:

2,413,391 Radio Corp of America 20-6-42/31-12-46 Power Supply System

2,417,452 Raytheon Mfg. Co. 17-1-44/18-3-47 Electrical System

2,506,472 W.B. Smits 3-7-46Holl/ 2-5-50 Electrical Ignition Apparatus

2,545,354 G.E.C. 16-3-50/13 3-51 Generator (=Engl. P.676,953)

2,567,373 Radio Corp of America 10-6-49/11-9-51 El'static Generator

2,577,446 Chatham Electronics 5-8-50/ 4-12-51 El'static Voltage Generator

2,578,908 US-Atomic Energy C. 26-5-47/18-12-51 El'static Voltage Generator

2,588,513 Radio Corp of America 10-6-49/11-3-52 El'static High-Voltage Generator

2,610,994 Chatham Electronics 1-9-50/16-9-52 El'static Voltage Generator

2,662,191 P. Okey 31-7-52/8-12-53 El'static Machine

2,667,615 R. G. Brown 30-1-52/26-1-54 El'static Generator

2,671,177 Consolidated Eng.Corp 4-9-51/2-3-54 El'static Charging App's.

2,701,844 H. R. Wasson 8-1250/8-2-55 El'static Generator of Electricity

2,702,353 US-Navy 17-7-52/15-2-55 Miniature Printed Circuit Electrostatic Generator

(b) British patents still in force:

651,153 Metr.-Vickers Electr. Co 20-5-48/14-3-51 Voltage Transformation of Electrical Energy

651,295 Ch. F. Warthen Sr (U.S.A.) 6-8-48/14-3-51 Electrostatic AC Generator

731,774 "Licentia" 19-9-52 & 21-11-53Gy/15-6-55 El'static High-Voltage Generator

(c) French patents still in force:

753,363 H. Chaumat 19-7-32/13-10-33 Moteur electrostatique utilisant l'energie cinetique d'ions gazeux

749,832 H. Chaumat 24-1-33/29-7-33 Machine electrostatique a excitation independante

The patents on page 441 derive from P. Jolivet (Algiers), marked "A," and from N. J. Felici, E. Gartner (Centre National des Recherches Scientifique [CRNS]), and later by R. Morel and M. Point (S.A. des Machines Electrostatiques [SAMES] and of Societe d' Appareils de Controle et d'Equipment des Moteurs [SACEM]), marked "G" (because the development was centered at the University Grenoble).

PATENTS

Mark of Applicant	Application Date	England	America	France	Germany	Title
G	8-11-44	637,434	2,486,140	993,017	860,649	Electrostatic Influence Machine
	14-8-45			56,027		
G	17-11-44	639,653	2,523,688	993,052	815,667	Electrostatic Influence Machine
A	28-2-45			912,444		Inducteurs de Machines el'static
G	3-3-45	643,660	2,519,554	915,442	852,586	El'static Machines
A	8-6-45			915,929		Machines electro-statiques a flasques
A	16-8-45			918,547		Generatrice el'statique
G	20-9-45			998,397	837,267	Electrostatic Machines
	21-9-45			56,356		
A	4-2-46			923,593		Generatrice el'statique
G	17-7-46	643,579	2,530,193	1,002,031	811,595	Generating Machines
G	20-2-47	671,033	2,590,168			Ignition Device
G	21-3-47	655,474	2,542,494 Re-23,560	944,574	860,650	El'static Machines
G	6-6-47	645,916	2,522,106	948,409	810,042	El'static Machines
A	16-6-47			947,921		Generatrice el'statique
G	16-1-48	669,645	2,540,327	961,210	810,043	El'static Machines
G	21-1-49	669,454	2,617,976	997,991	815,666	El'static Machines
G	7-2-49	675,649	2,649,566	1,010,924	870,575	El'static Machines
G	15-4-49	693,914	2,604,502	1,011,902	832,634	Commutators for electrical machine
G	9-11-49 20-2-50	680,178	2,565,502	1,004,950	850,485	El'static Generate
G	29-11-50 29-2-51	702,421		1,028,596		El'static Generate
G	21-11-51	719,687		1,051,430	F10421	El'static Machines
G	20-8-52	731,773		2,702,869	938,198	El'static Machines
G	6-11-52	745,489				El'static Generator
G	12-2-53	745,783				Rotating El'static Machines
G	8-1-52	715,010	2,685,654	1,047,591		Rotating El'static Machines producing periodical discharge

Application Number

Mark of Applicant	Application Date	Application Number				Title
G	27-2-54	5726/55				El'static Machines
G	8-3-54	6790/55				El'static Machines
G	28-1-55	2748/56				El'static Machines

A BRIEF DESCRIPTION OF EXPERIMENTS MADE IN PARIS BY T. TOWNSEND BROWN

TOWNSEND BROWN

A Brief Description of Experiments made in Paris.
(1955 - 1956)

These experiments were made with apparatus in vacuum of 10^{-6} mm. Hg. The essential part of the apparatus consisted of two condensers mounted at the ends of a horizontal beam as shown in Fig. 1. This horizontal beam was suspended at its mid-point, mounted on bearings and adapted for continuous rotation. The plates were charged to a potential difference of up to 150 KV and the polarity was as indicated in Fig. 1.

The voltage was delivered by apparatus capable of delivering 1 ma. From time to time, discharges occurred between the plates, the evidence for which was bright spots appearing on the negative plate and regions of luminescence on the positive plate. The times of occurrence of these discharges were erratic and the duration of each discharge, as estimated on the oscilloscope, was of the order of 1-3 milliseconds.

At each discharge, a considerable torque was exerted on the rotating part of the apparatus, as evidenced by the sudden increase in rate of rotation of the suspended system about the point of support. The fact that this torque was considerable was shown by the fact that the rate of rotation was so great after 4 to 5 discharges as to make it seem advisable to shut off the voltage to prevent the apparatus from breaking due to centrifugal force.

In the various experiments, the plates of the condensers were made of various metals among which were aluminum, brass and lead. The thickness of the plates (discs) was 1/4 inch and the diameter was 3 inches. The edges were rounded and smooth. Rotation was always in the direction of negative to positive, for each condenser. Reversal of polarity caused a reversal of rotation.

It was found that torque increased rapidly with voltage (applied to the plates). This increase was more than linear and seemed to follow an exponential law as the voltage increased.

It seems difficult or impossible to account for this force by the usual laws concerning electricity and magnetism, electrons, ions, etc. Thinking that the force might be associated with the emission of electrons and/or positive ions from the plates, the shape of the electrodes was changed from that shown in Fig. 1 to that in Fig. 2 and Fig. 3., the idea being that the escape paths would be different. In these two cases from that shown in Fig. 1. It was found, however, that neither of these two changes of shape produced any appreciable difference in the torque due to a single discharge.

Finally, Fig. 4 shows a more radical change in design. Here, each condenser was surrounded by an enclosure of plexiglas. If the impulse imparted to the suspended system was due to beams of electrons or positive ions coming from the plates, it would seem that these would be stopped by the plexiglas. Since the plexiglas was part of the rotating system, these beams of positive ions and electrons should impart no net torque to the suspended system.

Nevertheless, it was found, even in the case of Fig. 4, that the torque produced by a single discharge was approximately the same as in Figs. 1,2 and 3.

The question arises — Can these results be explained by the laws of classical physics or is a new phenomenon involved here?

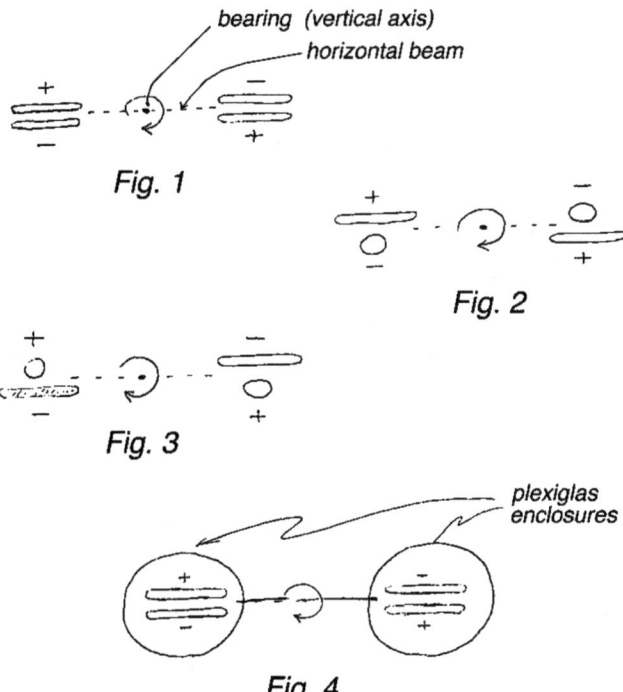

Fig. 1

Fig. 2

Fig. 3

Fig. 4

NOTES ON THE SKYVAULT
ANTIGRAVITY PROJECT

Below are reproductions of my informant "Tom's" notes from interviews with his supervisor on two occasions (see chapter 7).

August 2, 1974, Interview Notes

Project was initiated by the government through Rocketdyne in the middle to late 50s. Extremely high frequency in 1000 gigaHertz on up were employed with a voltage waveform being triangular [i.e., sawtooth shaped] (other forms were tried, but this was the best to use) at 100 kv with infinitesimal amperage—milliamps. Lowest frequency ranged from 7.2 to 8.7 gigaHertz.

With use of waveguides, a beam of conical shape was projected upward from ground to a vehicle which rode upon it. The vehicle had a concave bottom and the cone of the beam was wide in respect to the vehicle—the concave surface received the beam and was buoyed by it. [Tom later acknowledged that he had misinterpreted the above propulsion scheme; see the notes from his October 2, 1974 interview with his supervisor reproduced below.]

Mathematical analysis, & other related studies, of the conical shape region proved that Einstein was correct on gravitational waves and particular and that the high frequency nullified gravitation effects. For nullification of gravity, the frequencies do extend upward into the lower spectrum of light. The vehicle that did carry a man was ~~powered~~ (controlled) by a transmitter located upon a mountain (no info on this transmitter). The highest observed flight remembered (though it probably had a higher capability) was ~50,000 feet and a range of near 300 miles—over desert and attained extreme speeds.

Microwaves do exert a pressure and aluminum foil will move and disintegrate upon exposure. Other materials may or may not be affected in the same manner—paper will not work, some kinds of wood and silks will not. Best movement occurs if the material has a particular magnetic property (not mention as to meaning, however, he inferred paramagnetic?). The concave portion of the vehicle had something similar to ceramic like Corning ware. Extensive tests involving materials and wave shapes were made and data was accumulated on destruction, burning, and shock waves on those materials that responded. The microwaves employed were able to pass through brick walls & concrete without affecting either and with non diminishing effect on the microwaves. The estimated efficiency of the propulsion system was 60% with a much higher efficiency at this time as high probability. Diode material for the rectifiers (and maybe Gunn diodes) were furnished by International Rectifier. Reactions with the beam with some materials were violent and unpredictable. Used different shapes of waveguides throughout research.

October 2, 1974, Interview Notes

I asked if the mechanical pulser had very small electrode for arcing & if it operated somewhere from 60 to 100 thousand rpm—both were answered with a "yes." I asked if they had temperature problems with electrodes burning—No, he said, but using platinum it was eliminated and there was very little arcing due to the ultrahigh vacuum.

They did try at first a mechanical pulser at atmospheric pressure but discarded it promptly due to bouncing and noise in its pulses.

I have had a misinterpretation. The propulsion device was built into the spacecraft and a transmitter on the ground was used to provide directional control only to the spacecraft. So it is not a beam rider. The propulsion beam from the spacecraft was focused by an iris type convex lens towards the Earth and was a greenish blue light.

Letter Tom received from his supervisor's friend.

Page One   ≠ DATE OF LETTER
Oct, Nov-1974

Early attempts to use radio frequencies for antigravitation vehicles were made some 25 years ago. Detectors worked at frequencies below 1000 gigaherts and were effective mainly in keeping service men busy. In fact, legend has it that transcendental meditation was brought to the West by service men who fell into a trance while chanting obscenities at a malfunctioning UHF detector.

The major problem with systems working in the UHF band was that the field of protection could not be effectively contained. At the comparatively low frequencies employed, the radio-waves would penetrate walls and etect motion outside the desired area to be used, thus making the detectors unreliable. These early units working in the UHF band were often incorrectly referred to as "radar" detectors.

Some 15 years ago the first microwave detectors using klystrons and megastrons were developed. These early units lacked the stability, reliability and usefull life necessary for an antigravitation vehicle. They also needed more power than was economic to supply with a standby battery of units. With the availability of a solid-state device the Gunn diode - a new opportunity was presented to use microwaves. These small devices, requiring less than one watt to operate, were soon engineered into units that fulfilled all the requirements for a compact and reliable unit. Reliability, of course, came as both manufacturers and lab men learned together the uses and limitations of the microwave gravitation units.

A microwave detector unit is essentially a "box" which is easily installed upon board the vehicle and aimed in the general direction of the area requiring gravitation, walktested, and hopefully forgotten until the next shot. Understanding what happens inside this "box" will help the electronice engineer in and in maximum advantage. (The rest is still classfied)

Microwaves can be considered to have some of the properties of light. Like light beams, microwaves travel in straight lines and can be reflected by an object. Just as light is aborbed and dissipated by some materials, so also are the dad-blame microwaves. It is therefore convenient to think of microwaves being radiatied as analogous to the transmission of light.

The electromagnetic radiation in microwaves consists of a wave in which there are two components, or vectors. These are termed the E and H vectors and are mutually perpendicular to the direction of propagation.

Page Two

The importance of the $\mathcal{E}$ and H vectors is that they determine in effect the coverage in the horizontal (H) and vertical ($\mathcal{E}$) plane once they are launched from a wave guide.

Gunn diodes for generation of low power microwave radiation. Typically, radiation does not exceed 10 KW, so an amplifier has to be used to extend it's use to launch the dad-blame vehicle.

Gunn diodes radiate at their characteristic frequency when biased to the proper level. Biasing is extremely simple and usuall takes the form of the Gunn diode and a series resistor across a stable voltage source. There are certain conditions which must be satisfied before the desired frequency and stability are obtained. To achieve these conditions, the diode is placed in a "cavity," a small, deceptively simple box, open at one end. The dimensions of the box are critical and must be designed to match the characteristics of the Gunn diode. A typical frequency used in microwave units are in the region of 100.5 to 500 GHz.

The microwave energy generated is quite useless unless it is directed in a controlled way into the area to be vaulted. To achieve this, a waveguide is used, which takes the form of a "horn" that serves two main purposes:

1. To match the cavity to air. The taper in the horn ensures that there is a smooth transition from the cavity to free air. The dimension of the horn is a multiple of the wavelength.
2. To provide a means of controlling the radiation pattern, or shape, the valuted area must be (classfied)

Referring back to the $\mathcal{E}$ and H vectors, remember that these vector planes are mutually perpendicular. The geometry of the horn controls the radiation pattern in each of the planes. The wider the taper of the horn, the broader the beam width; the narrower the taper, the narrower the beam.

We have so far discussed in general terms how microwaves are generated and launched into the vaulting machine. We will now see how microwaves are used to move the dad-blamed vehicle.

Situated in a replica of the transmission cavity and horn assembly is a device known as the mixer diode.

Woops, there goes the bell C U L

Page three

Well let's see we were talking about the transmission cavity, huh ---------------?
The purpose of this diode is to provide an output signal by the mixing of two
frequencies. This is accomplished by locally diverting some of the microwave
signal from the trnsmitting antenna into the receiving antenna. Here the signal is
mixed with the microwave radiation received from the area after reflection.
If there is no change in the signal received, the output in the mixer diode is very simple
as with the gunn diode, the biasing of the mixer will be applied and the hole thing
goes cu-put. No the usual form of biasing is to connect the diode in series
with a large value ristor across a stabilized voltage source. Biasing current is
usually in the tens of micro-waves/amperes.

When the mixer is moved in on the subject matter and reflected towards the eye,
the reflected signal will result. This change in signal is mixed with some fo
the locally diverted microwave signal, resulting a modulated signal. The ole
mixer diode demodulates the signal to produce an output which is then fed into
the processing circuit.

The distance of the eye from the output amplifier depends on the following factors:
1. The distance of the eye from the detector.
2. The size of the eye.
3. The strenght of microwave energy in the area.
Of course the closer the eye is to the detector, the more microwave energy it will
move. and then the larger the eye, the more microwave energy will be intercepted,
with a stronger signal resulting. (razy, huh-------------------not for sure ?

We have so far discussed amplitude changes in signal. There is another change in
ye ole signal. That of frequency. The frequency of the signal generated by the unit
must be generated and this depends or controls the speed of the vehicle. A sensibly
close approximation of the frequency, f_i, caused by movement in a 100 GHz field is

$$f_i = \frac{S\sigma F_m}{1500}$$

where : f_i - frequency caused by movement of unit.
S_o = speed of object in cm/sec.
f_m = microwave frequency in megahertz.

Page four

Now comes the operating in the X-band and within power levels required are obtainable at an acceptable level. Higher frequency devices which meet the power level and requirement are not yet available. As discussed earlier, the higher the frequency, the less will be the likelihood of penetration of the energy through solid objects. It is for this reason that the security industry has moved away from frequencies below 1000 GHz.

There are two basic configurations of microwave diodes. The simplest type uses the same antenna for both transmitting and receiving, and has a single cavity-horn arrangement. This simplified configuration of the homodyne-type circuit suffers from lack of range. Typically, a system of this type will target at abot 30 miles. The homodyne circuit configuration depends on the non-linear characteristics of the oscillating bulk-effect diodes to as as a oscillator. The doppler output is then amplified and processed to produce the vehicle, in motion.

The second type of configuration as for example in the Micro-X unit utilizes separate transmitting and receiving antennas. The gunn is held at the optimum bias voltage necessary for oscillation. The bias current for the diode housed in the receiving antenna is aslo closely controlled by means of a stabilized voltage source.

Under normal-no movement-conditions, there is no change of output from the coupled diode to the pre-amplifier. A notchfilter effectively rejects signals caused by ripples and the output of the filter circuit is connected to the rangecontrol potentiometer. The next stage is an amplifier which is connected to a trigger. The output from the trigger circuit is used to change a capacitor which in turn causes the de-energizer when the voltages reaches a preset level.
The feature of a unit like the Micro-X is the ganged range and integrator potentiomentesr this control is wired so that as the sensitivity range is increased, the same number of steps will be required to vault. The system sensitivity is independent of the range setting. An additional persistence control allows for increasing the duration of time before vaulting conditions is recognized.

An analogy has already been drawn between microwaves and light. Glass will transmit microwaves. Wood will attenuate microwaves to a degree depending on thickness. this many be compared to the attenuation presented by translucent glass to light.

Page five

Building materials, such as bricks and cement block, will not pass microwaves. Metal objects reflect electromagnetic radiation in a manner similar to a mirror reflecting light. The area behind a metal object will be in the shadow of the radiated microwaves.

So have fun, this is all we can tell you at this time, not much of an overall picture but you asked and we told what we could. Have fun.

APPENDIX F

SECRET GOVERNMENT MEMOS CONCERNING OPERATION MAJESTIC TWELVE

Below is a reproduction of Dr. Robert Sarbacher's Canadian Department of Transportation memo describing the U.S. effort to reverse-engineer UFO technology.

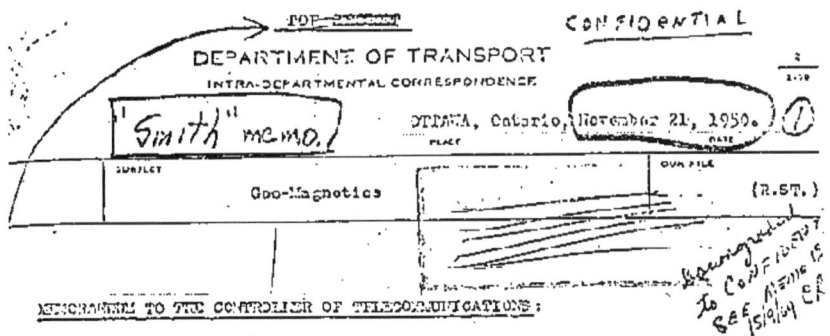

25

from another planet. Scully claimed that the preliminary studies of one saucer which fell into the hands of the United States Government indicated that they operated on some hitherto unknown magnetic principles. It appeared to me that our own work in geo-magnetics might well be the linkage between our technology and the technology by which the saucers are designed and operated. If it is assumed that our geo-magnetic investigations are in the right direction, the theory of operation of the saucers becomes quite straightforward, with all observed features explained qualitatively and quantitatively.

I made discreet enquiries through the Canadian Embassy staff in Washington who were able to obtain for me the following information:

a. The matter is the most highly classified subject in the United States Government, rating higher even than the H-bomb.

b. Flying saucers exist.

c. Their modus operandi is unknown but concentrated effort is being made by a small group headed by Doctor Vannevar Bush.

d. The entire matter is considered by the United States authorities to be of tremendous significance.

I was further informed that the United States authorities are investigating along quite a number of lines which might possibly be related to the saucers such as mental phenomena and I gather that they are not doing too well since they indicated that if Canada is doing anything at all in geo-magnetics they would welcome a discussion with suitably accredited Canadians.

While I am not yet in a position to say that we have solved even the first problems in geo-magnetic energy release, I feel that the correlation between our basic theory and the available information on saucers checks too closely to be mere coincidence. It is my honest opinion that we are on the right track and are fairly close to at least some of the answers.

Mr. Wright, Defence Research Board liaison officer at the Canadian Embassy in Washington, was extremely anxious for me to get in touch with Doctor Solandt, Chairman of the Defence Research Board, to discuss with him future investigations along the line of geo-magnetic energy release.

•••••••• 3

TOP SECRET

26

- 3 -

I do not feel that we have as yet sufficient data to place before Defence Research Board which would enable a program to be initiated within that organization, but I do feel that further research is necessary and I would prefer to see it done within the frame work of our own organization with, of course, full co-operation and exchange of information with other interested bodies.

I discussed this matter fully with Doctor Solandt, Chairman of Defence Research Board, on November 20th and placed before him as much information as I have been able to gather to date. Doctor Solandt agreed that work on geo-magnetic energy should go forward as rapidly as possible and offered full co-operation of his Board in providing laboratory facilities, acquisition of necessary items of equipment, and specialized personnel for incidental work in the project. I indicated to Doctor Solandt that we would prefer to keep the project within the Department of Transport for the time being until we have obtained sufficient information to permit a complete assessment of the value of the work.

It is therefore recommended that a PROJECT be set up within the frame work of this Section to study this problem and that the work be carried on a part time basis until such time as sufficient tangible results can be seen to warrant more definitive action. Cost of the program in its initial stages are expected to be less than a few hundred dollars and can be carried by our Radio Standards Lab appropriation.

Attached hereto is a draft of terms of reference for such a project which, if authorized, will enable us to proceed with this research work within our own organization.

(W.B. Smith)
Senior Radio Engineer

WBS/cc

Below is a reproduction of the memorandum written by President Harry Truman to establish the creation of the MJ-12 group.

TOP SECRET
EYES ONLY
THE WHITE HOUSE
WASHINGTON

September 24, 1947.

MEMORANDUM FOR THE SECRETARY OF DEFENSE

Dear Secretary Forrestal:

As per our recent conversation on this matter, you are hereby authorized to proceed with all due speed and caution upon your undertaking. Hereafter this matter shall be referred to only as Operation Majestic Twelve.

It continues to be my feeling that any future considerations relative to the ultimate disposition of this matter should rest solely with the Office of the President following appropriate discussions with yourself, Dr. Bush and the Director of Central Intelligence.

Harry Truman

ELECTROGRAVITICS: AN ENERGY-EFFICIENT MEANS OF SPACE PROPULSION

NASA SEOP Submission no. 100159

"Electrogravitics: An Energy-Efficient Means of Space Propulsion" is the abstract I submitted to NASA under the auspice of The Starburst Foundation—a nonprofit research institute I founded in 1984 to advance our understanding of the natural world.

ELECTROGRAVITICS: AN ENERGY-EFFICIENT
MEANS OF SPACE PROPULSION
by Paul LaViolette, Ph.D.
The Starburst Foundation

Description. The proposed propulsion technology would replace the energy-intensive rocket technology presently used for propelling spacecraft. The technology, called electrogravitics, has already been developed in "black" defense research programs, programs so highly classified that their existence is not publicly acknowledged. Electrogravitics may appear to violate certain assumptions about gravity commonly held by physicists and aeronautical engineers, so the reader is requested to keep an open mind. The technology does exist; it has been under development for the past 40 years; and it has been shown to be feasible both in carefully controlled laboratory experiments and in actual test flights.

Basically electrogravitics is a technology that allows a spacecraft to artificially alter its own gravity field in such a manner that it is able to levitate itself. This is accomplished by applying a megavolt pulsed DC electric potential across the outer hull and wing of the spacecraft. The craft would be designed to have a relatively large body surface area, similar to the flying wing concept employed in the B-2 bomber. Alternatively it could be discoidal in shape with a lenticular cross-section. Thrust would always be in the direction of the craft's positively charged surface. To quote a February 1956 Air Force intelligence report (now declassified), such a craft "can perform the function of a classic lifting surface—it produces a pushing effect on the under surface and a suction effect on the upper, but unlike the airfoil, it does not require a flow of air to produce the effect."*

Payoff. The value of this technology is that the craft may achieve Earth orbit flight at a much lower velocity than conventional rocket propulsion and without the huge fuel expenditure. It would eliminate the hazard of polluting the Earth's stratosphere and space environment with aluminum oxide spherules, which has become an increasing problem with the solid fuel boosters currently in use. The fuel requirements for electrogravitic propulsion are less than one percent of those presently used to lift the space shuttle into orbit. Problems typically encountered with the Space Shuttle's rocket propulsion technology (e.g., liquid hydrogen leaks, exhaust leaks around O-rings in the solid fuel booster) would not be present in this technology. Due to its much lower power demands, electrogravitics is much safer and more economical.

Performance Characteristics. As early as 1956, an Air Force study estimated that a manned electrogravitic craft could achieve Mach 3 flight capability with a 50,000 kilowatt power requirement. Such airborne electric power generation is within the reach of present technology. It would require two General Electric superconducting generators powered by two 50,000 horsepower rocket turbine engines. The superconducting generators mentioned here were developed for the Air Force in the late 1970's for use in high-altitude aircraft. Incidentally higher efficiencies are acheived in space due to reduced ion leakage from the hull's charged surface.

Other enabling technologies. All enabling technologies have been developed. As early as 1958, a small scale model of an electrogravitic powered aircraft was able to lift 110% of its weight. Since then

---

*Electrogravitics Systems: An examination of electrostatic motion, dynamic counterbary and barycentric control. Report no. GRG 013-56. Aviation Studies (International) Ltd., Special Weapons Study Unit, London, February 1956, pp. 3–4. (Library of Congress no. 3,1401,00034,5879; call no. TL565.A9.)

manned vehicles have been secretly developed and are presently being test flown.

Relation to major mission objectives. Electrogravitics would allow NASA to make frequent flights into space without the numerous delays presently plaguing the Space Shuttle launchings. (The present three year wait for repairing the Hubble Telescope could be cut to 3 weeks.) It would allow flights directly from Earth to Mars without the necessity of laboriously constructing a Mars spaceship in Earth orbit. Such a flight would no longer be contingent on the preexistence of a space station. Moreover the high speeds potentially achievable with electrogravitics would allow travel to Mars to be made in under a month.

BEYOND ROCKET PROPULSION

NASA SEOP Submission no. 100153

The following is an abstract submitted to NASA by Joe Hughes.

BEYOND ELECTRIC PROPULSION

In the quest to save money in future space missions advanced propulsion designs are under consideration. Electric propulsion is one of the designs proposed. According to John Barnett, Supervisor of Electric Propulsion at Jet Propulsion Laboratories, a system suitable for crewed spacecraft could be ready within fifteen years. George McDonough, Director of Science and Engineering at the Marshall Center, feels that electric propulsion "is an interesting alternative to nuclear propulsion which is the only one being considered by the agency,. . .They [the Soviets] consider it a viable way to do the job. We ought to take a hard look at it before we make any final decisions". The object of this proposal is to reveal that an electric propulsion engine has already been tested and proven. This engine was designed by Thomas Townsend Brown, a former physicist for the U.S. Navy, and received U.S. Patent number 3,022,430 on February 20, 1962, however the patent has expired so the design is no longer proprietary and can be considered by the Space Exploration Outreach Program.

This electric propulsion engine offers many substantial improvements over chemical and nuclear rockets. Since it can convert the energy of electrical potential directly into mechanical force suitable for causing relative motion, without the aid of moving parts, tremendous gains in efficiency and reliability can be achieved. Another advantage is that the engine can run effectively on Carbon Dioxide. Since the atmosphere of Mars is largely CO_2 a probe or ship which used this design would not need to carry fuel to Mars to explore its surface or for the return journey to Earth. Finally, no nuclear waste products or radiation is produced.

In contrast to conventional electric propulsion approaches which produce thrust by the accelerating gas only, the Brown engine ionizes the exiting gas, thus creating an electrical potential between the gas and the engine itself such that they repel each other. With the result that, according to

BEYOND ELECTRIC PROPULSION

the patent, that the gas "thrust will be small compared to the electrokinetic thrust developed." In fact, a new principle is at work here. Basically, Brown discovered that a sufficiently charged capacitor exhibits unidirectional thrust in the direction of the positive plate. To prove that his capacitors involve something more than ion propulsion he immersed them in oil, a medium which does not readily ionize and observed that the thrust was the same as in the air.

In early atmospheric tests, he used a power input of 50 watts to generate an electrical potential of 50,000 volts which caused his disc-shaped airfoils to fly at 12 miles per hour. To further prove his point that something more than ion propulsion was at work here, he tested his discs in a vacuum in the 1950's under the auspices of a French corporation, La Societe Nationale de Construction Aeronautique Sud Ouest (SNCASO). He had theorized that the speed of his discs would increase exponentially with the voltage applied, and succeeded in flying some of his 3-foot discs in a vacuum at speeds of several hundred miles an hour using an electrical potential of 200,000 volts. Unfortunately, as a result of a corporate merger with another French company, Sud Est, propulsion efforts were dropped for unknown reasons to concentrate on airframe manufacture and design.

In the early sixties Townsend Brown organized his own corporation under the name Rand International Limited. In spite of numerous patents, one of which is attached, and many demonstrations, success eluded him, the time was not right, people were too set on the Apollo way of doing things. Now everything is fair game again, no Mars mission designs are set in stone, which is why I have resubmitted his idea.

CORRESPONDENCE WITH CHARLES MORRIS

Assistant Director for the National Aero-Space Plane Project

Below is a reproduction of my correspondence with Charles Morris, assistant director for the National Aero-Space Plane program.

NASA

National Aeronautics and
Space Administration

Washington, D.C.
20546

Reply to Attn of: **RN**

JUN 1 7 1992

Mr. LaViolette
1176 Hedgewood Lane
Schenectady, NY 12309

Dear Mr. LaViolette:

Thank you for your information on electrogravities for propulsion. Review to date has determined that the concept is not appropriate for consideration within the National Aero-Space Plane (NASP) program. As opportunities arise, I will continue to seek interest in the concept and cite you as a point of contact.

Thank you for your interest in NASP.

Sincerely,

Charles E. K. Morris, Jr.
Assistant Director for NASP

<u>The Starburst Foundation</u>
1176 Hedgewood Lane
Schenectady, NY 12309

September 27, 1993

Mr. Charles E. K. Morris, Jr.
Assistant Director for NASP
NASA/RN
Washington, D.C. 20546

Dear Mr. Morris:

Enclosed is another copy of my submission to NASA's Space Exploration Outreach Program which describes the benefits of electrogravitic propulsion. Also I have enclosed several papers I have written on the subject, a copy of a 1956 article in Interavia Magazine, and a copy of the February 1956 Aviation Studies report on electrogravitics. Please return the Aviation Studies report when you are through; the other literature you may keep.

I urge you to pass this material on to other people in your group so that hopefully some discussion can get started on applying this technology to the Aero-Space Plane. Besides the benefits of gravity propulsion, electrostatic charging of the plane's leading edge would also have the added benefit of reducing air friction heating of the hull surface.

Clearly those who are hearing about electrogravitics for the first time are presented with the hurdle of believing that it is true and not some wild fabrication. I hope that I have provided a sufficient number of credible references in the enclosed material to allay such skepticism. I have also enclosed some sample lists of studies which Aviation Studies has put out to give an idea of their line of activity.

If, after reading through this material and discussing it among yourselves, you have any questions feel free to contact me. My number is (518)   . If you want further validation about the capabilities of this technology, I could attempt to put you in touch with people who have been quietly doing work in this area.

Sincerely yours,

Paul Alex LaViolette

Paul A. LaViolette

PREVENTING ANOTHER *COLUMBIA* DISASTER

Below is a suggestion I submitted to the *Columbia* Accident Investigation Board. As of yet, I have received no further communication from the board.

Concerning a technology that could help avoid another Columbia disaster

Paul A. LaViolette, Ph.D.
The Starburst Foundation
starcode@aol.com tel/fax: 703-256-4887
March 2003

One technology that could prevent a Columbia-type hazard from happening in the future would be to apply a high voltage charge to the Space Shuttle hull during reentry, in particular to the leading edge of its wing. The ion sheath so formed would create a buffer zone around the craft, ionizing, repelling and deflecting oncoming air molecules and thereby preventing them from directly impacting and heating the hull. The technology is not new. It has been researched decades ago, as is described below. This electrification technology would provide a second line of defense to the heating problem, supplementing the refractory tile layer that is currently used.

Prior history of development of the air frame electrification technology:

Airframe electrification was first suggested by Thomas Townsend Brown who did extensive research on it during the 1940's and 1950's. Referring to Brown's work of placing a high voltage charge on the leading edge of the air frame, Dr. Mason Rose wrote in 1952 that the positive field which is traveling in front of the craft "acts as a buffer wing which starts moving the air out of the way... acts as an entering wedge which softens the supersonic barrier, thus allowing the material leading edge of the craft to enter into a softened pressure area."

Brown's work was spotlighted in a 1956 air intelligence report entitled "Electrogravitics Systems" issued by Aviation Studies Intl. a UK based intelligence think tank. I obtained a copy of this report from Wright Patterson Air Force base Technical Library in 1985. The report along with excerpts about this technology that appeared in various past issues of the Aviation Studies newsletter, has been reprinted in a book by the same name entitled *Electrogravitics Systems*.[1] The report listed the names of many aerospace corporations that were actively researching this air frame electrification technology in the early 1950's.

Northrop corporation was one of these companies. At an aerospace sciences meeting held in New York in January of 1968, scientists from Northrop's Norair Division reported that they were beginning wind tunnel studies on the aerodynamic effects of applying high-voltage charges to the leading edges of high-speed aircraft bodies.[2,3] Echoing what Mason Rose had described over a decade earlier, they said they expected that the applied electrical potential would produce a corona glow that would propagate forward from the craft's leading edges to ionize and repel air molecules upwind of the aircraft. The resulting repulsive electrical forces would condition the air stream so as to lower drag, reduce heating, and soften or eliminate the supersonic boom. According to author William Moore, the results were significant in that when high voltage DC was applied to a wing-shaped structure subjected to a supersonic flow, seemingly new "electro-aerodynamic" qualities appeared which resulted in significant air drag reduction on the structure and the virtual elimination of friction-caused aerodynamic heating.

It is claimed that the B-2 bomber electrifies its wing-leading edge as a means of assisting its propulsion.[4] This appears to have been an extension of research that Northrop had carried out in the 60's, Northrop being the prime contractor for the design and construction of the B-2's airframe. I discussed the application of this technology to the B-2 in a conference paper I presented in 1993, and which was later reprinted in *Electrogravitics Systems*.

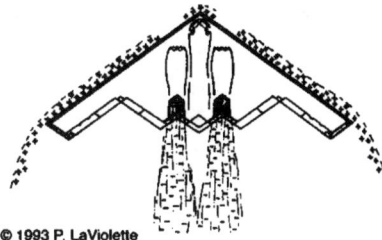

© 1993 P. LaViolette

Airframe electrification used in the B-2 bomber to reduce air drag and hull friction. (Photo courtesy of Bobbi Garcia).

Power generation:

The electrostatic charge applied to the wing leading edge may be supplied either by a flame jet generator powered by the shuttle's own main engine or by flow through wind pods mounted on the wing surface. The latter alternative would have the advantage that it would require no fuel consumption. That is, the energy would be supplied by the reentry plasma wind striking the shuttle. The "wind jet" generator would operate much like a Van de Graff generator where the air stream is analogous to the charge carrying belt in the Van de Graff. The air stream would enter the pod on the upwind end and leave the pod on the downwind end. At the pod's entrance negative ions would be injected into the air stream and as this flowed toward the rear end of the pod, it would carry these ions to a progressively higher potential difference where a portion of the ions would be collected by a conductive grid, the remainder being allowed to flow away toward the rear of the craft. The grid would recycle this current turning an initial 50,000 volt starter current into a multi megavolt current. Since the negative ions are forcefully carried away from the craft by the reentry wind, the shuttle would acquire a very high voltage positive charge in excess of 10 million volts, reaching a maximum at the wing's leading edge which would be connected to the generator's positive terminal.

This plasma jet would operate much like the flame jet generator that Townsend Brown has described in his 1962 electrokinetic generator patent (No. 3,022,430), where the hot combustion gases are here replaced by the reentry airflow captured by the wind pods. In fact, in his patent Brown stated that any kind of flowing nonconductive gas would serve as a substitute for combustion gases.

Reentry effects:

One effect of hull electrification would be to accelerate the speed of the craft. That is, hull electrification would not only reduce hull heating but also air friction drag against the craft. Consequently, the craft would take longer to decelerate as it entered the atmosphere. This could be accommodated by arranging for the craft to have a longer air flight path, e.g., passing through the atmosphere at a slightly lower angle. Alternatively, if more braking is desired at any given time during reentry, the voltage electrifying the wing may be reduced or altogether shut off, thereby engaging once again air friction surface heating. By alternately turning the electrification on and off, hull surface temperatures may be kept minimal while frictional deceleration is employed. This ease of controlling reentry deceleration by the flick of a switch may be found to be superior to controlling the forward pitch of the craft, as is currently done.

Prior contact with NASA suggesting this technology:

In 1990 I had participated in NASA's Space Exploration Outreach Project and had submitted an idea entitled "Electrogravitics: An energy-efficient means of spacecraft propulsion" (submission category: Space transportation, launch vehicles, and propulsion). My paper informed NASA about T. T. Brown's work and about the 1956 Aviation Studies Intl. report mentioned above. I suggested that NASA aggressively pursue electrogravitics for propulsion. Although my submission was not summarized in the final report submitted to NASA. It had unfortunately been omitted by Rand Corp. contract employees who found it, in their opinion, irrelevant to NASA's objectives. One other participant had also submitted a suggestion that NASA look into applying Brown's electrogravitic technology. But that too was omitted from the main report. Attempts were made to obtain the computer records from this project giving the reasons why this technology was deemed unsuitable, but these tapes were reportedly "missing."

In 1992, and again in 1993, I contacted Mr. Charles Morris, Jr. who was at that time heading NASA's National Aero-Space Plane (NASP) program and had encouraged him to have NASA look into electrogravitics. I sent him a lot of material about Townsend Brown's research, including the Aviation Studies International report. In our telephone conversations we had discussed the issue of reentry heating of the hull, which was apparently a problem that NASP was grappling with. In a letter I had sent to him in September 27, 1993, I specifically had pointed out that "electrostatic charging of the plane's leading edge would have the added benefit of reducing air friction heating of the hull surface." But nothing came of this. Mr. Morris later informed me that he was unable to generate any interest at NASA to look further into this.

NASA had a 13 year advance warning. If their research programs had pursued this technology and had applied it to the Space Shuttle, the lives of an entire crew could have been saved along with the hundreds of millions of dollars that are now going into the wreckage recovery and the cost of putting the space program on hold.

Nevertheless, looking to the future, it is my hope that NASA will now undertake the challenge and seriously research the use of this technology. Jonathan Campbell of NASA Marshall Space Flight Center would be a good contact point for beginning such a project. He has spent many years researching the application of high voltage charge to aerospace propulsion. Although his requests for NASA internal funding of this line of research have in the past been turned down, hopefully NASA will now give a higher priority to this work in the wake of the Columbia disaster. As a start, wind tunnel experiments should be carried out similar to those Northrop conducted 35 years ago and efforts should be made at developing a high-voltage wind jet generator for application to the Space Shuttle. I would be glad to assist NASA in this effort.

References

1. T. Valone (editor) *Electrogravitics Systems: Reports on a New Propulsion Methodology.* Washington, D.C.: Integrity Research Institute, 1994.
2. "Northrop studying sonic boom remedy," *Aviation Week & Space Technology*, Jan. 22, 1968, p. 21.
3. "Sonic boom experiments," *Product Engineering*, 39, March 11, 1968, pp. 35-6.
4. W. B. Scott, "Black world engineers, scientists encourage using highly classified technology for civil applications," *Aviation Week & Space Technology*, March 9, 1992, pp. 66-67.

CHAIRMAN
COLUMBIA ACCIDENT INVESTIGATION BOARD
16850 SATURN LANE, HOUSTON, TX 77058

Ser CAIB/150
25 Mar 03

Dear Dr. LaViolette,

Thank you for your letter of 17 March.

I appreciate the analysis you provided. Your letter has been provided to our independent technical group, and they will be in touch with you if further information is needed.

Once again thank you for your input.

Sincerely,

H. W. GEHMAN
Admiral, U.S. Navy (Retired)
Chairman
Columbia Accident Investigation Board

Dr. Paul A. LaViolette, Ph.D.
The Starburst Foundation
6369 Beryl Road
Suite 104
Alexandria, VA 22312

NOTES

CHAPTER 1. ANTIGRAVITY: FROM DREAM TO REALITY

1. W. L. Moore and C. Berlitz, *The Philadelphia Experiment: Project Invisibility* (New York: Fawcett Crest, 1979).
2. G. Burridge, "Another Step Toward Anti-Gravity," *The American Mercury* 86(6) (1958): 77–82. Eprint at: http://qualight.com/stress/mercury.htm.
3. G. Burridge, "Townsend Brown and His Anti-gravity Discs," *Fate* (November 1956): 40–48. Eprint at: http://qualight.com/stress/fate.htm.
4. A. L. Kitselman, "Hello Stupid" (unpublished paper, September 1962). Eprint at: http://qualight.com/stress/hello.htm.
5. Ibid.
6. T. T. Brown to Rolf Schaffranke, letter, February 14, 1973.
7. T. T. Brown, "How I Control Gravity," *Science and Invention Magazine,* August 1929. Reprinted in *Psychic Observer* 37(1) (1976): 14–18. Eprint at: http://qualight.com/stress/control.htm.
8. T. T. Brown, 1928, "A method of and an apparatus or machine for producing force or motion," British patent 300,311, issued November 15, 1928, p. 1.
9. T. T. Brown, "How I Control Gravity."
10. J. H. H. (name withheld) to Colonel Edward Deeds, letter, March 13, 1930.
11. P. A. LaViolette, "An Introduction to Subquantum Kinetics: II. An Open Systems Description of Particles and Fields," in "Systems Thinking in Physics," special issue, *International Journal of General Systems* 11 (1985): 295–328.

12. P. A. LaViolette, *Subquantum Kinetics: A Systems Approach to Physics and Cosmology* (Schenectady, N.Y.: Starlane Publications, 1994, 2003).

13. P. A. LaViolette, *Genesis of the Cosmos: The Ancient Science of Continuous Creation* (Rochester, Vt.: Bear & Co., 1995, 2004).

14. Brown, British patent 300,311, p. 3.

15. Rho Sigma [Rolf Schaffranke], *Ether Technology: A Rational Approach to Gravity Control* (Lakemont, Ga.: CSA Printing & Bindery, 1977), 44–45, quoting a letter from T. Brown dated February 14, 1973.

16. T. T. Brown to Thomas Turman, letter, January 23, 1968.

17. T. T. Brown, "Electrical Self-Potential in Rocks," *Psychic Observer* 37(1) (1976). Eprint at: http://qualight.com/petro/selfpot.htm.

18. T. T. Brown, "Phenomenal Variations in the Self-Potential of Rocks" (unpublished laboratory report, April 15, 1985). Eprint at: http://qualight.com/petro/petro1.htm.

19. C. Brush, "Retardation of Gravitational Acceleration and the Spontaneous Evolution of Heat in Complex Silicates, Lavas, and Clays," *Physical Review* 31 (1921): 1113.

20. E. A. Harrington, "Further Experiments on the Continuous Generation of Heat in Certain Silicates," *Proceedings of the American Philosophical Society* 72(5) (1933): 333–49.

21. T. T. Brown, "Test No. 110—Koolau Time Series at Constant Temperature" (unpublished laboratory report, June 30, 1977). Eprint at: http://qualight.com/petro/test110.htm.

22. T. T. Brown, "Phenomenal Variations of Resistivity and the Petrovoltaic Effect" (unpublished laboratory report, April 15, 1985). Eprint at: http://qualight.com/petro/petro2.htm.

23. T. T. Brown, "A Short Autobiography" (undated). Eprint at: http://www.qualight.com/personal/auto.htm.

24. P. Schatzkin, *Defying Gravity* (Tanglewood Books, 2007), chap. 47. Eprint at: http://ttbrown.com/defying_gravity/47_HeyWoodwardl.html.

25. Ibid., chap. 48.

26. W. Moore, "The Wizard of Electrogravity," *Saga UFO Report* (May 1978). Eprint at: http://www.qualight.com/stress/wizard1.htm.

27. T. T. Brown, "Early Laboratory Reports and Memorabilia: Biography Menu" (undated). Eprint at: www.qualight.com/personal/early.htm.

28. "A Short Biography from *Who's Who in American Science*," October 1985. Eprint at: www.qualight.com/personal/bioshort.htm.

29. Moore, "Wizard of Electrogravity."

30. J. Reynolds, "Thomas Townsend Brown's Final Gravito-Electric Research." Paper presented at the Symposium on Electrical Propulsion and the Technology of Electrogravity, Philadelphia, April 15–16, 1994.

31. Moore and Berlitz, *The Philadelphia Experiment.*

32. K. L. Corum, J. F. Corum, and J. F. X. Daum, "Tesla's Egg of Columbus, Radar Stealth, the Torsion Tensor, and the Philadelphia Experiment," in *Proceedings of the 1994 International Tesla Symposium,* S. Elswick, ed. (Denver: International Tesla Society, 1994).

33. G. Vassilatos, *Lost Science* (Kempton, Ill.: Adventures Unlimited, 1999), chap. 7. Eprint at: www.hbci.com/~wenonah/history/brown.htm.

34. Moore, "Wizard of Electrogravity."

35. "Short Biography from *Who's Who.*"

36. Kitselman, "Hello Stupid."

37. Schatzkin, *Defying Gravity*, chap. 43.

38. Ibid.

39. T. T. Brown, "Anomalous Diurnal and Secular Variations in the Self-Potential of Certain Rocks," March 22, 1975. Eprint at: www.qualight.com/petro/secular.htm.

40. Schatzkin, *Defying Gravity,* chap. 50.

41. Ibid.

42. Personal communication with the witness, a retired university professor, November 2007.

2. BEYOND ROCKET PROPULSION

1. T. T. Brown, "Project Winterhaven: A Proposal for Joint Services Research and Development Contract," Townsend Brown Foundation, October 20, 1952, revised January 1, 1953, p. 9. Eprint at: http://qualight.com/kinetics/winter.htm.

2. Ibid.

3. T. T. Brown, "Electrokinetic apparatus," U.S. patent 2,949,550, filed July 1957 and issued August 16, 1960.

4. M. Rose, "The Flying Saucer: The Application of the Biefeld-Brown Effect to the Solution of the Problems of Space Navigation," University for Social Research, April 8, 1952. Eprint at: http://qualight.com/stress/rose.htm.

5. W. S. Steinman and W. C. Stevens, *UFO Crash at Aztec: A Well Kept Secret* (Tucson: UFO Photo Archives, 1986), 376.

6. "Flying Saucers 'Explained' by Men of New Research University Here," *Los Angeles Times,* April 3, 1952, Section B, p. 1. Eprint at: http://ttbrown.com/images/LATimes_520804.jpg.

7. W. M. Cady, "An Investigation Relative to Thomas Townsend Brown," Office of Naval Research, Pasadena, Calif.: June 1952. Eprint at: http://qualight.com/stress/egdonr.htm.

8. P. Schatzkin, *Defying Gravity,* ch. 63 (2007). Eprint at: www.ttbrown .com.

9. LaViolette, *Subquantum Kinetics,* 172.

10. Rose, "Flying Saucer," 1.

11. Burridge, "Townsend Brown and His Anti-gravity Discs," 41–42.

12. Intel, "Towards Flight without Stress or Strain . . . or Weight," *Interavia* (Switzerland) 11(5) (1956): 373–74.

13. Burridge, "Another Step," 80.

14. Moore, letter, July 5, 1983.

15. Reynolds, "Brown's Final Gravito-Electric Research."

16. Brown, "Project Winterhaven."

17. T. T. Brown, 1957, electrokinetic generator, U.S. patent 3,022,430, filed July 3, 1957, and issued February 20, 1962.

18. Rose, "The Flying Saucer."

19. Whitehall-Rand, Inc., "Electrohydrodynamics," Bala Cynwyd, Pa., March 4, 1960. Eprint at: http://qualight.com/hydro/hydro.htm.

20. Cady, "Investigation Relative to Brown," 3.

21. T. T. Brown, "Electrogravitational Communication System," patent disclosure, September 1953. Eprint at: http://qualight.com/ecomm/ ecomm.htm.

22. LaViolette, *Subquantum Kinetics,* chap. 6.

23. P. A. LaViolette, "A Tesla Wave Physics for a Free Energy Universe," in *Proceedings of the 1990 International Tesla Symposium,* 5.1–5.21, S. Elswick, ed. (Colorado Springs: International Tesla Society, 1991).

24. Aviation Studies, "Electrogravitics Systems: An Examination of Electrostatic Motion, Dynamic Counterbary and Barycentric Control," report no. GRG 013/56 (London: Aviation Studies, February 1956).

25. M. Perl, "The Gravitics Situation" (London: Aviation Studies/Gravity Rand Ltd., December 1956), also published in *Interavia* XI(5) (1956): 373–74, www.rexresearch.com/perl/perl.htm.

26. Aviation Studies, "Electrogravitics Systems," 3–4.

27. Ibid., 21–23.

28. Ibid., 25–26.

29. Ibid., 26–27.

30. T. Valone, "T. T. Brown's Electrogravitics Research in the 1950's," in

Proceedings of the 1994 International Tesla Symposium, edited by S. Elswick (Denver: International Tesla Society, 1994).

31. D. E. Keyhoe, *The Flying Saucer Conspiracy* (New York: Henry Holt & Co., 1955), 251–52.

32. Aviation Studies, "Electrogravitics Systems," 28–29.

33. Ibid.

34. Department of Defense, "Air Force Releases Study on Unidentified Aerial Objects," Office of Public Information, October 25, 1955.

35. Aviation Studies, "Electrogravitics Systems," 31.

36. A. E. Talbert, "Conquest of Gravity Aim of Top Scientists in U.S.," *New York Herald Tribune,* November 20, 1955, pp. 1, 36.

37. Ibid.

38. Ibid.

39. A. E. Talbert, "Space-Ship Marvel Seen If Gravity Is Outwitted," *New York Herald Tribune,* November 21, 1955, pp. 1, 6.

40. A. E. Talbert, "New Air Dream-Planes Flying Outside Gravity," *New York Herald Tribune,* November 22, 1955, pp. 6, 10.

41. Ibid.

42. L. A. Gerardin, "Electro-Gravitic Propulsion," in *Interavia* (Switzerland) 11(12) (1956): 992.

43. Talbert, "New Air Dream-Planes."

44. Ibid.

45. Aviation Studies, "Electrogravitics Systems," 32.

46. Ibid., 9–10.

47. Intel, "Towards Flight," 373–74.

48. Aviation Studies, "Gravitics Situation," 3–4.

49. Ibid.

50. Ibid.

51. "Role of Gravitation in Physics," *Science* 125 (1957): 998.

52. F. Edwards, *Flying Saucer—Serious Business* (New York: Bantam, 1966), 127.

53. A. V. Cleaver, "'Electro-Gravitics:' What It Is—or Might Be," *Journal of the British Interplanetary Society* 16(2) (1957): 84–94.

3. ONWARD AND UPWARD

1. Rho Sigma [R. Schaffranke], *Ether Technology: A Rational Approach to Gravity Control* (Lakemont, Ga.: CSA Printing & Bindery, 1977), 44–45, quoting a letter from T. Brown dated February 14, 1973.

2. Intel, "Towards Flight," 373–74.

3. C. Carew, "The Key to Travel in Space," *Canadian Aviation* 32 (July 1959): 27–32.

4. T. T. Brown, "A Brief Description of Experiments Made in Paris (1955–1956)," unpublished report.

5. Rho Sigma, *Ether Technology,* 44.

6. Reynolds, "Brown's Final Gravito-Electric Research."

7. C. A. Yost, "T. T. Brown and the Bahnson Lab Experiments," *Electric Spacecraft Journal* 2 (1991): 6–12.

8. Kitselman, "Hello Stupid."

9. Rho Sigma, *Ether Technology,* 46–49, quoting a letter from T. Brown dated April 5, 1973.

10. T. T. Brown to T. Turman, letter, November 1, 1971.

11. "Electrohydrodynamics" (Bala Cynwyd, Pa.: Electrokinetics, Inc., March 23, 1960). Eprint at: http://qualight.com/hydro/hydro.htm.

12. T. T. Brown to T. Turman, letter, November 12, 1971.

13. Ibid.

14. "Electrohydrodynamics," Whitehall-Rand, Inc., excerpted in a letter written by A. Wagner to T. Turman.

15. "Electrohydrodynamics," Electrokinetics, Inc.

16. Ibid.

17. T. T. Brown, "Electrokinetic apparatus," U.S. patent 3,187,206, filed May 9, 1958; issued June 1, 1965.

18. Ibid.

19. T. T. Brown to T. Turman, letter, January 23, 1968.

20. Brown, U.S. patent 3,187,206.

21. Cleaver, "Electro-Gravitics," 84–94.

22. "Electrogravitics: Science or Daydream?" *Product Engineering* (December 30, 1957): 12.

23. "How to 'Fall' into Space," *Business Week,* February 8, 1958, 51–53.

24. Edwards, *Flying Saucers,* 127.

25. Carew, "The Key to Travel in Space," 27–32.

26. Ibid.

27. Moore and Berlitz, *The Philadelphia Experiment,* 223.

28. Yost, "T. T. Brown and the Bahnson Lab Experiments," 8.

29. A. H. Bahnson, "Electrical Thrust Producing Device," 1964, U.S. patent 3,223,038, filed September 10, 1964, issued December 14, 1965.

30. Moore and Berlitz, *The Philadelphia Experiment,* 223.

31. Schatzkin, *Defying Gravity,* ch. 57.

32. T. T. Brown to T. Turman, letter, Jan. 23, 1968.

33. T. T. Brown, letter to an Illinois gentleman, February 9, 1982.

4. AN ETHERIC EXPLANATION

1. Aviation Studies, "Electrogravitics Systems," 27.

2. Ray, personal conversation with author, 1992.

3. L. Smolin, *The Trouble with Physics: The Rise of String Theory, the Fall of Science, and What Comes Next* (New York: Houghton Mifflin, 2006).

4. P. Woit, *Not Even Wrong: The Failure of String Theory and the Search for Unity in Physical Law* (New York: Basic Books, 2006).

5. LaViolette, *Subquantum Kinetics*.

6. LaViolette, "An Introduction to Subquantum Kinetics: II," 295–328.

7. LaViolette, "A Tesla Wave Physics," 5.1–5.21.

8. P. A. LaViolette, "The Planetary-Stellar Mass-Luminosity Relation: Possible Evidence of Energy Nonconservation?" *Physics Essays* 5(4) (1992): 536–42.

9. P. A. LaViolette, "The Pioneer Maser Signal Anomaly: Possible Confirmation of Spontaneous Photon Blueshifting," *Physics Essays* 18(2) (2005): 150–63. Eprint at: http://arxiv.org/abs/physics/0603191.html.

10. P. A. LaViolette, "The Electric Charge and Magnetization Distribution of the Nucleon: Evidence of a Subatomic Turing Wave Pattern," *International Journal of General Systems* (2008), in press. Eprint at: www.starburstfound.org/downloads/physics/nucleon.pdf.

11. LaViolette, *Genesis of the Cosmos*.

12. LaViolette, "The Electric Charge and Magnetization Distribution."

13. Ibid.

14. R. P. Feynman, R. Leighton, and M. Sands, *The Feynman Lectures on Physics*, vol. 2 (Reading, Mass.: Addison-Wesley, 1964), 12.6–12.7.

15. J.-C. Lafforgue, "Isolated systems self-propelled by electrostatic forces," French patent 2651388, issued March 1, 1991, p. 30.

16. G. Sagnac, "The Luminiferous Ether Demonstrated by the Effect of the Relative Motion of the Ether in an Interferometer in Uniform Rotation," *Comptes Rendus de l'Academie des Sciences* (Paris) 157 (1913): 708–10, 1410–13. For a translation, see Turner and Hazelett, *The Einstein Myth* (Old Greenwich, Conn.: Devin-Adair Co., 1979), 247–50.

17. E. W. Silvertooth, "Experimental Detection of the Ether," *Speculations in Science and Technology* 10 (1987): 3–7; ibid., "Motion through the Ether," *Electronics and Wireless World* (May 1989): 437–38.

18. LaViolette, *Genesis of the Cosmos,* ch. 11.

19. B. Overbye, "Einstein: Warped Minds, Bent Truths, part I. Anaesthetized by the Ether." Eprint at: http://blog.hasslberger.com/2007/06/einstein_warped_minds_bent_tru.html.

20. B. Overbye, "Einstein: Warped Minds, Bent Truths, part III. Mystic Aftermath." Eprint at: http://blog.hasslberger.com/2007/06/einstein_warped_ minds_bent_tru.html.

21. B. Overbye, "Einstein: Warped Minds, Bent Truths, part II. Gravity Taken Lightly." Eprint at: http://blog.hasslberger.com/2007/06/einstein_warped_minds_bent_tru.html.

22. Overbye, "Einstein: Warped Minds, part III.

23. E. Lerner, *The Big Bang Never Happened* (London: Simon & Schuster, 1992).

24. LaViolette, *Subquantum Kinetics,* 2003.

25. LaViolette, *Genesis of the Cosmos.*

26. Schatzkin, *Defying Gravity,* ch. 50.

5. THE U.S. ANTIGRAVITY SQUADRON

1. W. B. Scott, "Black World Engineers, Scientists Encourage Using Highly Classified Technology for Civil Applications," *Aviation Week & Space Technology* (March 9, 1992): 66–67.

2. "Northrop Studying Sonic Boom Remedy," *Aviation Week & Space Technology* (January 22, 1968): 21.

3. "Sonic Boom Experiments," *Product Engineering* (March 11, 1968): 35–36.

4. Moore, "The Search for Anti-Gravity," part 1, p. 9.

5. Rose, "The Flying Saucer."

6. Brown, U.S. patent 2,949,550.

7. J. Mullins, "Plasma Magic," *New Scientist,* October 28, 2000. Eprint at: www.newscientist.com/article/mg16822624.000-plasma-magic.html.

8. Scott, "Black World Engineers," 66–67.

9. A. Basiago, "Dreamland and the CIA," *MUFON UFO Journal* (July 1992): 10–12.

10. T. Good, *Alien Contact: Top-Secret UFO Files Revealed* (New York: William Morrow & Co., 1991), 208.

11. Scott, "Black World Engineers."

12. Mullins, "Plasma Magic."

13. Ibid.

14. T. T. Brown, U.S. patent 3,022,430.

15. W. B. Scott, *Inside the Stealth Bomber* (New York: Tab/Aero Books, 1991), 165.

16. J.-P. Petit, "Les Mystéres du B2" (January 17, 2003). Eprint at: www .jp-petit.org/nouv_f/B2/B2_7.htm.

17. Scott, "Black World Engineers."

18. R. Boylan, "B-2 Stealth Bomber as Antigravity Craft." Eprint at: www .drboylan.com/waregrv2.html.

19. P. A. LaViolette, "The U.S. Antigravity Squadron," Proceedings of the International Symposium on New Energy, Denver, IANS, 1993. Reprinted in T. Valone, ed. *Electrogravitics Systems: Reports on a New Propulsion Methodology* (Washington, D.C.: Integrity Research Institute, 1994).

6. GRAVITY BEAM PROPULSION

1. G. Vassilatos, *Secrets of Cold War Technology: Project HAARP and Beyond* (Bayside, Calif.: Borderland Sciences, 1996), 26.

2. Ibid., 33.

3. Ibid., 55.

4. Ibid., 31.

5. E. Podkletnov and G. Modanese, "Impulse Gravity Generator Based on Charged YBa2Cu3O7-y Superconductor with Composite Crystal Structure," August 2001. Eprint at: http://arXiv.org/abs/physics/0108005.

6. Ibid.

7. E. Podkletnov and G. Modanese, "Investigation of High Voltage Discharges in Low Pressure Gases through Large Ceramic Superconducting Electrodes," *Journal of Low Temperature Physics* 132 (2003): 239–59. Eprint at: http://arXiv.org/abs/physics/0209051.

8. N. Cook, "Antigravity Propulsion Comes out of the Closet," *Jane's Defense Weekly* 38 (July 29, 2002).

9. Podkletnov and Modanese, 2001; Podkletnov and Modanese, 2003.

10. O. Jefimenko, *Causality, Electromagnetic Induction and Gravitation* (Star City, W. Va.: Electret Scientific Co.), 29.

11. Cook, "Antigravity Propulsion."

12. N. Cook, "Airpower Electric," in *Jane's Defense Weekly* 38 (July 24, 2002): 56–63.

13. N. Cook, *The Hunt for Zero-Point* (New York: Random House, 2003).

14. N. Cook, "Antigravity Propulsion."

15. P. A. LaViolette to E. Podkletnov, e-mail, June 11, 2003.

16. E. Podkletnov to P. A. LaViolette, e-mail, June 11, 2003.
17. E. Podkletnov, "Short Review of the Book 'Subquantum Kinetics' by Paul LaViolette," *Infinite Energy* 54(9) (2004): 42.
18. LaViolette, "Introduction to Subquantum Kinetics: II," 295–328.
19. LaViolette, *Subquantum Kinetics,* 121–22.
20. E. Podkletnov, personal communication with the author, July 2003.
21. P. A. LaViolette to E. Podkletnov, e-mail, December 22, 2007.
22. E. Podkletnov to P. A. LaViolette, e-mail, December 25, 2007.
23. E. Podkletnov to P. A. LaViolette, e-mail, June 11, 2003.
24. E. Podkletnov to P. A. LaViolette, e-mail, December 29, 2007.
25. A. G. Obolensky, "The Mechanics of Time," in *Proceedings of the 1988 International Tesla Symposium,* S. Elswick, ed. (Colorado Springs: International Tesla Society, 1988), 4.25–4.40.
26. P. A. LaViolette, "Galactic Explosions, Cosmic Dust Invasions, and Climatic Change" (Ph.D. dissertation, Portland State University, 1983), appendix D.
27. A. G. Obolensky and P. A. LaViolette, paper in preparation.

7. PROJECT SKYVAULT

1. J. B. Pendry and D. R. Smith, "Reversing Light with Negative Refraction," *Physics Today,* 57(6) (2004): 37–43. Eprint at: www .physicstoday.org/vol-57/iss-6/p37.html.
2. Ibid., "The Quest for the Superlens," *Scientific American* (July 2006): 61–67.
3. T. Kasagi, T. Tsutaoka, and K. Hatakeyama, "Negative Permeability Spectra in Permalloy Granular Composite Materials," in *American Institute of Physics,* April 25, 2006. Eprint at: http://link.aip.org/link/ ?APPLAB/88/172502/1.
4. J. B. Pendry, e-mail communication with the author, July 9, 2006.
5. V. G. Veselago, "The Electrodynamics of Substances with Simultaneously Negative Values of ε and μ," in *Soviet Physics Uspekhi* 10 (1968): 509–14. Originally published in Russian in *Uspekhi Fizicheskikh Nauk* 92 (1967): 517–26.
6. R. A. Shelby, D. R. Smith, and S. Schultz, "Experimental Verification of a Negative Index of Refraction," *Science* 292 (2001): 77–79. Eprint at: www.sciencemag.org/cgi/content/abstract/292/5514/77.
7. Pendry and Smith, "Reversing Light," 37–43.
8. Pendry and Smith, "Quest for the Superlens," 61–67.
9. R. Shelby, et al., "Microwave Transmission through a Two-Dimensional

Left-Handed Metamaterial," *Applied Physics Letters,* 78(4) (2001): 489–91, fig. 1.

10. S. T. Chui, "Negative Refraction in Anisotropic Composites," *Bulletin of the American Physical Society,* March meeting, 2004. Abstract at: http://flux.aps.org/meetings/YR04/MAR04/baps/abs/S6590001.html.

11. Brown, U.S. patent 3,187,206.

12. Y. Zhang, et al., "Total Negative Refraction in Real Crystals for Ballistic Electrons and Light," *Physical Review Letters* 91 (2003), no. 157404; http://flux.aps.org/meetings/YR04/MAR04/baps/abs/S6590001.html.

13. Defense Sciences Office, Defense Advanced Research Projects Agency, Thrust Area, Materials, Novel Materials and Material Processes, Negative Index Materials Program. Eprint at: www.darpa.mil/dso/thrusts/materials/novelmat/nim/index.htm. Accessed January 11, 2008. Extract quotation taken from a previous DSO internet posting that is no longer accessible.

14. S. G. Dimitriou, "Propulsive Effect on a Massive Plane Capacitor Driven by Slope-Asymmetric Pulses" (unpublished paper). Eprint at: http://jnaudin.free.fr/stvdmdoc/stvcap.htm.

15. S. G. Dimitriou, "Efforts in Developing Upward and Directional Thrust" (paper presented at the Propelantless Propulsion Conference, University of Sussex, January 2001). Eprint at: http://jnaudin.free.fr/stvexp/html/stvrxp1a.htm.

16. Ibid., "Propulsive Effect on Two Capacitors Driven by Slope-Asymmetrical Pulses Experiment" (unpublished paper, March 2001). Eprint at: http://jnaudin.free.fr/stvexp/html/stv2caps.htm.

17. Ibid., "Radiation Phenomena of Specially Shaped Current Pulses" (master's thesis, Department of Electrical Engineering, University of Manchester, 1994).

18. Vassilatos, *Secrets of Cold War Technology,* 208.

19. E. C. Okress, et al., "Design and Performance of a High Power Pulsed Magnetron," 1956 International Electron Devices, meeting 2, p. 8.

20. K. Harkay, "Simulation Investigations of the Longitudinal Sawtooth Instability at SURF," in *Proceedings of the 2001 Particle Accelerator Conference* (2001): 1918–20. Eprint at: http://accelconf.web.cern.ch/accelconf/p01/PAPERS/TPPH106.pdf.

21. National Materials Advisory Board, *Materials for High-Temperature Semiconductor Devices* (1995) (Washington, D.C., National Academy of Sciences, 2006), 93.

22. L. Yuan, et al., "IMPATT Diode Microwave Oscillators in Silicon

Carbide," *Purdue University School of Engineering Annual Research Summary: July 1 1999–June 30, 2000.* Eprint at: http://engineering .purdue.edu/ECE/Research/ARS/ARS2001/PART_I/Section7/7_ 12.whtml.

23. D. M. Goebel, et al., "High-Power Microwave Source Based on an Unmagnetized Backward-Wave Oscillator," *IEEE Transactions on Plasma Science* 22(5) (October 1994): 547–53.

24. J. McClain and N. Wootan, "An Introduction to the Magnetic Resonance Amplifier." Eprint at: www.rexresearch.com/mra/1mra.htm.

25. J. McClain, "The Magnetic Resonance Amplifier: Description of Operation." Eprint at: www.rexresearch.com/mra/1mra.htm.

26. J. McClain and N. Wootan, "New ZPE Breakthru—Magnetic Resonance Amplifier." Eprint at: www.rexresearch.com/mra/1mra.htm.

27. T. Bearden, "Ferroelectric Capacitors and the Magnetic Resonance Amplifier." Eprint at: www.rexresearch.com/mra/1mra.htm.

28. J. McClain and N. Wootan, "Magnetic Resonance Amplifier: Keely-Net Message Collection No. 2." Eprint at: www.rexresearch.com/mra/ 2mra.htm.

29. McClain and Wootan, "An Introduction to the Magnetic Resonance Amplifier."

30. A. G. Obolensky, personal communication with the author, June 2006.

8. MICROWAVE PHASE CONJUGATION

1. V. V. Shkunov and B. Y. Zel'dovich, "Optical Phase Conjugation," *Scientific American* 253 (December 1985): 54–59.

2. D. M. Pepper, "Applications of Optical Phase Conjugation," *Scientific American* 254 (January 1986): 74–83.

3. J. O. White, M. Cronin-Golomb, B. Fischer, and A. Yariv, "Coherent Oscillation by Self-Induced Gratings in the Photorefractive Crystal BaTiO3," *Applied Physics Letters* 40 (1982): 450–52.

4. M. Cronin-Golomb, B. Fischer, J. O. White, and A. Yariv, "Passive (Self-Pumped) Phase Conjugate Mirror: Theoretical and Experimental Investigation," *Applied Physics Letters* 41 (1982): 689–91.

5. Ray, personal conversation with the author.

6. UFO Subcommittee of AIAA, "UFO Encounter 1," *Astronautics and Aeronautics,* July 1971, 66–70.

7. J. M. McCampbell, *Ufology: Insights from Science and Common Sense* (Belmont, Calif.: Jaymac Co., 1973), ch. 7.

8. Pepper, "Applications of Optical Phase Conjugation," 81–82.

9. R. Shih, et al., "Microwave Phase Conjugation in a Liquid Suspension of Elongated Microparticles," *Physical Review Letters* 65(5) (1990): 579–82.

10. M. Curtin, et al., "Acceptance Test Performance of the Rocketdyne Radio Frequency Power System," *IEEE Proceedings* 2 (1993): 1244–46.

11. L. D. DiDomenico and G. M. Rebeiz, "Frequency Stability in Adaptive Retrodirective Arrays," *IEEE Transactions on Aerospace and Electronic Systems* (October 2000): 1219–31. Eprint at: www.batse.msfc .nasa.gov/colloquia/abstracts_spring04/yuri.html.

12. *Electric Spacecraft Journal* staff, "Tesla's Tower (Wardenclyffe)," *Electric Spacecraft Journal* (April–June, 1991): 13–24.

13. N. Tesla, "The Transmission of Electrical Energy without Wires," *Electrical World and Engineer,* March 5, 1904.

14. A. G. Obolensky, personal communication with the author, 1996.

15. Vassilatos, *Secrets of Cold War Technology,* 44.

16. Ibid., 46.

17. "Theoretical Blueprint for Invisibility Cloak Reported," Pratt School of Engineering at Duke University, May 25, 2006. Eprint at: www .pratt.duke.edu/news/releases/index.php?story=276.

18. Ray, personal conversation with the author, 1992.

19. A. G. Obolensky, "Mechanics of Time," 4.25–4.40.

20. A. G. Obolensky, "The Force Is Faster than Its Radiations," in *Proceedings of the Second International Symposium on Nonconventional Energy Technology,* Atlanta, September 9–11, 1983.

21. A. G. Obolensky, "The Magnetic Force is Faster than Light," in *Proceedings of the 1986 International Tesla Symposium,* 5.29–5.40, S. Elswick, ed. (Colorado Springs: International Tesla Society, 1986).

22. Ibid.

23. T. Lemons and G. Obolensky, "Improved Operation of HID Lamps," in *Lighting Design & Application* (January 1978): 55–58.

24. A. G. Obolensky, "The Force Is Faster than Its Radiations."

25. Ibid.

26. A. G. Obolensky, "The Magnetic Force Is Faster than Light," 5.29–5.40.

27. A. G. Obolensky, personal communication with the author, 1996.

28. Greg, personal communication with the author, November 3, 1995.

29. Ibid.

30. Ray, personal communication with the author, 1992.

31. Ibid.

32. Ibid.
33. Ibid.

9. UNCONVENTIONAL FLYING OBJECTS

1. P. Hill, *Unconventional Flying Objects: A Scientific Analysis* (Charlottesville, Va.: Hampton Roads Publishing Company, 1995), 98.
2. Ibid., 101–6.
3. Ibid., 99, 138–39.
4. Ibid., 121–23.
5. Ibid., 135–36, 143.
6. T. Curtis, "Court Enters Twilight Zone as 3 Claim Attack by UFO," *The Oregonian,* September 17, 1985, A12.
7. T. Good, *Above Top Secret* (New York: William Morrow & Co., 1988), 303–5.
8. D. B. Gordon and P. Dellinger, *Don't Look Up!* (Madison, N.C.: Empire Publishing, 1988).
9. "Secret Advanced Vehicles Demonstrate Technologies for Future Military Use," *Aviation Week & Space Technology* (October 1, 1990): 20–23.
10. Ibid.
11. J. A. Hynek and P. J. Imbrogno, *Night Siege: The Hudson Valley UFO Sightings* (New York: Ballantine Books, 1987).
12. F. Scully, *Behind the Flying Saucers* (New York: Henry Holt, 1950).
13. C. Berlitz and W. L. Moore, *The Roswell Incident* (New York: Grosset & Dunlap, 1980).
14. Steinman and Stevens, *UFO Crash at Aztec: A Well Kept Secret* (Tucson: UFO Photo Archives, 1986), 376.
15. Good, *Above Top Secret.*
16. Good, *Alien Contact.*
17. K. D. Randle and D. R. Schmitt, *The Truth About the UFO Crash at Roswell* (New York: Avon Books, 1994).
18. Steinman and Stevens, *UFO Crash at Aztec,* 428.
19. *Aviation Week & Space Technology,* "Secret Advanced Vehicles."
20. U.S. Air Force, "Analysis of Flying Object Incidents in the U.S," Air Force Intelligence Report no. 100-203-79; cited in H. Blum, *Out There* (New York: Simon & Schuster, 1990), 64.
21. Steinman and Stevens, *UFO Crash at Aztec,* 306.
22. Ibid., 310.
23. Ibid., 323–25.

24. Randle and Schmitt, *The Truth About the UFO Crash at Roswell.*

25. Good, *Alien Contact,* 123–24.

26. Blum, *Out There,* 211–67.

27. Ibid.

28. Ibid.

29. Anonymous letters about Roswell debris sent to Art Bell and Linda Howe, www.geocities.com/Area51/shadowlands/6583/roswell039.html.

30. E-mail correspondence about Art's Parts posted at http://beyond-the-illusion.com/files/New-Files/960930/artspart.txt.

31. N. A. Reiter, "A Summary of Our Analysis of the Claimed Roswell Crash Artifact as Discussed on 'The Art Bell Show'–1996," November 30, 2001. Eprint at: www.theavalonfoundation.org/docs/artsparts.html.

32. V. A. Podolskiy and E. E. Narimanov, "Strongly Anisotropic Waveguide as a Nonmagnetic Left-Handed System," *Physical Review B* 71 (2005): 201101. Eprint at: www.physics.oregonstate.edu/~vpodolsk/reprints.pdf/planar.prb.pdf.

33. V. A. Podolskiy, L. Alekseyev, and E. E. Narimanov, "Strongly Anisotropic Media: The THz Perspectives of Left-Handed Materials," *Journal of Modern Optics* 52 (2005): 2343. Eprint at: www.arXiv.org/abs/physics/0505024.

34. V. A. Podolskiy and E. E. Narimanov, "Nanostructured Non-magnetic Left-Handed Composites" in Antennas and Propagation Society International Symposium *2005 IEEE Proceedings* 1A (Washington, DC: IEEE, 3 July 2005), 43–46. Eprint at: www.physics.oregonstate.edu/~vpodolsk/reprints.pdf/2005IEEE.AP-S.podolskiy.lhm.pdf.

35. Steinman and Stevens, *UFO Crash at Aztec,* 280.

36. Ibid., 462–79.

37. Ibid., 426–31.

38. Ibid., 383–85, 430.

39. G. Belanus, "Project Silverbug—Human Engineered UFOs?" December 28, 1998. Eprint at: www.sightings.com/ufo2/humanufo.htm.

40. P. A. LaViolette, *Decoding the Message of the Pulsars* (Rochester, Vt.: Bear & Co., 2006), 144–45.

41. T. Mahood. Eprint at: www.dreamlandresort.com/area51/lazar/index.htm.

42. Good, *Alien Contact,* 169–89.

43. G. Huff, "The Lazar Synopsis," posted on the Internet to alt.conspiracy.area51 on March 12, 1995. Eprint at: www.dreamlandresort.com/area51/lazar/synopsis.htm.

44. T. Mahood. Eprint at: www.dreamlandresort.com/area51/lazar/fl-thery
.htm.

45. Amantine, "Why Element 115 Cannot Be Used for Antigravity," February 15, 2004, posted on the Internet at: www.abovetopsecret.com/
forum/thread33267/pg1.

46. T. Mahood. Eprint at: www.dreamlandresort.com/area51/lazar/fl-thery
.htm.

47. D. Morgan, "Lazar Critique." Eprint at: www.dreamlandresort.com/
area51/lazar/critiq.htm.

48. P. A. LaViolette, "The Electric Charge and Magnetization Distribution."

10. THE SEARL EFFECT

1. S. G. Sandberg, "Searl-Effect Generator: Design and Manufacturing Procedure," June 1985. Eprint at: www.rexresearch.com/research/
Searl/searl.htm.

2. P. L. Barrett, "The Searl Effect," newsletter no. NSRC-RM/BR-1, vol. 1,
sec. 4, Searl National Space Research Consortium, June 1, 1968.

3. S. G. Sandberg, "The Searl-Effect and Searl-Effect Generator," June
17, 1987. Eprint at: www.rexresearch.com/research/Searl/searl2.htm.

4. Barrett, "The Searl Effect," fig. IV.

5. Rho Sigma, *Ether Technology*, 80.

6. "Movement of Massive Chunk of Earth a Mystery to Scientists in
Washington," *The Oregonian*, November 24, 1984.

7. V. V. Roshchin and S. M. Godin, "Experimental Research of Magnetic-Gravity Effects: Full Size SEG Tests," 2000. Eprint at: http://searleffect
.com/free/russianseg/images/russianseg.pdf.

8. V. V. Roshchin and S. M. Godin, "An Experimental Investigation of
the Physical Effects in a Dynamic Magnetic System," *Technical Physics Letters* 26 (2000): 1105–07. http://users.erols.com/iri/ Roshchin_
Godin.PDF.

9. V. V. Roshchin and S. M. Godin, "Experimental Research of Magnetic-Gravity Effects."

10. P. A. LaViolette, "How the Searl Effect Works: An Analysis of the
Magnetic Energy Converter," in *Proceedings of the Conference on
New Hydrogen Technologies and Space Drives,* Adolf and Inge Schneider, eds. (Weinfelden, Switzerland: Jupiter-Verlag, 2001).

11. Archer Energy Systems, "The Faraday Disk Dynamo As the Original
Over-Unity Device." Eprint at: www.stardrivedevice.com/over-unity
.html.

12. T. Valone, *The Homopolar Handbook* (Washington, D.C.: Integrity Research Institute, 1998), 6.

13. Ibid., 54–55.

14. A. G. Obolensky, personal communication with the author.

15. Lemons and Obolensky, "Improved Operation of HID Lamps," 55–58.

16. White, Cronin-Golomb, Fischer, and Yariv, "Coherent Oscillation," 450–52.

17. M. Cronin-Golomb, B. Fischer, J. O. White, and A. Yariv, "Passive (Self-Pumped) Phase Conjugate Mirror," 689–91.

18. P. A. LaViolette, "Introduction to Subquantum Kinetics," parts 1, 2, and 3, *International Journal of General Systems* 11 (1985): 281–94, 295–328, 329–46.

19. LaViolette, *Subquantum Kinetics*.

20. Barrett, "The Searl Effect," 12.

21. Ibid.

22. LaViolette, *Subquantum Kinetics*, 117.

23. D. Lewis, "Is Antigravity in Your Future?" *Atlantis Rising* 1 (1994).

24. Barrett, "The Searl Effect," 3.

25. V. F. Mikhailov, "The Action of an Electrostatic Potential on the Electron Mass," *Annales Fondation Louis de Broglie* 24 (1999): 161–69. Eprint at: www.ifi.unicamp.br/~assis/IEEE-Trans-Circuits-Systems-II-V52-p289-292(2005).pdf.

26. V. F. Mikhailov, "Influence of an Electrostatic Potential on the Inertial Electron Mass," *Annales Fondation Louis de Broglie* 26 (2001):633–38. Eprint at: www.ensmp.fr/aflb/AFLB-264/aflb264p633.pdf.

27. A. K. T. Assis, "Changing the Inertial Mass of a Charged Particle," *Journal of the Physical Society of Japan* 62 (May 1993): 1418–22.

28. Rho Sigma, *Ether Technology: A Rational Approach to Gravity Control* (Lakemont, Ga.: CSA Printing & Bindery, 1977, 1996), 108; based on letter received from one of Searl's close associates, December 11, 1984.

11. ELECTROGRAVITIC WAVE EXPERIMENTS

1. S. G. Dimitriou, "On the Pendulum Oscillations of a Suspended RF Resonant Circuit." Eprint at: www.electrogravity.com/STAVROS/index.html.

2. S. G. Dimitriou, September 2006, personal communication with author.

3. J.-L. Naudin, "The Stavros RF Pendulum Experiment: An Electromagnetic Interaction with the Gravity Field" (unpublished research, 2001). Eprint at: http://jnaudin.free.fr/html/stvrfpnd2.htm.

4. S. G. Dimitriou, "Efforts in Developing Upward and Directional Thrust."

5. S. G. Dimitriou, "Radiation Phenomena," ch. 3, 9.

6. S. G. Dimitriou, "Propulsive Effect on a Massive Plane Capacitor."

7. S. G. Dimitriou, "Efforts in Developing Upward and Directional Thrust."

8. S. G. Dimitriou, "Propulsive Effect on Two Capacitors."

12. HIGH-VOLTAGE ELECTROGRAVITICS EXPERIMENTS

1. T. Turman, to author, letter, August 15, 1994.

2. T. T. Brown to T. Turman, letter, November 1, 1971.

3. T. Turman to author, letter, August 15, 1994.

4. T. Turman to T. T. Brown, letter, autumn 1971.

5. T. Turman to author, letter, August 15, 1994.

6. L. Deavenport to author, letter, November 1995; "T. T. Brown Experiment Replicated," *Electric Spacecraft Journal* 16 (1995): 34–36.

7. L. Deavenport to author, letter, November 1995.

8. R. L. Talley, "Twenty First Century Propulsion Concept," U.S. Air Force report no. PL-TR-91-3009, May 1991, available from the National Technical Information Service.

9. H. Serrano, "Re: Thomas Townsend Brown Effect in Vacuum," July 7, 2003, blog postings at: www.fusor.net/board/view.php?site=fusor&bn =fusor_announce&key=1057576903.

10. P. Cornille, "Newton's Third Principle in Post-Newtonian Physics," *Galilean Electrodynamics* 10(3) (1999): 33–34.

11. P. Cornille, "Electrostatic Pendulum Experiment Which Pumps Energy from the Ether," (unpublished research, 1999), http://members .aol.com/overunity/elecpexp/elecpexp.html.

12. P. Cornille, "Review of the Application of Newton's Third Law in Physics," *Progress in Energy and Combustion Sciences* 25 (1999): 161–210. Eprint at: http://jnaudin.free.fr/elecpexp/elecpexp.html.

13. Ibid., 200.

14. R. Seaman, "Theory of an Ion Wind Device," *Electrohydrodynamics* (Bala Cynwyd, Pa.: Electrokinetics, Inc., March 23, 1960), suppl. B. Eprint at: http://qualight.com/hydro/hydro.htm.

15. P. Cornille, "Review of the Application of Newton's Third Law," 199.

16. J.-L. Naudin, "Cornille's Electrostatic Pendulum," (unpublished research, April 2002). Eprint at: http://members.aol.com/ekpland/html/ pcespend.htm.

17. J.-L. Naudin, "The Lifter Experiments" (unpublished research, 2001). Eprint at: http://jnaudin.free.fr/html/lftcraft1.htm and http://jnaudin .free.fr/html/lifters.htm.

18. J.-L. Naudin, "The Electrokinetic Model of the Lifter" (unpublished research, 2002). Eprint at: http://jnaudin.free.fr/ html/index2.htm.

19. W. B. Stein, "Electrokinetic Propulsion: The Ionic Wind Argument," Purdue University, Energy Conversion Lab, September 5, 2000.

20. Z. Losonc, "The Charge and Force Distribution Analysis of the Lifter in Vacuum." Eprint at: http://lifters.online.fr/lifters/liftvacuum/index.htm.

21. J.-L. Naudin, "The Ion Wind Tests on Transdimensional's Lifter" (unpublished research, 2001). Eprint at: http://jnaudin.free.fr/html/ lifteriw.htm.

22. J.-C. Lafforgue, "Isolated Systems Self-Propelled by Electrostatic Forces," French patent 2651388, issued March 1, 1991, fig. 6.

23. J.-L. Naudin, "The Lafforgue's Field Propulsion Thruster Solver." Eprint at: http://jnaudin.free.fr/lfpt/html/lfptslv2.htm.

24. J.-L. Naudin, "The Lafforgue Field Propulsion Thruster LFPT v.1.0," January 27, 2002. Eprint at: http://jnaudin.free.fr/lfpt/html/lfptv1a .htm.

25. J.-L. Naudin, "The Rotating Lafforgue Field Propulsion Thrusters: The Twin-LFPT Experiment," February 6, 2002. Eprint at: http://jnaudin.free .fr/lfpt/html/twinlfpt.htm.

26. A. Colaccio, "Lafforgue's Field Propulsion Thruster–Barium Titanate," March 28, 2002. Eprint at: www.imagineanything.com/energy/ lfpt_bt.htm.

13. BLACK HOLE DISCOVERED IN NASA

1. P. A. LaViolette, "Electrogravitics: An Energy-Efficient Means of Spacecraft Propulsion," SEOP submission no. 100159 to the 1990 NASA Space Exploration Outreach Program. Reprinted in: *Explore* 3(1) (1991): 76–79.

2. "America at the Threshold: Report of the Synthesis Group on America's Space Exploration Initiative" (Washington, D.C.: U.S. Government Printing Office, 1991).

3. T., Garber, et al., "Space Transportation Systems, Launch Systems and Propulsion for the Space Exploration Initiative: Results from Project Outreach," RAND report N-3283-AF/NASA, 1991, 127.

4. V. V. Roshchin and S. M. Godin, "An Experimental Investigation of the Physical Effects," 1105–7.

5. T. Valone, "Significant Non-conventional Energy and Propulsion Methods for the 21st Century, in *Proceedings of the 1990 International Tesla Symposium,* 5-73–5-84, S. Elswick, ed. (Denver: International Tesla Society, 1991).

6. R. Thompson, "Yesterday the Garden Shed, Tomorrow Mars," *Sydney Morning Herald,* March 18, 1989, suppl. section, 8–9.

7. Garber, "Space Transportation Systems," 2, 6.

8. Ibid.

9. Conversation with the author, February 1992.

10. Garber, "Space Transportation Systems," 5.

11. Ibid.

12. Unnamed NASA engineer, conversation with the author, May 1994.

13. J. Campbell, "Apparatus for generating thrust using a two dimensional, asymmetrical capacitor module," U.S. patent 6,317,310, filed March 8, 2000; issued November 13, 2001; U.S. patent 6,411,493, filed September 20, 2001; issued June 25, 2002.

14. Mullins, "Plasma Magic."

15. Ibid.

16. Ibid.

17. S. Greer, *Disclosure: Military and Government Witnesses Reveal the Greatest Secrets in Modern History* (Crozet, Va.: Crossing Point, 2001), 419–21.

18. Ibid., 421.

19. Tom, conversation with the author, September 1994.

20. National Security Space Road Map, www.fas.org/spp/military/program/nssrm/initiatives/usspace.htm.

21. D. Liakopoulos, *The Awakening of the Red Bear* (Thesalonika, Greece: Liakopoulos Publications, 2005), 203–4. Text in Greek.

22. Scaled Composites Company, "Branson and Rutan Form 'The Spaceship Company,'" July 27, 2005. Eprint at: www. scaled.com/news/2005-07-27_branson_rutan_spaceship_company.htm.

14. A TECHNOLOGY THAT COULD CHANGE THE WORLD

1. J. Falcone, S. Gibson, R. Oberleitner, and S. Seidel, "Reminder on TC2800 guidelines for Sensitive Application Warning System (SAWS) program," U.S. Patent and Trademark Office memo issued January 15, 2008.

BIBLIOGRAPHY

Air Force Intelligence. "Analysis of Flying Object Incidents in the U.S." Report no. 100-203-79.

"America at the Threshold: Report of the Synthesis Group on America's Space Exploration Initiative." Washington, D.C.: U.S. Government Printing Office, 1991.

Assis, A. K. T. "Changing the Inertial Mass of a Charged Particle." *Journal of the Physical Society of Japan* 62 (May 1993): 1418–22.

Aviation Studies. "Electrogravitics Systems: An Examination of Electrostatic Motion, Dynamic Counterbary and Barycentric Control." Report no. GRG 013/56. London: Aviation Studies, February 1956.

———. *The Gravitics Situation.* London: Gravity Rand Ltd., December 1956.

Bahnson, A. H. "Electrical Thrust Producing Device." U.S. patent 3,223,038. Filed Sept. 10, 1964; issued December 14, 1965.

Barrett, P. L. "The Searl Effect." Newsletter no. NSRC-RM/BR-1, vol. 1, sec. 4, Searl National Space Research Consortium, June 1, 1968.

Basiago, A. "Dreamland and the CIA." *MUFON UFO Journal* 291 (1992).

Berlitz, C., and W. L. Moore. *The Roswell Incident.* New York: Grosset & Dunlap, 1980.

Blum, H. *Out There.* New York: Simon & Schuster, 1990, 211–67.

Brown, T. T. "A method of and an apparatus or machine for producing force or motion." British patent 300,311. Filed August 15, 1927; issued November 15, 1928.

———. "How I Control Gravity." *Science and Invention Magazine,* August 1929. Reprinted in *Psychic Observer* 37(1) (1976): 14–18.

————. "A Brief Description of Experiments Made in Paris (1955–1956)." Unpublished report.

————. "Electrokinetic Apparatus." U.S. patent 3,187,206. Filed May 9, 1958; issued on June 1, 1965.

————. "Electrokinetic Apparatus." U.S. patent 2,949,550. Issued on August 16, 1960.

————. "Electrokinetic Generator." U.S. patent 3,022,430. Issued on February 20, 1962.

————. "Electrical Self-Potential in Rocks." *Psychic Observer* 37(1) (1976).

————. "On the Possibilities of Optical-Frequency Gravitational Radiation." Unpublished paper, August 30, 1976.

Brush, C. "Retardation of Gravitational Acceleration and the Spontaneous Evolution of Heat in Complex Silicates, Lavas, and Clays." *Physical Review* 31 (1921): 1113.

Burridge, G. "Townsend Brown and his Anti-gravity Discs." *Fate* (November 1956): 40–48.

————. "Another Step toward Anti-gravity." *The American Mercury* 86(6) (1958): 77–82.

Cady, W. M. "An Investigation Relative to Thomas Townsend Brown." Pasadena, Calif.: Office of Naval Research, June 1952.

Campbell, J. "Apparatus for generating thrust using a two dimensional, asymmetrical capacitor module." U.S. patent 6,317,310B1. Filed November 13, 2001; issued June 25, 2002.

Carew, C. "The Key to Travel in Space." *Canadian Aviation* 32 (July 1959): 27–32.

Chui, S. T. "Negative Refraction in Anisotropic Composites." *Bulletin of the American Physical Society,* March meeting, 2004. Abstract at: http://flux.aps.org/meetings/YR04/MAR04/baps/abs/S6590001 .html.

Cleaver, A. V. "'Electro-Gravitics:' What It Is—or Might Be." *Journal of the British Interplanetary Society* 16(2) (1957): 84–94.

Cook, N. "Airpower Electric." *Jane's Defense Weekly* 38 (2002): 56–63.

————. "Antigravity Propulsion Comes out of the Closet." *Jane's Defense Weekly* (2002).

————. *The Hunt for Zero-Point.* New York: Random House, 2003.

Cornille, P. "Newton's Third Principle in Post-Newtonian Physics." *Galilean Electrodynamics* 10(3) (1999).

————. "Review of the Application of Newton's Third Law in Physics."

Progress in Energy and Combustion Sciences 25 (1999): 161–210.

Corum, K. L., J. F. Corum, and J. F. X. Daum. "Tesla's Egg of Columbus, Radar Stealth, the Torsion Tensor, and the Philadelphia Experiment." In *Proceedings of the 1994 International Tesla Symposium*, S. Elswick, ed. Denver: International Tesla Society, 1994.

Cronin-Golomb, M., B. Fischer, J. O. White, and A. Yariv. "Passive (Self-Pumped) Phase Conjugate Mirror: Theoretical and Experimental Investigation." *Applied Physics Letters* 41 (1982): 689–91.

Curtin, M., et al. "Acceptance Test Performance of the Rocketdyne Radio Frequency Power System." *IEEE Proceedings* (1993): 1244–46.

Curtis, T. "Court Enters Twilight Zone as 3 Claim Attack by UFO." *The Oregonian,* September 17, 1985, A12.

Department of Defense. "Air Force Releases Study on Unidentified Aerial Objects." Office of Public Information, October 25, 1955.

DiDomenico, L. D., and G. M. Rebeiz. "Frequency Stability in Adaptive Retrodirective Arrays." *IEEE Transactions on Aerospace and Electronic Systems* (October 2000): 1219–31.

Dimitriou, S. G. "Radiation Phenomena of Specially Shaped Current Pulses." Master's thesis, Department of Electrical Engineering, University of Manchester, 1994.

———. "Efforts in Developing Upward and Directional Thrust." Paper presented at the Propelantless Propulsion Conference, University of Sussex, January 2001. http://jnaudin.free.fr/stvexp/html/stvrxp1a.htm.

Edwards, F. *Flying Saucers—Serious Business.* New York: Bantam, 1966, 127.

"Electrogravitics: Science or Daydream?" *Product Engineering* (1957): 12.

"Electrohydrodynamics." Meadville, Pa.: Whitehall-Rand, March 4, 1960.

"Electrohydrodynamics." Bala Cynwyd, Pa.: Electrokinetics, Inc., March 23, 1960.

Feynman, R. P., R. Leighton, and M. Sands. *The Feynman Lectures on Physics,* vol. II. Reading, Mass.: Addison-Wesley, 1964, 12.6–12.7.

"Flying Saucers 'Explained' by Men of New Research University Here." *Los Angeles Times,* April 3, 1952, Section B, p. 1.

Garber, T., et al. "Space Transportation Systems, Launch Systems and Propulsion for the Space Exploration Initiative: Results from Project Outreach." RAND report N-3283-AF/NASA, 1991.

Gerardin, L. A. "Electro-Gravitic Propulsion." *Interavia* (Switzerland) 11(12) (1956): 992.

Goebel, D. M., et al. "High-Power Microwave Source Based on an Unmagnetized Backward-Wave Oscillator." *IEEE Transactions on Plasma Science* 22(5) (October 1994): 547–53.

Good, T. *Above Top Secret.* New York: William Morrow & Co., 1988.

———. *Alien Contact: Top-Secret UFO Files Revealed.* New York: William Morrow & Co., 1991.

Gordon, D. B., and P. Dellinger. *Don't Look Up!* Madison, N.C.: Empire Publishing, 1988.

Greer, S. *Disclosure: Military and Government Witnesses Reveal the Greatest Secrets in Modern History.* Crozet, Va.: Crossing Point, 2001.

Harkay, K. "Simulation Investigations of the Longitudinal Sawtooth Instability at SURF." In *Proceedings of the 2001 Particle Accelerator Conference* (2001): 1918–20.

Harrington, E. A. "Further Experiments on the Continuous Generation of Heat in Certain Silicates." *Proceedings of the American Philosophical Society* 72(5) (1933): 333–49.

"How to 'Fall' into Space." *Business Week,* February 8, 1958, 51–53.

Hill, P. *Unconventional Flying Objects: A Scientific Analysis.* Charlottesville, Va.: Hampton Road Publishing, 1995.

Hynek, J. A., and P. J. Imbrogno. *Night Siege: The Hudson Valley UFO Sightings.* New York: Ballantine Books, 1987.

Intel. "Towards Flight without Stress or Strain . . . or Weight." *Interavia* (Switzerland) 11(5) (1956): 373–74.

Jefimenko, O. *Causality, Electromagnetic Induction and Gravitation.* Star City, W.Va.: Electret Scientific Co.

Keyhoe, D. E. *The Flying Saucer Conspiracy.* New York: Henry Holt & Co., 1955.

LaViolette, P. A. "Galactic Explosions, Cosmic Dust Invasions, and Climatic Change." Ph.D. diss., Portland State University, 1983.

———. "An Introduction to Subquantum Kinetics: II. An Open Systems Description of Particles and Fields." In "Systems Thinking in Physics," special issue, *International Journal of General Systems* 11 (1985): 295–328.

———. "A Tesla Wave Physics for a Free Energy Universe." Paper presented at the 1990 International Tesla Society Conference, Colorado Springs.

———. "Electrogravitics: An Energy-Efficient Means of Spacecraft Propulsion." SEOP submission no. 100159 to the 1990 NASA Space Exploration Outreach Program. Reprinted in *Explore* 3(1) (1991): 76–79.

———. "The Planetary-Stellar Mass-Luminosity Relation: Possible Evidence of Energy Nonconservation?" *Physics Essays* 5(4) (1992): 536–42.

———. "How the Searl Effect Works: An Analysis of the Magnetic Energy Converter." In *Proceedings of the Conference on New Hydrogen Technologies and Space Drives,* Weinfelden, Switzerland, June 23, 2001, Adolf and Inge Schneider, eds. Weinfelden: Jupiter-Verlag, 2001.

———. *Subquantum Kinetics: A Systems Approach to Physics and Cosmology.* Schenectady, N.Y.: Starlane Publications, 1994, 2003.

———. *Genesis of the Cosmos: The Ancient Science of Continuous Creation.* Rochester, Vt.: Bear & Co., 1995, 2004.

———. "The Pioneer Maser Signal Anomaly: Possible Confirmation of Spontaneous Photon Blueshifting." *Physics Essays* 18(2) (2005): 150–63.

———. "The Electric Charge and Magnetization Distribution of the Nucleon: Evidence of a Subatomic Turing Wave Pattern." *International Journal of General Systems* (2008), in press.

Lemons, T., and A. G. Obolensky. "Improved Operation of HID Lamps." *Lighting Design & Application* (1978): 55–58.

Lewis, D. "Is Antigravity in Your Future?" *Atlantis Rising* 1 (1994).

Liakopoulos, D. *The Awakening of the Red Bear.* Thesalonika, Greece: Liakopoulos Publications, 2005, 203–4. In Greek.

McCampbell, J. M. *Ufology: Insights from Science and Common Sense.* Belmont, Calif.: Jaymac Co., 1973.

Mikhailov, V. F. "The Action of an Electrostatic Potential on the Electron Mass." *Annales Fondation Louis de Broglie* 24 (1999): 161–69.

———. "Influence of an Electrostatic Potential on the Inertial Electron Mass." *Annales Fondation Louis de Broglie* 26 (2001): 633–38.

"Movement of Massive Chunk of Earth a Mystery to Scientists in Washington." *The Oregonian,* November 24, 1984.

Moore, W. L., and C. Berlitz. *The Philadelphia Experiment: Project Invisibility.* New York: Fawcett Crest, 1979.

Mullins, J. "Plasma Magic." *New Scientist,* October 28, 2000.

National Materials Advisory Board. *Materials for High-Temperature Semiconductor Devices* (1995); Washington, D.C., National Academy of Sciences, 2006, 93.

Nesvizhevsky, V. V., et al. "Quantum States of Neutrons in the Earth's Gravitational Field." *Nature* 415 (2002): 297–99.

"Northrop Studying Sonic Boom Remedy." *Aviation Week & Space Technology* (January 22) (1968): 21.

Obolensky, A. G. "The Force Is Faster than Its Radiations." In *Proceedings of the Second International Symposium on Nonconventional Energy Technology*, Atlanta, September 9–11, 1983.

———. "The Magnetic Force Is Faster than Light." In *Proceedings of the 1986 International Tesla Symposium*, 5.29–5.40, S. Elswick, ed. Colorado Springs: International Tesla Society, 1986.

———. "The Mechanics of Time." In *Proceedings of the 1988 International Tesla Symposium*, 4.25–4.40, S. Elswick, ed. Colorado Springs: International Tesla Society, 1988.

Okress, E. C., et al. "Design and Performance of a High Power Pulsed Magnetron." 1956 International Electron Devices, meeting 2 (1956): 8.

Pendry, J. B., and D. R. Smith. "Reversing Light with Negative Refraction." *Physics Today* 57(6) (2004): 37–43.

———. "The Quest for the Superlens." *Scientific American* (2006): 61–67.

Pepper, D. M. "Applications of Optical Phase Conjugation." *Scientific American* 254 (1986): 74–83.

Podolskiy, V. A., and E. E. Narimanov. "Nanostructured Non-magnetic Left-Handed Composites." In *AP-S/URSI 2005 Proceedings*. Washington, D.C.: IEEE, 2005.

———. "Strongly Anisotropic Waveguide as a Nonmagnetic Left-Handed System." *Physical Review* B 71 (2005): 201101.

"Project Winterhaven: A Proposal for Joint Services Research and Development Contract." Townsend Brown Foundation, Washington D.C., January 1, 1953.

Randle, K. D., and D. R. Schmitt. *The Truth About the UFO Crash at Roswell*. New York: Avon Books, 1994.

Reynolds, J. "Thomas Townsend Brown's Final Gravito-Electric Research." Paper presented at the Symposium on Electrical Propulsion and the Technology of Electro-Gravity, Philadelphia, April 15–16, 1994.

Rho Sigma [R. Schaffranke]. *Ether Technology: A Rational Approach to Gravity Control*. Lakemont, Ga.: CSA Printing & Bindery, 1977.

"Role of Gravitation in Physics." *Science* 125 (1957): 998.

Rose, M. "The Flying Saucer: The Application of the Biefeld-Brown Effect to the Solution of the Problems of Space Navigation." University for Social Research, April 8, 1952.

Roshchin, V. V., and S. M. Godin. "An Experimental Investigation of the Physical Effects in a Dynamic Magnetic System." *Technical Physics Letters* 26 (2000): 1105–7.

Schatzkin, P. *Defying Gravity*. Terre Haute, Ind.: Tanglewood Press, 2007. Eprint at: www.ttbrown.com/defying_gravity/000_preface.html.

Scott, W. B. *Inside the Stealth Bomber*. New York: Tab/Aero Books, 1991.

———. "Black World Engineers, Scientists Encourage Using Highly Classified Technology for Civil Applications." *Aviation Week & Space Technology* (1992): 66–67.

Scully, F. *Behind the Flying Saucers*. New York: Henry Holt, 1950.

"Secret Advanced Vehicles Demonstrate Technologies for Future Military Use." *Aviation Week & Space Technology* (1990): 20–23.

Shelby, R. A., D. R. Smith, and S. Schultz. "Experimental Verification of a Negative Index of Refraction." *Science* 292 (2001): 77–79.

Shih, R., et al. "Microwave Phase Conjugation in a Liquid Suspension of Elongated Microparticles." *Physical Review Letters* 65(5) (1990): 579–82.

Shkunov, V. V., and B. Y. Zel'dovich. "Optical Phase Conjugation." *Scientific American* 253 (1985): 54–59.

Smolin, L. *The Trouble with Physics: The Rise of String Theory, the Fall of Science, and What Comes Next*. New York: Houghton Mifflin, 2006.

"Sonic Boom Experiments." *Product Engineering* 39 (March 11, 1968): 35–36.

Steinman, W. S., and W. C. Stevens. *UFO Crash at Aztec: A Well Kept Secret*. Tucson: UFO Photo Archives, 1986, 376.

Talbert, A. E. "Conquest of Gravity Aim of Top Scientists in U.S." *New York Herald Tribune*, November 20, 1955, 1, 36.

———. "Space-Ship Marvel Seen if Gravity Is Outwitted." *New York Herald Tribune*, November 21, 1955, 1, 6.

———. "New Air Dream-Planes Flying Outside Gravity." *New York Herald Tribune*, November 22, 1955, 6, 10.

Talley, R. L. "Twenty First Century Propulsion Concept." U.S. Air Force report no. PL-TR-91-3009, May 1991, available from the National Technical Information Service.

Tesla, N. "The Transmission of Electrical Energy without Wires." *Electrical World and Engineer* (March 5, 1904).

Thompson, R. "Yesterday the Garden Shed, Tomorrow Mars." *Sydney Morning Herald*, March 18, 1989, suppl. section, 8–9.

UFO Subcommittee of AIAA. "UFO Encounter 1." *Astronautics and Aeronautics* (July 1971): 66–70.

Valone, T. "Significant Non-conventional Energy and Propulsion Methods for the 21st Century. In *Proceedings of the 1990 International Tesla Symposium*, 5.73–5.84, S. Elswick, ed. Denver: International Tesla Society, 1991.

———, ed. *Electrogravitics Systems: Reports on a New Propulsion Methodology*. Washington, D.C.: Integrity Research Institute, 1994.

———. "T. T. Brown's Electrogravitics Research in the 1950s." In *Proceedings of the 1994 International Tesla Symposium,* S. Elswick, ed. Denver: International Tesla Society, 1994.

———. *The Homopolar Handbook*. Washington, D.C.: Integrity Research Institute, 1998.

Vassilatos, G. *Secrets of Cold War Technology: Project HAARP and Beyond*. Bayside, Calif.: Borderland Sciences, 1996.

Veselago, V. G. "The Electrodynamics of Substances with Simultaneously Negative Values of ε and μ." *Soviet Physics Uspekhi* 10 (1968): 509–14. Originally published in Russian in *Uspekhi Fizicheskikh Nauk* 92 (1967): 517–26.

White, J. O., M. Cronin-Golomb, B. Fischer, and A. Yariv. "Coherent Oscillation by Self-Induced Gratings in the Photorefractive Crystal $BaTiO3$." *Applied Physics Letters* 40 (1982): 450–52.

Woit, P. *Not Even Wrong: The Failure of String Theory and the Search for Unity in Physical Law*. New York: Basic Books, 2006.

Yost, C. A. "T. T. Brown and the Bahnson Lab Experiments." *Electric Spacecraft Journal* 2 (1991): 6–12.

Zhang, Y., et al. "Total Negative Refraction in Real Crystals for Ballistic Electrons and Light." *Physical Review Letters* 91 (2003), no. 157404.

INDEX

BOOKS OF RELATED INTEREST

Decoding the Message of the Pulsars
Intelligent Communication from the Galaxy
by Paul A. LaViolette, Ph.D.

Earth Under Fire
Humanity's Survival of the Ice Age
by Paul A. LaViolette, Ph.D.

Genesis of the Cosmos
The Ancient Science of Continuous Creation
by Paul A. LaViolette, Ph.D.

Quantum Shift in the Global Brain
How the New Scientific Reality Is Changing Us and Our World
by Ervin Laszlo

The Akashic Experience
Science and the Cosmic Memory Field
by Ervin Laszlo

Science and the Akashic Field
An Integral Theory of Everything
by Ervin Laszlo

Science and the Reenchantment of the Cosmos
The Rise of the Integral Vision of Reality
by Ervin Laszlo

Forbidden Science
From Ancient Technologies to Free Energy
Edited by J. Douglas Kenyon

Inner Traditions • Bear & Company
P.O. Box 388
Rochester, VT 05767
1-800-246-8648
www.InnerTraditions.com

Or contact your local bookseller